BALKEMA – Proceedings and Monographs
in Engineering, Water and Earth Sciences

PROCEEDINGS OF THE FIFTH INTERNATIONAL WORKSHOP ON LIFE-CYCLE COST ANALYSIS AND DESIGN OF CIVIL INFRASTRUCTURE SYSTEMS, SEOUL, KOREA, OCTOBER 16–18, 2006

Life-Cycle Cost and Performance of Civil Infrastructure Systems

Editors

Hyo-Nam Cho
Hanyang University, Ansan, Korea

Dan M. Frangopol
Lehigh University, Bethlehem, PA, USA

Alfredo H-S. Ang
University of California, Irvine, CA, USA

Assistant Editor

Jung Sik Kong
Korea University, Seoul, Korea

Taylor & Francis
Taylor & Francis Group

LONDON / LEIDEN / NEW YORK / PHILADELPHIA / SINGAPORE

Taylor & Francis is an imprint of the Taylor & Francis Group, an informa business

© 2007 Taylor & Francis Group, London, UK

Typeset by Charon Tec Ltd (A Macmillan Company), Chennai, India

Published by: Taylor & Francis/Balkema
                P.O. Box 447, 2300 AK Leiden, The Netherlands

ISBN 13: 978-0-415-41356-5 (hbk)

Table of contents

Life-Cycle Cost and Performance of Civil Infrastructure Systems – Cho, Frangopol & Ang (eds)
© 2007 Taylor & Francis Group, London, ISBN 978-0-415-41356-5

Preface

The civil infrastructure systems of the world represent a huge investment for both governments and taxpayers. The life-cycle benefits of this investment must be maximized to ensure that the needs of our society are optimally met, taking into consideration safety, economy, and sustainability requirements. The Life-Cycle Cost (LCC) analysis and design of civil infrastructure systems plays an important role in maximizing these benefits. For this reason, the International Association for Bridge Maintenance and Safety (IABMAS) decided to initiate a series of International Workshops on Life-Cycle Cost Analysis and Design of Civil Infrastructure Systems. The First, Second, Third and Fourth IABMAS Workshops on Life-Cycle Cost Analysis and Design of Civil Infrastructure Systems were held, respectively, in Honolulu, Hawaii, USA (August 7-8, 2000), Ube, Yamaguchi, Japan (September 27-29, 2001), Lausanne, Switzerland (March 24-26, 2003), and Cocoa Beach, Florida, USA (May 8-11, 2005). Following this tradition, it was decided to hold the Fifth IABMAS Workshop on Life-Cycle Cost Analysis and Design of Civil Infrastructure Systems (LCC05) in Seoul, Korea (October 16-18, 2006).

Most of the papers presented at the LCC05 Workshop are contained in this book. Included herein are the keynote lectures and technical contributions. This book should serve as a valuable reference on the recent developments on life-cycle performance of deteriorating structures and on the state-of-the-art and research and application needs in this field.

The joint event was chaired by Hyo-Nam Cho (Korea) and Dan M. Frangopol (USA). The International Scientific Committee was chaired by Alfredo H-S. Ang (USA) and the Secretariat was lead by Jung Sik Kong (Korea).

The Editors wish to express their sincere gratitude to the work by all members of the Organizing Committee and to all the sponsors. Thanks are especially due to Dr. Jung Sik Kong for his excellent and continuous efforts. Thanks are also due to the authors of the various papers for the excellent contributions they made to the workshop. Without each of them, this international workshop would not have been so successful. The Editors believe that this book will help further advance the current state-of-knowledge in the field of life-cycle performance assessment, design and management of civil infrastructure systems.

The Editors,

Hyo-Nam Cho, Dan M. Frangopol and Alfredo H-S. Ang
February 2007

Workshop Organization

Organizing Association

International Association for Bridge Maintenance and Safety (IABMAS)
(http://www.iabmas.org/)

Workshop Chairs

Dan M. Frangopol, Lehigh University, Bethlehem, PA, USA
Hyo-Nam Cho, Hanyang University, Ansan, Korea

Steering Committee

Dan M. Frangopol, USA (Chair)
Hitoshi Furuta, Japan (Co-Chair)
Alfredo H-S. Ang, USA (Ex-Officio)
Eugen Bruehwiler, Switzerland
Hyo-Nam Cho, Korea (Ex-Officio)
Michael H. Faber, Switzerland
Ayaho Miyamoto, Japan
Andrzej S. Nowak, USA

International Scientific Committee

Alfredo H-S. Ang, USA (Chair)
Jim Beck, USA
Fabio Biondini, Italy
Christian Bucher, Germany
Harald Budelmann, Germany
Joan R. Casas, Spain
Moe Cheung, Hong Kong
Hyo-Nam Cho, Korea (Ex-Officio)
Marios K. Chryssanthopoulos, UK
Paulo J. S. Cruz, Portugal
Armen Der Kiureghian, USA
Allen Estes, USA
Luis Esteva, Mexico
Dan M. Frangopol, USA (Ex-Officio)
Mircea Grigoriu, USA
Jun Kanda, Japan
Sang-Hyo Kim, Korea
Anne Kiremedjian, USA
H.E. Klatter, The Netherlands
Xila Liu, China
Chin-Hsiung Loh, Taiwan
Marc Maes, Canada
Sami F. Masri, USA
Robert Melchers, Australia
Andrzej Nowak, USA
Udo Peil, Germany
Jorge Riera, Brazil
John D. Sorensen, Denmark
Mark G. Stewart, Australia
Man-Chung Tang, USA
Wilson Tang, Hong Kong
Palle Thoft-Christensen, Denmark
Eiichi Watanabe, Japan
Y.K. Wen, USA
Chung-Bang Yun, Korea

Local Organizing Committee

Hyo-Nam Cho, Korea (Chair)
Yoon Koog Hwang, Korea
Jong Kwon Lim, Korea
Jung Sik Kong, Korea (Secretary)
Dae Hong Min, Korea
Kyung Hoon Park, Korea
Yong Su Kim, Korea

Organized by

IABMAS (International Association for Bridge Maintenance and Safety)

Sponsored by

KICT (Korea Institute of Construction and Technology)
KSSC (Korean Society of Steel Construction)
KALCEM (Korean Association for Life-cycle Engineering and Management)

Sponsors from Industry

DAELIM Industrial Company Ltd.
DAEWOO Engineering & Construction. Co. Ltd.
GS Engineering & Construction
HYUNDAI Development Company
HYUNDAI Engineering & Construction. Co. Ltd.
KOREA Development Corporation
LOTTE Engineering & Construction
NAMKWANG Engineering & Construction
POSCO Engineering & Construction. Co. Ltd
SAMSUNG Engineering & Construction
SHINSUNG Engineering & Construction. Co. Ltd
SK Engineering & Construction. Co. Ltd
TAEYOUNG Corp.

Photo of Participants

History of LCC Workshops

Year	Venue	Chairs
2000	*Honolulu, Hawaii*	*D.M. Frangopol and H. Furuta*
2001	*Ube, Japan*	*A. Miyamoto and D.M. Frangopol*
2003	*Lausanne, Switzerland*	*E. Bruehwiler, M.H. Faber, and D.M. Frangopol*
2005	*Cocoa Beach, FL, USA*	*A.S. Nowak and D.M. Frangopol*
2006	*Seoul, Korea*	*H. N. Cho and D.M. Frangopol*

Keynote papers

A structural monitoring system for rc-/pc structures

H. Budelmann & K. Hariri
Technical University Braunschweig, Germany

ABSTRACT: A comprehensive analysis and assessment system for structural monitoring of concrete structures is under way within several projects of the collaborative research center CRC 477 "Structural Monitoring" at the Technical University Braunschweig, Germany. Main components of the system are a reliability-based system assessment tool, prognostic models of degradation and new sensoring techniques. The system assessment tool is based on the reliability method; its task is to find out weak points of a structure being first responsible for a possible failure. If prognostic models of degradation are combined with monitoring an improved accuracy of prognosis can be achieved. To get relevant monitoring data from a structure a new generation of sensors is being developed. These three parts of a structural monitoring system are presented in the contribution. The methods mentioned before are tested and demonstrated at a real scale pc-bridge, serving as an experimental structure. This structure provides typical structural situations and degradation processes in a realistic manner. The object is to examine under conditions close to real structures the combined use of reliability analysis in order to find weakpoints and to decide on the places where sensors should be applied on one side, and on the other side the practical operability and usefulness of new sensors and adaptive prognostic models for concrete degradation as presented above. The experimental bridge structure is presented and the first results of the probabilistic system analysis are sketched finally.

1 INTRODUCTION

Innovative management systems for civil structures allocate maintenance resources to deteriorating structures. A life-cycle cost analysis (LCA) provides a convenient means for decision makers to compare alternative management solutions using economic terms. The fundamental goal is to improve the condition, safety, and long-term performance of a structure or of a group of structures with reduced life-cycle costs. Most of the existing methodologies are based on life-cycle cost minimization. Several models have been developed in the last decade, e.g. (Frangopol et al. 2005, Nishijima et al. 2006).

Since civil structures are quite complex and the number of influences to be included is large, probabilistic treatment becomes indispensable to accounting for various sources of uncertainty in modeling time-dependent variation of structural performance (condition, safety, life cycle cost). Generally the usefulness of a LCA depends on the quality of information on structures concerning their current state and the development of further degradation (Liu et al. 2005). In particular, the actual structural capacity under deterioration should be accurately modelled.

Reinforced and, in particular, prestressed concrete structures of civil infrastructure are affected by load, environment and several chemical attacks as well. Their long-term safe, efficient and economical use requires current knowledge of remaining service life. Modern methods of structural monitoring achieve important contributions for the prediction of the realistic life time and the prolongation of the service life of civil engineering structures. A comprehensive monitoring system is under way at the TU Braunschweig. It is precondition for and part of a life-cycle management tool. Figure 1 points up the framework how to integrate such a structural health monitoring system into life-cycle management (Messervey et al. 2006).

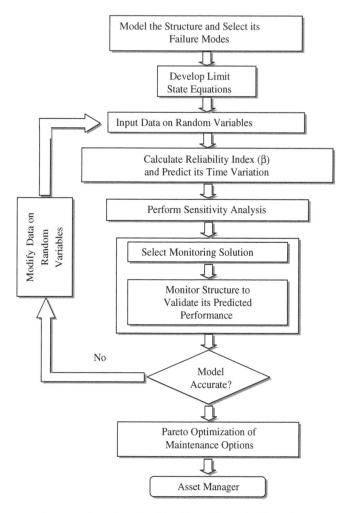

Figure 1. Framework for integration of structural health monitoring in life cycle management (Messervey et al. 2006).

Within a collaborative research project new sensor techniques, prognostic models of degradation and a reliability-based system assessment tool have been developed to an advanced stage. The actual task is to bring them together in order to provide an analysis and assessment system for structural monitoring of concrete structures which can be used for LCA in future. The main components of the system are introduced in this paper: a reliability-based system assessment tool (chapter 3), prognostic models of degradation (chapter 4) and new sensing techniques (chapter 5). In chapter 6 the experimental full scale bridge structure and its application for system analysis is presented.

2 DEGRADATION MODELING OF CONCRETE STRUCTURES

Any structure holds an initial resistance inventory against degradation, recapitulating expressed by the initial value of performance p_0. This initial performance is the result of planning, construction, quality control etc., including individual deficiencies. During its service life any structure degrades. The degradation process to be expected, covering different deterioration processes, must

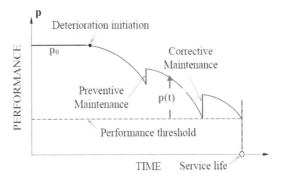

Figure 2.   Degradation of performance of a structure during service life (based on Frangopol et al. 2005).

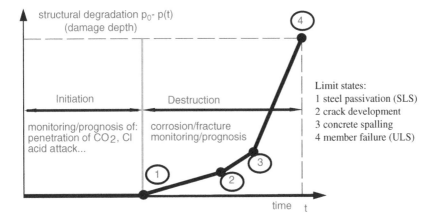

Figure 3.   Development of Corrosion Induced Deterioration and Limit States (based on Tuutti 1982).

be estimated in advance for the purpose of any life-cycle management. Monitoring is needed to win current condition data to achieve improvement of prognosis. The forecasted gradient of degradation is the basis for any maintenance decision. Figure 2 shows a principal example of structural degradation.

More than 50% of maintenance and repairing measures at concrete bridges in Germany are necessary because of corrosion damages. Regarding deterioration processes induced by corrosion of reinforcement it is well known that it is developing in at least two overriding periods of time (Tuutti 1982), Figure 3. The deterioration processes generate a cumulative loss of resistance $p_0 - p(t)$. The first period is called incubation period or initiation phase. This period ends at the point of time when depassivation of reinforcement is starting.

At this point the serviceability limit state (SLS) is reached. Several deterioration processes may proceed during the first period: such as dissolving acid attack with the subsequence of reducing concrete cover, sulfate attack with the subsequence of destroying concrete cover, carbon dioxide attack with the subsequence of carbonation of concrete cover and loosing alkalinity or chloride ingress.

Time development of incubation can be monitored in-situ by help of suited sensors within the first period in the reinforcing steel neighbourhood, e.g. in the concrete cover or inside grouted ducts. The corrosion affecting parameters would be e.g. moisture, salt-content (chloride, sulfates etc.), pH value (carbonation) and temperature. Alarm values resp. threshold values of corrosion-relevant parameters must a priori be specified for the evaluation of the corrosion danger.

In the second phase, the damage period, corrosion of reinforcement is developing, leading to a final failure of the structure (ultimate limit state, ULS). Today there are only limited monitoring methods available to observe type, place, intensity and extend of corrosion procedure or of corrosion induced damages. First direct measuring procedures are being developing, i.e. (Hariri et al. 2003).

The time development of the processes, mentioned above, generally can be calculated by help of numerical models. A number of forecasting models for the processes of the first phase (initiation period) has been developed and is increasingly going into practice. However, forecasting models for the second phase (destruction period), taking into account the different consequences of steel corrosion on concrete members (for example crack development, concrete spalling, failure of reinforcing element or of rc-member) still are in a early phase of development.

3 PROBABILISTIC SYSTEM ANALYSIS OF CONCRETE STRUCTURES

A framework for reliability-based system assessment based on data from structural health monitoring (SHM) is developed within CRC 477 (Klinzmann et al. 2006). Its main objective is the optimization of SHM measures by help of probabilistic methods. The methodology is able to identify relevant parameters, to weigh the critical areas and to determine the actual safety level of structures. It concentrates on the individual assessment of a structure based on results from SHM. The basis of the framework is a probabilistic model of the structure (PROBILAS), which can be formulated after a thorough anamnesis. In the anamnesis the engineer identifies typical weak points and failure paths of the structure and includes them into the model. Then the assessment process proposed by the framework can be started. The overall procedure is shown in Figure 4.

The framework utilizes first and second order reliability methods (FORM/SORM) as calculation procedure for system reliability analyses. Before the probabilistic model can be used for calculation, it should be calibrated using data from inspection and/or quality control. Afterwards, the phase of continuous assessment of the structure is started. In this phase, the actual state of the structure as well as the expected performance in future are considered in the reliability analysis. Data from the SHM process is included if available. Based on the results of these analyses, further decisions

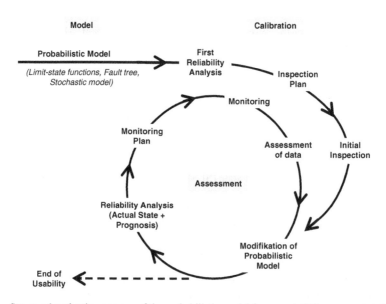

Figure 4.   Structural evaluation process of the probabilistic model framework (Klinzmann et al. 2006).

concerning inspection and monitoring can be made. The phase of continuous assessment ends when the reliability of the structure falls below a target reliability level.

The elements of the system analysis tool named PROBILAS, developed at the TU Braunschweig (Hosser et al. 2004), are shortly described below. The first step is to acquire all necessary input data describing the structure. Then, the structure is analyzed with methods of system theory. The reliability analysis starts with identifying the different sources of risk of the structure, afterwards the system discretisation in subsystems and components is following. All possible failure relevant events are shown in an event tree. And in a fault tree all possible causal sequences of components failure leading to the system failure are regarded. For each component in the fault tree an ultimate limit state has to be defined. The limit state equation is comparing the quantity of the resistance with the quantity of the action, both being functions of stochastically described parameters.

Now by help of reliability analysis the failure probability and the safety index can be calculated, first for each limit state equation, afterwards for the whole system using the logical model of the fault tree. On basis of the calculated values the failure path with the highest probability of occurrence can be found. Especially the parameters of the limit state equations within this failure path should be investigated further. For example the concentration of chemical agents within the concrete cover can be monitored directly using new sensor techniques, as described in this paper.

To illustrate how a weak point of a structure (in this example the "definitive path of failure") can be analyzed with the help of probabilistic assessment, in (Hosser et al. 2004) a simple example is presented. In this example prestressed slab strip elements are systematically corroded with a 5% sodium chloride solution. More details can be found in (Schnetgöke et al. 2005).

4 PROGNOSTIC MODELING OF DETERIORATION PROCESSES OF CONCRETE STRUCTURES

Forecasting models, describing the durability or the degradation processes resp. of a structure or of a group of structures, are indispensable components of a life-cycle management system. Forecasting models for networks of structures usually are physically simple probabilistic models. Such models cannot consider data from monitoring, decisions are derived from calculated values. Forecasting for single structures, however, can be based on physically sound models, describing the phenomena to be observed in an appropriate manner. The accuracy of the prognosis can be up-dated considerably by monitoring data.

Most durability models for concrete structures are based on design data and do not consider measured performance criteria of the structure. Such models must already consider the unknown actions during the life time at the time of structural design. Uncertainty is unavoidable. A reliable prognosis of durability and remaining service life during the utilization of a structure requires actual information about the behavior of the structure and a simulation model which is able to treat this information. An adaptive simulation model that improves its accuracy itself by the use of measurable data is needed. So the durability will become a property that can be monitored. There is another advantage of such an adaptive algorithm. Unlike a non-adaptive algorithm, the properties of the non-corroded concrete structure do not need to be known very accurately. So the expensive and time consuming determination of initial data for the simulation can be shortened.

Such an adaptive durability model for the monitoring of concrete structures is being developed. The adaptive model is based on the software system TRANSREAC (*trans*port and *reac*tion). TRANSREAC combines the calculation of chemical reactions by a thermodynamic algorithm, transport processes within a structure and corrosive effects (Schmidt-Döhl et al. 2004).

For testing and validation of the adaptive model reinforced concrete testing structures are used and exposed to real climatic conditions as well as to acid, sulfate, chloride and ammonium solutions. The structures are prestressed to ensure realistic stress situations and cracks. These testing structures are realistic "building substitutes". One of the testing structures is shown in Figure 5. It is equipped with different sensors. The environmental climatic data are automatically recorded. The climatic data, the cement composition, the concrete mixing properties, the composition of the aggressive

Figure 5.   Concrete testing structure for chemical and weathering exposure.

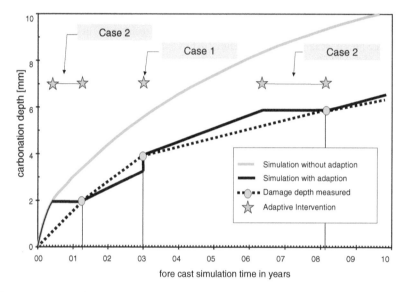

Figure 6.   Development of concrete carbonation, measured and calculated.

solutions and the sensor signals are input data for the adaptive simulation. The corrosion behavior of these structures is simulated. The calculated and measured data are compared and the function of the adaptive algorithm is checked.

Fig. 6 shows an example for the adaptive simulation of concrete carbonation. The walls used for computation have been concreted in 1987. Since then they have been stored in outdoor climate being documented at regular intervals.

The figure shows the carbonation depth as a function of time. The dotted line connects the observed carbonation depths at certain points of time. The light line shows the calculation result for the case that the model parameters have not been up-dated by adaptive action. The solid black line shows the calculation corrected in an adaptive approach. It is evident from the graph that at the beginning the calculated carbonation development overestimates the real carbonation progress until the point of time when the first adaptive correction was performed. The adaptive algorithm reduces the velocity of any additional mass transport (case 2). At the second event of comparison the calculated development of carbonation depth is behind reality. So the adaptive intervention accelerates the reaction (case 1) until the calculated event is synchronized with the real event. Later

8

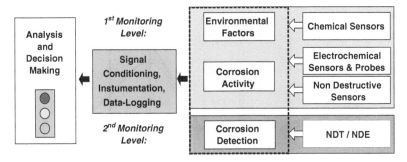

Figure 7.   Schematic for a corrosion monitoring strategy within the 1st and 2nd corrosion monitoring level.

again a decelerating correcting is necessary (case 2). It can be seen that with a limited number of adaptive interventions the forecast quality can be improved considerably.

Of course, the calculation program TRANSREAC can also be used as a probabilistic model by combining the main algorithm with a Monte-Carlo-simulation. This can be realized in a simple way since the distribution functions and the scatterings of the basic variables of TRANSREAC are rather well known .

5   SENSOR TECHNIQUES FOR THE FIRST AND SECOND MONITORING LEVEL

5.1   *General remarks*

Monitoring of a concrete structure covers both the initiation phase during which transport and possibly reaction processes occur in the concrete cover or inside the duct and the destruction phase being responsible for corrosion and cracking of reinforcing or prestressing steel elements. The phases have been defined in chapter 2. Since the mechanisms of deterioration in the first and second monitoring level are quite different, also the sensors and monitoring concepts differ. Figure 7 proposes a schematic for a corrosion monitoring strategy.

5.2   *Sensors for the initiation phase*

At the first monitoring level at which the preconditions for corrosion of the reinforcing or prestressing steel elements are proceeding in their vicinity (concrete cover or grouting mortar), following processes are of relevance for monitoring:

– Observation of ingressing substances or of condition alterations: e.g. moisture content, chloride-concentration, pH-value.
– Observation of changes of material properties: e.g. electrical conductivity, impedance.
– Observation of corrosion of "substitute sensors" in the vicinity: e.g. thin depth staged steel wires.

Several new sensor concepts have been developed in the CRC 477. Figure 8 shows a new fiber-optic chemo-sensor, qualified for the measuring of the moisture content or pH-value of concrete inside the concrete. It indicates the measured quantity by changing the wave length spectrum of the light being absorbed from the dye at the sensor top.

The dye within the sensor shown in Figure 8 is a so-called azo dye, which is acting as optical transducer with a characteristic absorption spectrum, dependent on the pH-value of the pore solution of cement stone. The dye is polymer-bound to prevent leeching. This material was chosen for its stability in basic range. There are other dyes being developing for humidity (Reichard's dye) or chloride content. But a dye which is sensitive against chloride ions and simultaneously resistant in an alkaline solution is not yet found sufficiently.

9

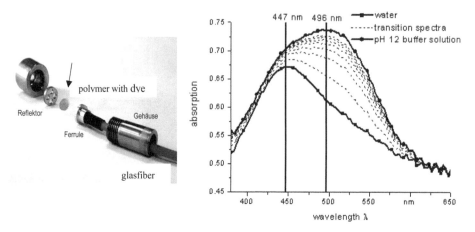

Figure 8. Fiber-optical chemo-sensor for measurement of pH-value.

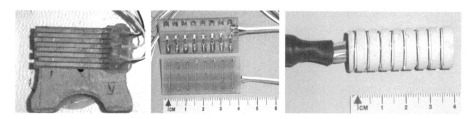

Figure 9. Different layouts of corrosion "substitute sensors". (*left*): Finger sensor made from copper fixed at a concrete spacer (upside down), (*middle*): Filament sensor mounted at a plastic board made with SMD-resistances as multiplier (series resistances) and (*right*): Sensor with prefabricated mortar slices for bore-hole instrumentation (Ø 20 mm) at existing structures.

The electrochemical parameter corrosion potential or current and concrete impedance can be measured in varying depths of rc- or pc-structures by means of galvanic probes or "watch dog sensors", (e.g. Schiegg 2002). This electrochemical sensing principle was implemented at miniature sensors by our research group. Several small size sensing elements (only a few cm's) made from a plastic circuit board were developed.

The "finger"-electrode sensor depicted in Figure 9a is consisting of 4 to 16 straight stripes of iron or other metals (electrodes) galvanized at a plastic board. Due to the finger design a good bond between the board and the concrete is achieved. This was verified by means of pull out tests. Because of slight swelling of the plastic board in concrete shrinkage of concrete can be compensated. Thus it is made sure that no gap between concrete and sensor could distort transport processes.

Another innovative type of corrosion sensor is the "filament" sensor. A prototype for application during construction is depicted in Figure 9b and for borehole instrumentation at existing structures in Figure 9c. 0,065…0,5 mm thin iron made wires are mounted at a plastic board at different depths within the concrete cover of the steel element. If the depassivation (chloride) front reaches the sensor the thin wires will corrode one by one very fast corresponding to the transport progress. In laboratory tests the time span until the rupture of the wire took only hours to a few days. By measuring the transition ohmic resistance the local rupture can be monitored easily. The erratic change of resistance due to corrosion of the wire amounts to 3 to 4 decades. This is normally a very clear signal even in very conductive wet concrete.

In Figure 10 a chart of measuring with two filament sensors according to Figure 9b) at prestressed concrete stripes is shown. The stripes are under a long-term bending constraint in a cracked condition and treated with NaCl-solution on top (see Figure 10 right). The steps of resistance alteration due

10

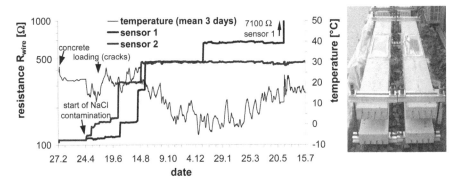

Figure 10. Measured data of 2 filament sensors due to corrosion, installed in chloride-treated pc-stripes.

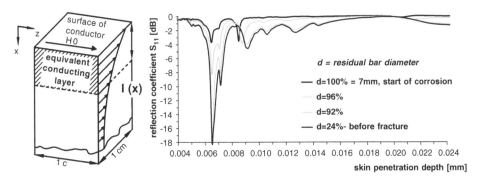

Figure 11. Skin-Effect for corrosion monitoring. (Lab test at a single wire in a concrete beam corroded by impressed current, with the amplitude of reflection coefficient- beam and bar length 3 m, free corrosion length 1 cm).

to corrosion of single level wires are observable. One can see certain differences influenced by the concrete material itself, acting as an additional conductor in parallel. After corroding of a wire its resistance is not infinite because the corrosion products and the adjacent concrete are slightly conductive. This impairs partially the measurement due to fluctuations of the signal.

5.3 Sensors for the destruction phase

Destruction phase means the corrosion process of a steel element itself. A direct monitoring of a corrosion process of steel members embedded in concrete or grouting mortar is a difficult task and not yet realized to practice.

A microwave reflection method seemed to be promising from lab tests performed at the CRC 477 (Holst et al. 2004). In this method the surface of the steel serves as a good conductor for alternate currents. With increasing frequency the primary current is concentrated towards the surface due to eddy currents. This phenomenon is designated as skin effect. The ousting process is illustrated in the left part of Figure 11. The skin phenomenon generally can be used for the interrogation of defects in the surface area such as corrosions notches, at which a part of the generated RF-wave is reflected back towards the signal source. The amplitude and the shift of the reflected signal include the information of the severity of damage at the discontinuity point.

In the right part of Figure 11 the measured reflection coefficient at a 3 m long concrete beam is depicted. At a certain point of the beam a single wire was corroded in an accelerated test. The integral information concerning the corrosion status correlates with the amplitude and the shift of the minima of the reflection parameter. On the other hand the measured signal is an interfered

11

Figure 12. Electromagnetic resonance measurement (ERM) for fracture detection, (*left*): Principle, (*middle*): Open ended wide-band permittivity sensor and (*right*): External adapter for feeding the RF-signal onto the tendon at a hollow slab.

signal depending also from tendon's environment and several other influencing parameters which can not be controlled in real structural members. Thus the method seems to be limited to well defined situations, such as at monostrands or ground anchors.

To a more promising development status a method for fracture detection is thriven (Holst et al. 2006): The fracture detection and localization of prestressed cables by means of electromagnetic resonance measurement (ERM). The basic idea of this method is to consider the tendon itself as an unshielded resonator located in concrete or grout as a lossy electromagnetic material. The steel is excited at one accessible end by electromagnetic waves of systematically varying frequency, as shown in Figure 12 left. By systematically scanning the reflection coefficient over a wide-band spectrum up to frequencies of 500 MHz by a vector network analyzer, resonance frequencies of the tendon are recorded. The length of a tendon l possibly shortened by a fracture is inverse proportional to the spacing Δf between two adjacent resonance (minimum of S_{11}) frequencies also depends on the filling material as shown in the following equation:

$$l = \frac{c_0}{2 \cdot \sqrt{\varepsilon_r} \cdot \Delta f},$$ (1)

with the vacuum speed of light c_0 and the (averaged) dielectric constant (permittivity) of the surrounding medium ε_r as a material parameter.

The dielectric constant depends on the frequency, the moisture as well as the concrete composition and varies between 4 and 10 for dry and between 10 and 20 for wet concrete or grout. The permittivity can be measured by locally RF-sensors made of semi-rigid cable as depicted in the middle position of Figure 12.

Figure 12 right shows the handheld sensor for externally coupling of the RF-signal into the tendon. It is made of a 10 cm long, standard semi-rigid cable Ø 0,25" with a microwave connector. The inner wire sticks out of the coaxial cable with about 3 mm and is pressed to the surface of the tendon.

The main advantages of the ERM are: Detection and localization of fracture, no patrolling of the tendon, only access at one point is required, location of tendon is irrelevant, every position (depth) of tendon and also ground anchors are measurable.

ERM-measurements at a 55 year old bowstring bridge in Hünxe, Germany before and after cutting the hanger were carried out for practical evaluation. The hanger consisted of four 7,65 m long prestressed bars (diameter 26 mm). Each tendon was wrapped with a bituminous textile bandage. The ordinary reinforcement and the tendons themselves were short-circuited. The measured reflection coefficients of the tendon bar No. 2 are depicted in Figure 13. There are clear resonances with varying magnitude of the spacing Δf because of the dispersion. Then bar No. 2 was cut. A significant alteration of resonance frequency before and after cutting/fracture of bar No. 2 is discernable. But the difference Δf_i does not linearly correspond with the inverse ratio of the bar lengths. This is due to the superposition of the electrical contact between the reinforcement of the arch and the tendon.

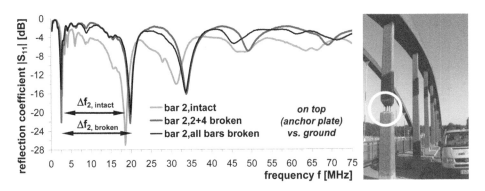

Figure 13. Bridge Hünxe: Reflection coefficient $|S_{11}|$ as function of frequency f for intact and broken prestressed bar no. 2 measured externally at tendon's end above and view with cutted hanger.

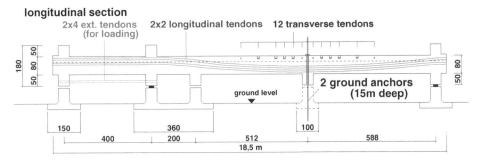

Figure 14. Longitudinal section of the experimental bridge.

6 SYSTEM TESTING AND VALIDATION AT A FULL SCALED SUBSTITUE STRUCTURE

6.1 Experimental full scale pc-bridge structure

A real scale prestressed concrete bridge was constructed, serving as an experimental structure. This structure provides typical structural situations in a realistic manner. The object is to examine under conditions close to real structures the combined use of reliability analysis in order to find weakpoints (critical failure paths) and to decide on the places where sensors should be applied on. Furthermore an object is to prove the practical operability and usefulness of new sensors and of adaptive prognostic models for concrete degradation.

The designed structure provides typical structural situations in a realistic manner. It consists of a double-T cross section with a height of 80 cm and has a total length of 17 m with a span of 13 m and a cantilever of 4 m as depicted in Figure 14. In longitudinal as well as in transverse direction the bridge is prestressed, with internal or external tendon profile, with or without bond. So prestressing techniques as known from conventional motorway bridges have been included. Traffic load is applied via hydraulic actuators respectively via additional prestressing an deadman. The design of the bridge was chosen with a view on high flexibility with regard to static sytem, mechanical and chemical loading and implementation of sensors respectively. So 4 different static systems can be realized by help of movable abutments.

The bridge structure is prepared to contain about 130 different monitoring sensors, commercially available ones as well as new sensors from SFB 477 as presented in chapter 5. About 50 sensors have been included directly at construction.

Figure 15. Experimental bridge structure, (*top*): bridge slab instrumented with commercial and novel developed corrosion sensors before concreting, (*bottom*): prestressing procedure.

The superstructure is built in concrete C20/25 with $270 \, kg/m^3$ cement CEM I and $60 \, kg/m^3$ limestone in order to realize a moderate concrete quality. The water-cement-ratio is 0.58. The achieved concrete cube-strength was 33.1 MPa while the modulus of elasticity was 27.1 GPa at an age of 28 days.

Within this structure several faults and weaknesses are included. Furthermore devices are supposed for chemical and mechanical attacks, leading to degradation processes. The weak points are hollow members and ungrouted parts of ducts with a local attack of chloride on the strand. Some of the tendons can be fractured by drilling and by means of electro-chemical corrosion via applied currents. Other factors regarding degradation of the superstructure are honeycombs, cracks in concrete to allow a faster penetration of degrading agents like chlorides or NH_4SCN.

Figure 15 gives an impression of the bridge structure during construction. More details on bridge construction and instrumentation can be seen in (Budelmann et al. 2006).

6.2 *Validation of the monitoring system at the experimental bridge structure*

The application of the monitoring system explained at the beginning, is currently been tested at the experimental bridge structure. A reliability analysis in order to find weakpoints by help of probabilistic modelling of the structure and to decide then on parameters to be monitored and on the places where sensors should be applied on, is just under way. Furthermore, the testing of practical operability and usefulness of new sensors and adaptive prognostic models for concrete degradation as presented in chapter 4 is running. Some first results of the probabilistic structural analysis, presented in (Schnetgöke et al. 2006, Hosser et al. 2006), are illustrated subsequently.

Initial point is to define the technical system in its components. In this case, the bridge is a simple one span construction with a one-sided cantilever. The cross section is a double T. It is prestressed in longitudinal and transverse direction. Then the possible failure combinations leading to a structural failure (top-event) have to be identified by help of a fault-tree-analysis. Possible failure positions

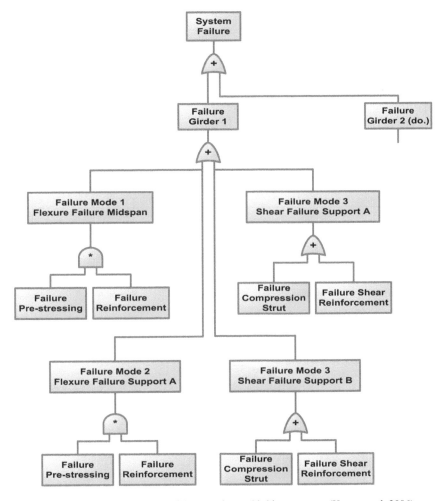

Figure 16.   Fault tree for the assessment of the experimental bridge structure (Hosser et al. 2006).

are the middle of the span and the supports A and B (see figure 14). The fault tree as a basis for the assessment is shown in Figure 16.

Afterwards for every components of the fault tree limit state equations and failure criteria have to be formulated. All parameters being included to these relations need to be described stochastically (density functions, stochastic parameters). Since the complete set of data never can be described sufficiently, a calibration procedure is carried out within a first reliability analysis using Bayesian updating procedures.

Having repeated this procedure for all selected parameters, the probabilistic model is calibrated and the reliability analysis can be performed. Figures 17 and 18 show some first results of the safety analysis of the structural subsystems (safety index β, Figure 17) and of the sensitivity of the structural components to contribute to the failure (sensitivity factors γ, Figure 18).

The predominant influence on the structural failure comes from the prestressing elements at the middle of the span. So, monitoring should concentrate first on the prestressing elements. A result like this is not surprising and could have been deduced from conventional engineering analysis too; it was just to show the procedure at an example. The results reflect the situation at time of analysis. Integrating the reliability analysis into a prognosis of degradation, possibly an adaptive prognosis,

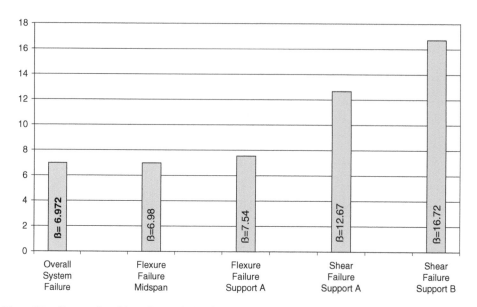

Figure 17.   First results of the safety analysis of structural subsystems, safety index ß (Hosser et al. 2006).

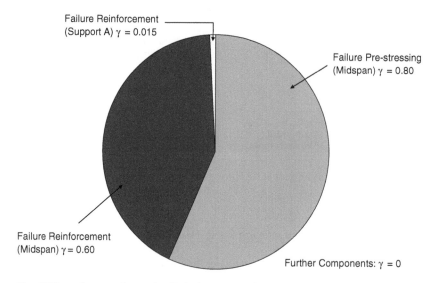

Figure 18.   Different layouts of corrosion "substitute sensors".

using monitoring data, the time development of the safety indices and of the sensitivity factors can be calculated.

7   CONCLUSIONS

Life-cycle management systems for groups of structures or single structures, especially if including costing aspects, need a precise prognosis of time dependent changes of performance. Especially rc- and pc-structures are subjected to different influences from load and environment, which lead to deterioration processes. A reliable prognosis of deterioration processes, e.g. of carbonation or chloride ingress or of subsequent steel corrosion, needs updating by monitoring.

A monitoring system for concrete structures is presented in this contribution. Main components of the system being presented are a reliability-based system assessment tool, prognostic models of degradation and new sensing techniques. The system assessment tool is based on the reliability method; its task is to find out weak points of a structure being first responsible for a possible failure. It is shown in the paper, how the components are linked together.

Future work will be focused on models for prognosis of the behavior of maintenance and repairing and on monitoring methods to observe their time dependent changes.

REFERENCES

Blumentritt, M., Brodersen, O., Flachsbarth, J. et al. 2006. "Novel Sensor Systems for Structural Health Monitoring". In *Proceedings of 3rd European Workshop on Structural Health Monitoring,* Granada, Spain, July 7–9.

Budelmann, H., Hariri, K., & Holst, A. 2006. "A Reale Scale PC Bridge for Testing and Validation of Monitoring Methods". 3rd.*Int. Conference IABMAS'06.* Porto, Portugal, July 16–19.

Frangopol, D.M., & Liu, M. 2005. "Life-cycle analysis and optimization of civil infrastructure under uncertainty," *Third Probabilistic Workshop on Technical Systems and Natural Hazards,* K. Bergmeister, D. Rickeenmann, & A. Strauss, eds. Vienna. November 24–25: 23–30.

Hariri, K., Holst, A., Wichmann, H.-J., & Budelmann, H. 2003. "Assessment of the State of Condition of Prestressed Concrete Structures with Innovative Measurement Techniques". In *Structural Health Monitoring Journal* Vol.2, No.2, June 2003, 179–185.

Holst, A., Wichmann, H.-J., Hariri, K. & Budelmann, H. 2006. "Monitoring of Tension Members of Civil Structures – New Concepts and Testing". In *Proceedings of 3rd European Workshop on Structural Health Monitoring,* Granada, Spain, July 7–9.

Hosser, D., Klinzmann, C., & Schnetgöke, R. 2004. "Optimisation of Structural Monitoring using Reliability-Based System Assessment". In *Proceedings of 2nd European Workshop on Structural Health Monitoring,* Munich, Germany, July 7–9.

Hosser, D., Klinzmann, C., & Schnetgöke, R. 2006. "A Framework for reliability-based system assessment based on structural health monitoring". In: *Structure and infrastructure engineering,* accepted for publication.

Klinzmann, C., Schnetgöke, R., & Hosser, D. 2006. "Framework for the optimization of structural health monitoring on a probabilistic basis". In *Proceedings of 3rd European Workshop on Structural Health Monitoring,* Granada, Spain, July 7–9.

Liu, M., & Frangopol, D.M. 2005. "Multiobjective maintenance planning optimization for deteriorating bridges considering condition, safety, and life-cycle cost," *Journal of Structural Engineering,* ASCE, 131(5): 833–842.

Messervey, T.B., Frangopol, D.M., & Estes, A.C., 2006. "Reliability-based life-cycle bridge management using structural health monitoring," 3rd. *Int. Conference IABMAS'06,* Porto, Portugal, July 16–19.

Nishijima, K., & Faber, M.H. 2006. A budget management approach for societal infrastructure projects.

Schiegg, Y. 2002. "Online-Monitoring zur Erfassung der Korrosion der Bewehrung von Stahlbetonbauten," *PhD-Dissertation, ETH Zurich.*

Schmidt-Döhl, F., Rigo, E. Bruder S., & Budelmann, H. 2004. "Chemical attack on mineral building materials, features and examples of the simulation program Transreac," presented at the *2nd International Conference Lifetime-Oriented Design Concepts,* Bochum, Germany, March 1–3.

Schnetgöke, R., Klinzmann, C., & Hosser, D., 2005. "Structural Health Monitoring of a Bridge Using Reliability Based System Assessment". *Proceedings of the 5th International Workshop on Structural Health Monitoring,* Stanford University, USA., September 12–14.

Schnetgöke, R., Klinzmann, C., & Hosser, D., 2006. "Zuverlässigkeitsorientierte Bewertung von Bauwerken auf Grundlage der Bauwerksüberwachung". *Beton- und Stahlbetonbau* 101, Heft 8, 585–595.

Tuutti, K. 1982. "Corrosion of steel in concrete". Stockholm: Swedish Cement and Concrete Research Institute. In: *CBI Research,* No. Fo 4:82.

Life-Cycle Cost and Performance of Civil Infrastructure Systems – Cho, Frangopol & Ang (eds)
© 2007 Taylor & Francis Group, London, ISBN 978-0-415-41356-5

Life-Cycle Cost analysis and management of reinforced/prestressed concrete structures

Moe M.S. Cheung, Xueqing Zhang & Kevin K.L. So
Department of Civil Engineering, The Hong Kong University of Science and Technology, Hong Kong

ABSTRACT: Infrastructure is essential to the social and economic development of a city and to the quality of life of the people living in the city. However, the increasing stocks of public infrastructure and serious deterioration of infrastructure systems in many cities present great financial, safety, technical and operational challenges to government organizations in charge of public infrastructure development and management. It is recognized that an integrated whole life-cycle design and management strategy needs to be taken in order to meet these challenges. In this paper, the deterioration mechanisms of reinforced/prestressed concrete structures are examined as a first step towards the development of a methodology for the whole life-cycle design of reinforced/prestressed concrete structures. Then, some existing life-cycle cost analysis models are reviewed. Finally, an integer-programming model based on the Markov decision process is developed, in which Markov chains are used to model the change of condition index and integer-programming employed to optimize annual management actions and annual budget allocation, subject to various types of constraints. This Markov-based optimization model would provide a useful tool for life-cycle cost analysis and management in a network of many reinforced/prestressed concrete structures.

1 INTRODUCTION

Infrastructure is essential to social and economic development in developing as well as developed countries. The importance of developing efficient transportation infrastructure is well recognized worldwide. For example, an efficient transportation infrastructure system is required to speed up the development of remote areas, to meet the demands arising from various economic and social activities, to create prosperity, and to enhance the well-being of the people.

Reinforced/prestressed concrete is widely used in modern infrastructure development in the 20th century. It is commonly known that the combination of concrete and steel reinforcement is an optimal one not only because of the mechanical performance but also from the point of view of long-term performance. Concrete, is a material that is much more durable than steel. The encasement of steel in concrete provides the steel with a protective environment and allows it to function effectively as reinforcement. Theoretically, this combination should be highly durable, as the concrete cover over the steel provides a physical protection to the steel, and eliminates steel corrosion problems that occur quickly in bare steel structures. Unfortunately, this highly desirable durability is not always achievable in practice due to corrosion of steel in concrete, which is a commonly encountered cause of deterioration in many concrete structures in recent decade.

In this paper, the deterioration mechanisms of reinforced/prestressed concrete structures are examined as a first step towards the development of a methodology for the whole life-cycle design of reinforced/prestressed concrete structures. Then, some existing life-cycle cost analysis models are reviewed. Finally, an integer-programming model based on the Markov decision process is developed, in which Markov chains are used to model the change of condition index and integer-programming employed to optimize annual management actions and annual budget allocation, subject to various types of constraints. It is believed that this Markov-based optimization model would provide a useful tool for life-cycle cost analysis and management in a network of many reinforced/prestressed concrete structures.

2 CORROSION INDUCED PROBLEMS IN REINFORCED/PRESTRESSED CONCRETE STRUCTURES

There are many causes for concrete structure deterioration. Common among these are reinforcement corrosion due to chloride attack or carbonation, freeze/thaw, cycling, alkali-silica reaction and poor quality of detailing, materials or workmanship. Among them, the corrosion of embedded steel in reinforced/prestressed concrete has become a major problem. Although corrosion in concrete may be by acid attack, hydrogen embrittlement or electrolysis due to 'stray' electrical current, the vast majority of steel corrosion in practice occurs as a result of electrochemical processes (Beamish et al. 1998).

2.1 Reinforced concrete

Good quality concrete provides reinforcing steel with sufficient corrosion protection by the high alkalinity ($pH > 12$) of the cement paste. When reinforcement is placed in concrete, the alkaline condition leads to the formation of a passive layer on the surface of the steel. However, if the pH value is reduced by the penetration of acidic gases such as carbon dioxide, or if chloride ions are present at the steel surface, the protective film may be disrupted, leading to corrosion of the reinforcement. The products of corrosion, rust, will absorb water and occupy a greater volume than the original steel. The forces generated by this expansive process can far exceed the tensile strength of the concrete, resulting in cracking and spalling of concrete cover. Not to mention, the corrosion rate of steel reinforcement will be accelerated since the induced cracks allow more moisture, carbon dioxide or chloride getting in easily rather than just penetrating the concrete matrix.

Corrosion can also cause structure distress resulting from the loss of both concrete and reinforcement section area and consequent loss of load capacity (Broomfield 1997). Such damage can be observed usually in critical locations such as parts of the structure where humidity is more readily maintained, or at the base of columns in contact with the soil where there is a greater tendency for accumulation of salts due to capillary rise. If repair actions are not taken at this early stage, the corrosion of the steel will proceed further, causing serve damage through delamination and spalling, as well as exposure of steel and reduction of its cross-section to an extent which may become a safety hazard. In some structures, such as bridges and marine structures, such damage can be quite critical.

2.2 Prestressed concrete

As with steel reinforcing bars, prestressing steel must be protected form attack by moisture permeating the surrounding concrete. Indeed, protection against corrosion of prestressing steels is more critical than in the case of non-prestressed steel. Such precaution is necessary since the strength of the prestressed concrete component is a function of the prestressing force, which in turn is a function of the prestressing tendon area. Reduction of the prestressing steel area due to corrosion can drastically reduce the nominal moment strength of the prestressed section, which can lead to premature failure of the structural system.

In pretensioned members this is prevented by having adequate cover surrounding the tendons and also by using concrete with a sufficiently low water/cement ratio, which is usually the case for the high-strength concretes used for prestressed concrete. In post-tensioned members, protection can be obtained by full grouting of the ducts after prestressing is completed or by greasing (Hurst 1998).

Another form of wire or strand deterioration is stress corrosion, which is characterized by the formation of microscopic cracks in the steel which lead to brittleness and failure. This type of reduction in strength can occur only under very high stress and though infrequent is difficult to prevent (Nawy 2006).

3 REINFORCEMENT CORROSION MECHANISM

The reinforcing steel in concrete will note corrode unless chemicals intrude to change the normally passive conditions under the alkaline environment in the concrete. The alkaline condition leads to the formation of a "passive" layer on the steel surface. A passive layer is a dense and impenetrable film, which if fully established and maintained, prevents further corrosion of the steel. The thin layer of oxide leads to a very slow of corrosion. However, the passivating environment is not always maintained. Two processes can breakdown the passivating environment in concrete.

3.1 *Carbonation*

For the carbonation attack, carbon dioxide will react with the hydroxides in the pore solution to form carbonates, plus water. As a result of this reaction, the *pH* decreases to about 8 and steel will corrode in an environment where *pH* is lower than 8. The penetration rate of the carbonated front into the hardened concrete depends upon the partial pressure of CO_2, concrete permeability, cement type, and cement content. Increasing concrete cover depth over steel, increasing density of the concrete and using higher amounts of cement in the concrete will all help to deter incidences of carbonation caused damage.

3.2 *Chloride attack*

For the case of chloride attack, when chloride ions contact the steel surface, the passive condition is disrupted and corrosion can occur. Chloride ions from deicing salt or from sea-spray in marine environments can diffuse through the concrete to reach the steel. Chloride ions also can be mixed with concrete inadvertently as part of the water or aggregate. In some cases chloride ion is present in admixtures such as calcium chloride, which is a set-accelerator.

Steel corroded by the anodic reaction, which involves iron turning into ferrous ion. There must be a simultaneous cathodic reaction and in concrete it is the reduction of oxygen. Once the passive layer breaks downs, area of rust will start appearing in the steel surface. The chemical reaction is the same whether corrosion occurs by chloride attack or carbonation. When steel in concrete corrodes it dissolves in the pore water to give up electron: $Fe \rightarrow Fe^{2+} + 2e^-$ (Anodic Reaction).

The two electrons $(2e^-)$ created in the anodic reaction must be consumed elsewhere on the steel surface to preserve electrical neutrality. In other words, it is not possible for large amounts of electrical charge to build up at certain place on the steel; another chemical reaction must consume the electrons. This is a reaction that consumes water and oxygen: $\frac{1}{2}O_2 + H_2O + 2e^- \rightarrow 2OH^-$ (Cathodic Reaction). Oxygen must be available for the corrosion reaction to proceed. Therefore, in some instances, like under water, the diffusion of oxygen through concrete to the reinforcing steel controls the reaction.

If iron simply dissolves in the pore water (but Fe^{2+} is soluble), it would not lead to cracking and spalling of the concrete. Several stages are involved for "rust" to form. The full corrosion process will generate the unhydrated ferric oxide Fe_2O_3, which has a volume of about twice that of the steel it replaces. When it becomes hydrated it swells even more and becomes porous. The volume increase at the steel and concrete interface is two to ten times (Broomfield 1997). This leads to the cracking and spalling of concrete cover observed as the usual consequence of the steel corrosion in concrete structures.

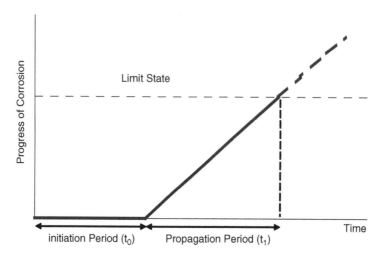

Figure 1.   Determination of service life with respect to corrosion of reinforcement (Sarja and Vesikari 1996).

4   CORROSION MODELS

The state of corrosion of steel in concrete may be expected to change as a function of time. Corrosion process has three distinct stages, namely, depassivation, propagation and final state (Schiessl 1988). Two limit states can be identified with regard to the service life as shown in Figure 1.

   Limit State 1: The service life ends when the steel is depassivated. Depassivation is the loss of oxide (passive) layer over the reinforcement, which is initially formed due to the high alkalinity of concrete. The process of depassivation takes an initiation period, t_0, which is the time from construction to the time of initiation of corrosion. Thus, the following formula for service life can be used (Sarja and Vesikari 1996):

$$t_L = t_0 \qquad (1)$$

where t_L = the service life
        t_0 = the initiation time of corrosion

   Limit State 2: This limit state is based on cracking of the concrete cover due to oxides generated during corrosion. In this case the service life includes a certain propagation period of corrosion during which the cross-sectional area of steel is progressively decreased, the bond between steel and concrete is reduced and the effective cross-sectional area of concrete is diminished due to spalling of the concrete cover. This approach is applied in cases where generalized corrosion is developing due to carbonation. During the propagation period, which begins at the moment of depassivation, the reinforcement corrosion is schematically assumed to be in a steady state as indicated by a straight line in Figure 1.

   The service life based on cracking of the concrete cover is defined as the sum of initiation time of corrosion and the time for cracking of the concrete cover to a given limit (Sarja and Vesikari 1996):

$$t_L = t_0 + t_1 \qquad (2)$$

where t_L = the service life
           t_0 = the initiation time of corrosion
           t_1 = the propagation time of corrosion

4.1   *Initiation – carbon dioxide induced corrosion*

Immediate after exposure of concrete to air, the concrete paste begins to form carbonates. These carbonates reduce the *pH* of the concrete and can damage the protective oxide film bound around the reinforcing steels. If the carbonates are found in conjunction with moisture and oxygen the reinforcing steel may corrode. The rate of carbonation is a function of the concrete quality, the relative humidity and the concentration of carbon dioxide (Cheung and Kyle 1996).

The carbonation reaction starts at the external surface and penetrates into the concrete producing a low pH front. The rate of carbonation decreases in time, as CO_2 has to diffuse through the pores of the already carbonated outer layer. The penetration in time of carbonation can be described by (Broomfield 1997):

$$d = K_c t^{1/n}$$
(3)

where d = the depth of carbonation (mm) at time t
K_c = the carbonation coefficient (mm/yrs$^{1/n}$)
t = time (yrs)
n = exponent value usually equal to 2

Often exponent n is approximately equal to 2 and, therefore, a parabolic trend can be considered. The initiation time of corrosion t_0 can be re-written and determined according to the formula as follows:

$$t_0 = \left(\frac{d}{K_c} \right)^2$$
(4)

Based on experimental data, the parameter K_c has values around 1.0 to 1.5 (mm/yrs$^{1/2}$) for good concrete, but can increase to 7.0 to 8.0 (mm/yrs$^{1/2}$) for poor concrete and industrial environment situations (Mangat 1991).

In fact, the carbonation coefficient K_c (mm/yrs$^{1/2}$) depends on a number of factors such as concrete strength, binding agents, cement content, relative humidity and temperature.

In dense and/or wet concrete, however, the reduction of the carbonation rate with time is stronger than that described by the parabolic formula, so that $n > 2$; in very impervious concrete the carbonation rate even tends to become negligible after a certain time (Bertolini et al. 2004).

To relate more precisely with the humidity, Parrott (1992) suggested that the depth of carbonation can be determined on the basis of the oxygen permeability of concrete.

$$d = \frac{64K^{0.4}t^n}{c^{0.5}}$$
(5)

where d = the depth of carbonation at time t
K = the oxygen permeability of concrete at 60% relative humidity
t = time
c = alkaline (CaO) content in the concrete cover
n = power exponent which is usually about 0.5 but decreases as the relative humidity increase above 70%.

Relevant parameters can be obtained from tables in Parrott (1992) and the initiation time for corrosion could be predicted empirically. However, since corrosion due to carbonation does not yet appear to be a major deterioration mechanism in Hong Kong, the effects of carbonation will continue to be monitored and the development and implementation of deterioration models for use with its predictive system will proceed as warranted.

4.2 *Initiation – chloride induced corrosion*

The common sources of chlorides are seawater and deicing salts. Admixed chloride is not considered here. Chloride penetration from the environment produces a profile in the concrete characterized by high chloride content near the external surface and decreasing contents at greater depths. The experience on both marine structure and highway structures exposed to de-icing salts or seawater-spray has shown that in general, these profiles can be approximately modeled one-dimensionally by means of Fick's second law:

$$\frac{\partial C}{\partial t} = D\frac{\partial^2 C}{\partial x^2}$$

(6)

where C is the concentration of the diffusing substance at a distance x from the surface at time t, and D is the diffusion coefficient of the process.

It is a non-stationary diffusion process. The equation is usually integrated under the assumptions that the concentration of the diffusing ion, measured on the surface of the concrete, is a constant in time and is equal to C_0 ($C = C_0$ for $x = 0$ and for any y), that the coefficient of diffusion D does not vary in time, that the concrete is homogeneous, so that D does not vary through the thickness of the concrete, and that it does not initially contain chlorides ($C = 0$ for $x > 0$ and $t = 0$). The solution thus obtained is (Bentur et al. 1997):

$$C(x,t) = C_o\left[1 - erf\left(\frac{x}{2\sqrt{tD}}\right)\right]$$

(7)

where
$C(x,t)$ = the chloride content at a distance x from the concrete surface at time
$\quad C_0$ = surface chloride content (% by mass of cement or concrete)
$\quad D$ = diffusion coefficient for chloride (m²/s)
$\quad t$ = time
$\quad erf$ = error function

The diffusion coefficient and the surface chloride content are calculated by fitting the experimental data and are often used to describe chloride profiles measured on real structures. By replacing the parameter to the cover thickness (c) of the concrete structure, the initiation time for corrosion is obtained from the following formula:

$$C_{th} = C_0\left[1 - erf\left(\frac{c}{2\sqrt{t_0 D}}\right)\right]$$

(8)

where
C_{th} = the critical chloride content (% by mass of cement or concrete)
C_0 = surface chloride content (% by mass of cement or concrete)
$\quad c$ = the concrete cover and
$\quad D$ = diffusion coefficient for chloride (m²/s)
$\quad t_0$ = the initiation time of corrosion
erf = error function

This formula can be simplified by using a parabola function and re-written in the following form for initiation time of corrosion t_0: (Sarja and Vesikari 1996)

$$t_0 = \frac{1}{12D}\left(\frac{c}{1 - (C_{th}/C_0)^{1/2}}\right)^2$$

(9)

Many standards require threshold value not higher than 0.4% of chloride by weight of cement for reinforced concrete and 0.2% for prestressed concrete. This corresponds approximately to 0.05–0.07 by weight of concrete (0.025–0.035 for prestressed concrete). For example, the European standard 206 (BS EN 206) restricts chloride contents to 0.2–0.4% by mass of cement for reinforced

Table 1. Typical surface chloride concentrations (Magnat 1991).

Structure	Environment	C_0,%
Bridge deck	Air zone	1.6
Bridge column	Splash zone	2.5
Bridge column	Tidal zone	5
Bridge deck	Deicing salt	1.6
Bridge column	Deicing salt	5

concrete and 0.1–0.2% for prestressed concrete (Concrete Structures Euro-Design Handbook 2004). In British Standard BS 8110, the maximum allowed chloride contents are 0.2–0.4% chloride ions by mass of cement for reinforced concrete and 0.1% for prestressed and heat-cured concrete.

Concerning values of C_0, field experience has shown this quantity to be time dependent at early ages but to tend toward a maximum after a number of years. For the sake of calculation it is usually considered constant. Normal values may be about 0.3–0.4 by weight of concretes. (Sarja and Vesikari 1996).

Typical surface chloride concentrations for bridge structures were recommended by Magnat (1991) in Table 1.

In general, the value of the diffusion coefficient for a particular concrete is considered as the rate determining parameter. The coefficient of diffusion D is roughly in the order of 10^{-7}–10^{-8} cm^2/s (Sarja and Vesikari 1996).

The value is often calculated by fitting the experimental data to equation. Equation (9) suggests that the lower the diffusion coefficient, the higher the resistance to chloride penetration.

Concrete properties, particularly water-to-cement ratio, have a direct influence on the diffusion coefficient. Rosowsky and Stewart (1998) summarize a computer-integrated knowledge system developed by Bentz, Clifton and Snyder to predict the diffusion coefficient for a given water-to-cement ratio. The least-square line of best fit of the predicted diffusion coefficients gives:

$$D = 10^{-10+4.66\frac{w}{c}} \qquad (10)$$

where D is the diffusion coefficient in cm^2/s and $\frac{w}{c}$ is the water-to-cement content ratio.

For a typical water-to-cement ratio of 0.5, for example, the diffusion coefficient is approximately 2.0×10^{-8} cm^2/s. Rosowsky and Stewart (1998) suggest that equation (10) is consistent with the statistical analysis of bridge diffusion coefficient conducted in the United States, where the estimated mean diffusion coefficient for each state ranged from 0.6×10^{-8} cm^2/s to 7.5×10^{-8} cm^2/s, with an overall mean of approximately 2.0×10^{-8} cm^2/s.

For existing structures, the water-to-cement ratio can be estimated backward from Bolomey's formula (Rosowky and Stewart 1998):

$$\frac{w}{c} = \frac{27}{\left(f'_{cy} + 13.5\right)\left(3.7\right)} \qquad (11)$$

where $\frac{w}{c}$ is the water-to-cement content ratio and f'_{cy} is the concrete compressive strength of a standard test cylinder in MPa.

However, the empirical formula developed by Rosowsky and Stewart (1998) did not take the temperature effect into account. Berke and Hicks (1992) developed the estimates of diffusion coefficient as a function of concrete mixture proportions and temperature in the Table 2. The values in the table were determined from fitting the equation (7) to actual chloride profile data.

Table 2. Estimates of effective diffusion coefficient of concrete mixture proportions and temperature (Berke and Hicks 1992).

Mixture	Effective diffusion coefficient (m²/s)			
	10°C	18°C	22°C	27°C
w/c = 0.5	5.3×10^{-12}	9.2×10^{-12}	12×10^{-12}	17×10^{-12}
w/c = 0.45	2.6×10^{-12}	4.6×10^{-12}	6×10^{-12}	8.3×10^{-12}
w/c = 0.40	1.3×10^{-12}	2.3×10^{-12}	3×10^{-12}	4.2×10^{-2}
w/c = 0.35 or 0.4 with fly ash	8.8×10^{-12}	1.5×10^{-12}	2×10^{-12}	2.8×10^{-12}

Clear (1976) proposed that the time prior to chloride induced may be empirically modeled as follows:

$$RT = \frac{129\left(\dfrac{S_i}{25.4}\right)^{1.22}}{K^{0.42}\left(\dfrac{w}{c}\right)} \tag{12}$$

where RT is the time (years) to onset of corrosion, S_i is the concrete cover (mm), K is the chloride concentration of water deposited on slab (ppm), and $\frac{w}{c}$ is the water-to-cement ratio.

4.3 Propagation

Corrosion begin when the passive film is destroyed as a result of falling pH due to carbonation or as a result of the chloride content rising above the threshold close to the reinforcement. The great need for volume cause tensile stress in concrete around the stell bar leading to cracking or spalling of the concrete cover. Propagation period may be quantified by Alonso and Andrade (1993):

$$t_1 = \frac{\Delta R_{\max}}{r} \tag{13}$$

where t_1 = the propagation time of corrosion (yrs)
  ΔR_{max} = the maximum loss of radius of the steel bar
  r = the rate of corrosion
  The propagation time leading to cracking occurs can be approximated by the following empirical equation suggested by Siemes et al. (1985).

$$t_1 = 80\frac{C}{Dr} \tag{14}$$

where C = the thickness of concrete cover (mm)
  D = the diameter of the rebar (mm)
  r = the rate of corrosion in concrete (μm/yr)
  The rate of corrosion or reinforcement can be evaluated by Siemes et al. (1985):

$$r = c_T r_0 \tag{15}$$

where c_T = the temperature coefficient
  r_0 = the rate of corrosion at $+20°C$
  Primary factors that affect the rate of corrosion in concrete at $+20°C$ are the relative humidity of air (or concrete) and the chloride content. Other factors such as the water-to-cement ration and the type of cement may also have some influence. The values of corrosion rate in anodic areas of reinforcement taken as approximate averages determined on the experimental data and the temperature coefficient determined on the basis of the findings were reported by Tuutti (1982).

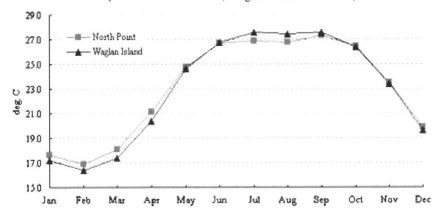

Figure 2.   Monthly mean sea surface temperature record in Hong Kong.
(Source: Hong Kong Observatory).

5   FACTORS AFFECTING THE CORROSION RATE – RELATIVE HUMIDITY AND TEMPERATURE

5.1   *Effect of temperature*

Temperature affects the corrosion rate directly. The rate of oxidation reaction is affected by the amount of heat energy available to drive the reaction. The concrete resistivity also reduces with increased temperature as ions become more mobile and chloride containing salts become more soluble. It affects the relative humidity in the concrete, lowering it when the temperature increases and therefore introducing an opposite effect.

As the low end of the temperature scale the pore water will freeze and corrosion stops as the ions can no longer move. It should be noted that this will happen well below ambient freezing point as the ions in the pore water depress the freezing point well below 0°C. However, this situation is rarely encountered in Hong Kong. The monthly sea surface temperature and the monthly air temperature in Hong Kong are shown in Figures 2–3 respectively.

5.2   *Effect of relative humidity*

Relative humidity is a factor in determining how much water in the pores enabling the corrosion reaction to be sustained. Chloride induced corrosion is believed to be a maximum when the relative humidity within concrete is around 90–95% (Tuutti 1982). For the carbonation, there is experimental evidence that the peak is around 95–100%. However, it is important to recognize that relative humidity in the pores is not simply related to atmosphere relative humidity; water splash, run off or capillary action, formation of dew, solar heat gain or other factors may intervene.

Increased water saturation will slow corrosion through oxygen starvation because once the pores are filled with water, oxygen cannot get in. Conversely, totally dry concrete cannot corrode. However, when the concrete is close to saturation the embedded steel can corrode rapidly without signs of cracking. This is due to the limited amount of oxygen available as the iron ions stay in solution without forming solid rust that expand and cracks the concrete. Neville (1995) suggested the optimum relative humidity for corrosion is 70–80%. In Figure 4, it is noted the relative humidity

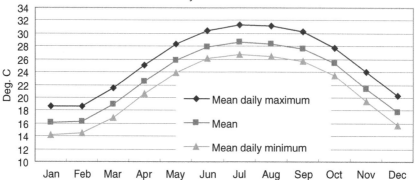

Figure 3.   Monthly air temperature record in Hong Kong.
(Source: Hong Kong Observatory).

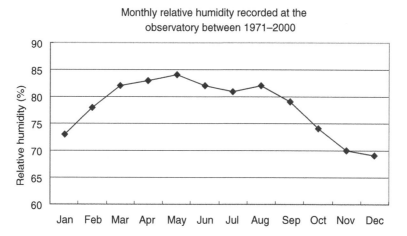

Figure 4.   Monthly relative humidity record in Hong Kong.
(Source: Hong Kong Observatory).

in Hong Kong falls within the optimum range for corrosion. The maximum and the minimum relative humidity are around 85% and 70% respectively.

5.3   *Empirical formulas application to Hong Kong conditions*

Although the aforesaid empirical equations are formulated based on the experimental and field data in Europe and U.S., it is still worthwhile to simulate the results by using existing information.

5.3.1   *Initiation – carbon dioxide induced corrosion*

As shown in Table 3, the initiation period in Hong Kong environment is estimated by applying the formula (5) and using the Parrott's experimental data and Hong Kong's humidity.

The average relative humidity of 80% is assumed around the year in Hong Kong. The cement type 4 according to the European standards is also assumed which drives the lowest CaO content in the concrete cover. Under the relative humidity 80%, parameters m, n, k_{60} and c can be obtained

Table 3. Estimated initiation period using Parrott's formula and Hong Kong's humidity.

Concrete Cover, d (m)	m	$k60$ ($10-16\,m^2$)	k ($10-16m^2$)	c (kg/m^3)	n	t_0 (yr)
50	0.564	46	25.94	240	0.415	17.6
60	0.564	46	25.94	240	0.415	27.3
70	0.564	46	25.94	240	0.415	39.5

Table 4. Estimated diffusion coefficients.

Water-to-cement Ratio, w/c	Diffusion coefficient, D (cm^2/s)
0.5	2E-08
0.6	6E-08
0.7	2E-07
0.8	5E-07
0.9	2E-06

Table 5. Estimated initiation time of corrosion of reinforced concrete under chloride attack.

	Berke and Hicks	Rosowky and Stewart				Berke and Hicks	Rosowky and Stewart
Cover, c (cm)	Diffusion coeff, $D1$ (cm^2/s)	Diffusion coeff, $D2$ (cm^2/s)	C_{th}	C_0		t_{01} (yr)	t_{02} (yr)
5	1.50E-07	2.00E-08	0.06	0.4		1.2	8.8
6	1.50E-07	2.00E-08	0.06	0.4		1.7	12.7
7	1.50E-07	2.00E-08	0.06	0.4		2.3	17.2

from Parrott's table. The value k can be calculated by the multiplication of m and k_{60}. Concrete cover thickness is randomly chosen for comparison only.

5.3.2 Initiation – chloride induced corrosion

The value of the diffusion coefficient D for a particular concrete is considered as the rate determining parameter. It is not a constant, but depends on the temperature, the nature of the diffusing substance, and the nature of material through which diffusion is occurring. For chloride ions, the effective diffusion constant may be sensitive to interactions between the diffusing substances and the hydrated cement constituting the wall of the pores through which diffusion takes place. The value is often calculated by fitting the experimental data to equation. This equation suggests that the lower the diffusion coefficient, the higher the resistance to chloride penetration.

By using formula (10), a table showing the relationship between water-to-cement ratio and diffusion coefficient is illustrated in Table 4. In contrast, under the assumptions of Hong Kong's average temperature of 25°C and water-to-cement ratio of 0.5, Berke and Hicks (1992) produced the effective diffusion coefficient with value approximately to $15 \times 10^{-12}\,m^2/s$ which is $1.5 \times 10^{-7}\,cm^2/s$.

By applying formula (9) and the different diffusion coefficients obtained from both Berke's table and Rosowky's formula, the initiation time of corrosion of reinforced concrete structures under chloride attack is estimated, as shown in Table 5. In this estimation, $C_{th} = 0.4\%$ of chloride by weight of cement for reinforced concrete. This corresponds approximately to 0.06 by weight of concrete and $C_0 = 0.4$ by weight of concrete.

Table 6. Estimated propagation period using Siemes's formula and Hong Kong's relative humidity and temperature.

Cover, C (mm)	Rebar Dia, D (mm)	Carbonated concrete		Chloride-contaminated concrete	
		r (μm/yr)	Propagation period, t_1 (yr)	r (μm/yr)	Propagation period, t_1 (yr)
50	12	3	111	78	4
	16	3	83	78	3
	24	3	56	78	2
60	12	3	133	78	5
	16	3	100	78	4
	24	3	67	78	3
70	12	3	156	78	6
	16	3	117	78	4
	24	3	78	78	3

5.3.3 Propagation period

By applying formula (14) and using Siemes' experimental data and Hong Kong's humidity and temperature, a table showing the propagation period is illustrated in Table 6. The maximum relative humidity of 85% is assumed around the year in Hong Kong. Under the relative humidity 85%, the rate of corrosion corresponding to both carbonated and chloride-contaminated concrete at +20°C obtained from Tuutti (1982) are 3 μm/yr and 78 μm/yr respectively. Cover thickness and rebar diameters are randomly chosen for comparison only.

It is found that the propagation period of corrosion process under chloride attack is much shorter than that under carbonation which means that the chloride attack in Hong Kong is more influential. One important thing should be noted is that the temperature effect has not been calibrated yet and the average temperature in Hong Kong is around 25°C which has +5°C discrepancy. However, it can still be presumed the rate of corrosion at 25°C should be faster than the value obtained by Tuutti (1982).

6 SOME EXISTING LIFE-CYCLE COST ANALYSIS MODELS

The prediction of both initial service life and the effects of maintenance and rehabilitation options present a considerable challenge. The decision making process must consider the impact of any repair and maintenance options upon the future performance of the structure. This is a very complex and indeterministic process.

Thoft-Christensen and Sorensen (1987) proposed a reliability-based methodology to optimize inspection, maintenance and repair of structure systems. Key design variables considered are inspection quality and frequency. The model minimizes total inspection and repair cost while maintaining system reliability to an acceptable level.

A service life prediction model was developed for bridge superstructures by Jiang and Sinha (1989) using the Markov chain technique and a third-order polynominal performance function. The performance function related the bridge age to an average condition rating. The model is based upon a subjective condition rating and it does not permit an enhanced assessment as a result of repair and rehabilitiation.

Mori and Ellingwood (1992) evaluated the time-dependent reliability of reinforced concrete structures. The sensitivity of structural reliability to three degradation models was evaluated. Linear, square-root and parabolic functions were used to represent corrosion, sulphate stack and diffusion-controlled deterioration mechanisms respectively. The focus of the work is upon nuclear power plants and the authors conclude that similar approaches could be used in evaluation of other civil structures where safety versus time is of concern.

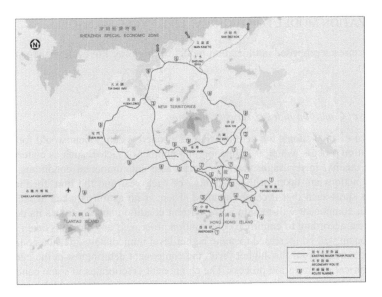

Figure 5. Major road networks in Hong Kong.
(Source: Hong Kong highways department).

Table 7. Roads and highway maintenance expenditure in Hong Kong (Mak 2005).

Year	Expenditure (HK$ million)	Approx. areas of roads and highways maintained (million sq. m) (year)
2004/05	784.8 (estimated)	22.3 (2004)
2003/04	738.8 (estimated)	22.2 (2003)
2002/03	801.1 (actual)	22.1 (2002)
2001/02	777.0 (actual)	21.9 (2001)

7 TRADITIONAL HIGHWAY MAINTENANCE IN HONG KONG

Hong Kong's roads have one of the highest vehicle densities in the world. In December 2005, there were over 540,000 licensed vehicles and only 1,955 km of roads −436 km on Hong Kong Island, 449 km in Kowloon and 1,070 km in the New Territories. Furthermore, there are 11 major road tunnels, 1,129 flyovers and bridges, 649 footbridges and 398 subways to keep people and goods on the move. Figure 5 shows the major road networks in Hong Kong.

The high vehicle density, combined with the difficult terrain and the dense building development, imposes a constant challenge to road builders to cope with the ever increasing transport demands. The traditional mode of highway maintenance is to employ contractors to carry out maintenance works while the Government staff is responsible for inspecting the public roads and highways, identifying and planning maintenance works. This mode follows a deterministic maintenance approach that is a highly labour intensive, consuming much manpower in site inspection, estimating, checking, measuring and so forth.

Hong Kong spends over HK$700 million each year to maintain its highway assets to upkeep the safety and serviceability level of the road system comprising over 1900 km of roads and the associated highway structures, roadside slopes and street furniture. The maintenance expenditure on the roads and highways network and the area of roads and highways maintained are shown in Table 5 (Mak 2005).

8 OPTIMIZATION MODEL FOR HIGHWAY LIFE CYCLE MANAGEMENT

This part discusses a proposed optimization model for highway life cycle management based on Zhang (2006).

8.1 *Change of Condition Index (CI) over life cycle*

In the absence of any actions, the deterioration of a concrete component would follow an empirical deterioration process, which is a function of a number of factors such as material properties, construction quality, load, usage, years in service and environmental conditions. Assume that management actions are taken at the beginning of each year and the time for these actions to implement is short such that it can be ignored. Then, if no action is taken at the beginning of some year, an component will remain at the current condition and continues its deterioration in the current year according to empirical deterioration process. If any action is taken, the condition of the component will be increased to some higher level immediately after the completion of the action. Then, starting from this higher level, the component deteriorates in the current year follows the empirical deterioration process. The component deteriorates to some condition level by the end of the current year (or the beginning of the following year), at which time a new round of management decision is to be made and an action taken on the component. The component either remains in the then current condition or goes to a higher level depending on what action is taken. In the following year, the component deteriorates following the previous deterioration mechanism starting from the point of condition at the beginning of the following year immediately after a management action is taken. This cycle repeats till the end of the planned long time horizon.

There is a wide range of variability and uncertainty related to the quality and property of materials, construction methods and technology, loading and usage of the facility, and the environment where the facility operates. This has a combined effect on the CI and renders the CI changing process of a component to a stochastic one, mainly including two aspects:

(1) The component may go to different conditions with varying probabilities immediately after a management action is taken; and
(2) During the current year after a management action is taken, the component may deteriorate to different conditions with varying probabilities by the end of the current year (the beginning of the following year).

8.2 *Markov decision process*

A Markov decision process consists of five aspects: decision epochs, states, actions, transition probabilities and rewards (Puterman 1994), which are discussed in the following in the context of a concrete structure.

8.2.1 *Decision epochs and periods*
Decision epochs are the points of time when decisions are made. For the management of concrete structures, assume that decisions are made at the beginning of each year of the planned time horizon of N years, and let T denote the set of decision epochs, then $T = \{1, 2, \ldots, N\}$.

8.2.2 *State and action sets*
At each decision epoch, each component occupies a state, the *CI* of the component. Let Ω denote the set of possible states for each component. Then, $\Omega = [\alpha, \beta]$, with α indicating the worst CI and β the best CI. Ω is a continuous set of states, which may be converted to a descriptive and discrete

state set Ω' as follows:

$$\Omega' = \{E, G, F, P, I\}$$

$$CI = \begin{cases} E & E_{\min} \leq CI \leq \beta \\ G & G_{\min} \leq CI < E_{\min} \\ F & F_{\min} \leq CI < G_{\min} \\ P & P_{\min} \leq CI < F_{\min} \\ I & \alpha \leq CI < P_{\min} \end{cases}$$

where $E =$ excellent; $G =$ good; $F =$ fair; $P =$ poor; and $I =$ insufficient; $E_{\min} =$ minimum numerical value of CI that belongs to category E; $G_{\min} =$ minimum numerical value of CI that belongs to category G; $F_{\min} =$ minimum numerical value of CI that belongs to category F; and $P_{\min} =$ minimum numerical value of CI that belongs to category P.

At the beginning of year t, a management action a is taken on an component that is at state $I_t \in \Omega$. Assumed that there are always four allowable actions (a_1 – replacement, a_2 – major rehabilitation, a_3 – minor rehabilitation, and a_4 – no action) no matter at what state an component is, then $A = \{a_1, a_2, a_3, a_4\}$ is the allowable set of management actions. Ω and A do not vary with time t.

8.2.3 *Rewards and Transition Probabilities*
For an component, as a result of choosing an action $a \in A$ in state $I_t \in \Omega$ at the beginning of year t,

(1) The asset manager receives a reward, R_{ta}, that is, the average performance of the component in year t;
(2) The component goes to state $J_{ta} \in \Omega$ with a probability $P_{J_{ta}I_{ta}}$ immediately after action a is taken. "No action" does not have any effect on the condition of an component, "replacement" will result in a new component with condition index of β, and "major rehabilitation" and "minor rehabilitation" will increase the condition index to a higher level depending on the current condition of the component.
(3) The component is at state $I_{t+1} \in \Omega$ with probability $Q_{I_{t+1}J_{ta}}(I_{t+1} \leq J_{ta}$ due to deterioration in year t) at the beginning of year $t + 1$ before any action is taken.

The reward R_{ta} is dependent on the current CI of the component, the effects of the management action and the projected future CI as indicated by its deterioration mechanism. This is calculated using the following equations:

$$R_{ta} = \frac{\sum_{J_{ta} \in \Omega} P_{J_{ta}I_t} \left(J_{ta} + \sum_{I_{t+1} \in \Omega} Q_{I_{t+1}J_{ta}} I_{t+1} \right)}{2}$$

$$\sum_{J_{ta} \in \Omega} P_{J_{ta}I_t} = 1$$

$$\sum_{I_{t+1} \in \Omega} Q_{I_{t+1}J_{ta}} = 1$$

For t = 1, 2, ..., N and a = 1, 2, 3 and 4

8.3 *Optimization model based on markov decision process*

8.3.1 *Methodology of the optimization model*
The performance of a highway network is dependent on the performance of all its components. The actual performance curve of a highway network can be derived from the actual performance of all these components. The objective of highway facilities management is to maximize the overall performance of the network.

Different costs are required for different management actions on different highway components. While "no action" spends no money, increasing costs are needed for "minor rehabilitation", "major rehabilitation" and "replacement." The aim of the optimization model is to maximize the overall performance of the highway network over a planning horizon by optimizing the set of annual management actions on all highway components, subject to various constraints, such as annual budget and the minimum performance requirements. This requires that the limited resources be allocated to alternative management actions cost-effectively. An integer-programming model based on the Markov decision process has been developed based on the methodology discussed above. Details of this model are presented in the following sections.

8.3.1.1 Decision variables

Y_{tIa}^c: a binary variable. $Y_{tIa}^c = 1$ if action a is taken when component c of the network is currently at a CI of I at the beginning of year t. $Y_{tIa}^c = 0$ if action a is not taken.

c: component index. $c = 1, 2, \ldots, m$, where m is the total number of components in the highway network.

I: condition index of component c at the beginning of year t before action a is taken, $I \in \Omega$.

a: an action to be taken, $a = 1$ (replacement), 2 (major rehabilitation), 3 (minor rehabilitation) and 4 (no action).

8.3.1.2 Objective Function

$$Max\ Z_t = \frac{1}{2}\sum_{c=1}^{m}\sum_{a=1}^{4}[W_c Y_{tIa}^c P_{J_{ta}I_t}^c (\sum_{J_{ta}^c \in \Omega} J_{ta}^c + \sum_{I_{t+1}^c \in \Omega} I_{t+1}^c Q_{I_{t+1}J_{ta}}^c)]$$

Z_t: Weighted overall performance level of the highway network in year t.

W_c: Weight of component c of the highway network, $\sum_{c=1}^{m_c} W_c = 1$.

J_t^c: Condition index of component c at the beginning of year t immediately after action a is taken, when the component is at condition I_t^c at the beginning of year t before any action is taken.

$P_{J_{ta}I_t}^c$: Probability of component c to go to condition J_{ta}^c immediately after action a is taken when it is in condition I_t^c at the beginning of year t, $\sum_{J_{ta}^c \in \Omega} P_{J_{ta}I_t}^c = 1$, for $a = 1, 2, 3$ and 4.

$Q_{I_{t+1}J_{ta}}^c$: Probability of component c to go to condition I_{t+1}^c at the beginning of year $t+1$ before any action is taken when it is in condition J_{ta}^c at the beginning of year t, $\sum_{I_{t+1}^c \in \Omega} Q_{I_{t+1}J_{ta}}^c = 1$, for $a = 1, 2, 3$ and 4.

8.4 Constraints

8.4.1 Network budget constraints

$$\sum_{c=1}^{m}\sum_{a=1}^{4}(C_{tIa}^c Y_{tIa}^c) \leq B_t^N$$

where C_{tIa}^c is the cost corresponding to action a when component c is at condition I in the beginning of year t; and B_t^N is the total budget available in year t.

8.4.2 Minimum acceptable performance constraints

Minimum performance levels may be required for the highway network as a whole and for individual components, depending on their relative importance, health, safety and environment requirements, and the economics.

(1) Minimum Performance Requirement for a Component

$$\frac{1}{2}\sum_{a=1}^{4}\sum_{J_{ta}^c \in \Omega} P_{J_{ta}I_t}^c (J_{ta}^c + \sum_{I_{t+1}^c \in \Omega} I_{t+1}^c Q_{I_{t+1}J_{ta}}^c)Y_{tIa}^c \geq M_c$$

for $c = 1, 2, \ldots, m$

where M_c is the required minimum level of performance allowable for component c.

(2) Minimum Performance Requirement for the Network

$$\frac{1}{2}\sum_{c=1}^{m}\sum_{a=1}^{4}\sum_{J_{ta}^c\in\Omega}W_c P_{J_{ta}^c I_t}^c (J_{ta}^c + \sum_{I_{t+1}^c\in\Omega}I_{t+1}^c Q_{I_{t+1}^c J_{ta}}^c)Y_{tla}^c \geq M_N$$

where M_N is the required minimum performance level for the network.

Only One Action Actually Taken for an Component

$$\sum_{a=1}^{4}Y_{tla}^c = 1$$

for $c = 1, 2, \ldots, m$; and $I \in \Omega$.

Binary Constraints
Y_{tla}^c is binary
for $c = 1, 2, \ldots, m$; $I \in \Omega$; and $a = 1, 2, 3, 4$.

Other Constraints
Other constraints may be added when necessary.

8.5 *Outputs of the optimization model*

The following information can be obtained based on the solutions of the optimization model discussed above.

8.5.1 *Optimal annual actions to take*

If $Y_{tla}^c = 1$, action a ($a = 1$, replacement; 2, major rehabilitation; 3, minor rehabilitation and 4, no action) is taken when component c is in condition I at the beginning of year t. If $Y_{tla}^c = 0$, action a is not taken.

8.5.2 *Optimal annual budget allocation*

Let C_t^c be the budget allocated to component c in the beginning of year t, then

$$C_t^c = \sum_{a=1}^{4}C_{tla}^c Y_{tla}^c$$

for $c = 1, 2, \ldots, m$; and $t = 1, \ldots, N$

8.5.3 *Other outputs*

Other useful information can be obtained from the proposed optimization model, including (1) the expected annual values of CI for each component; (2) the expected annual performance of the network as a whole; and (3) different scenarios and sensitivity analysis.

9 CONCLUSIONS

Literature review was conducted as a first step to understand the corrosion mechanism of reinforced/prestressed concrete structures. To develop a corrosion prediction model to carter for the Hong Kong environment, it is necessary to make appropriate adjustments for some parameters, including the surface chloride concentration, temperature and humidity. Further investigation on different empirical models under chloride attack and carbonation will be undertaken to identify the most appropriate empirical models for service life prediction. Nonetheless, an integer-programming model based on the Markov decision process has been developed, in which the performance of a highway network is measured by the condition index (CI) that is modeled by Markov chains. This mathematical model optimizes periodical management actions and consequently the allocation of limited periodical resources toward achieving the maximum overall performance of the network that is subject to various types of constraints.

The proposed optimization model requires substantial input data, which is often derived from in-service evaluation, historical data and expert opinions. Accuracy of inputs is critical to the successful application of this optimization model. The predicted deterioration process, the predicted change of the CI due to a management action, the predicted costs corresponding to alternative management actions, and the predicted transition probabilities regarding the change of the CI should be compared with the actual data obtained from in-service evaluation during the planning time horizon, and appropriate modifications and adjustments made to improve future predictions.

ACKNOWLEDGEMENT

The authors wish to acknowledge the contribution of the Research Grant Council RGC No. 610505 to the development of life cycle cost analysis and management of reinforced/prestressed concrete structures.

REFERENCES

Bentur, A., Diamond, S. and Berke, Neal S. (1997). *Steel Corrosion in Concrete- Fundamentals and Civil Engineering Practice*, 1st Edition, E&FN SPON.

Berke, N. S. and Hicks, M. C. (1992). "The life cycle of reinforced concrete decks and marine pile using laboratory diffusion and corrosion data", *Corrosion Forms and Control for Infrastructure*, ed. V. Chaker, ASTM STP 1137, Americal Society of Testing and Materials, Philadelphia, p.p. 207–31.

Bertolini, L., Elsener, B., Pedeferri, P. & Polder, R. (2004). *Corrosion of Steel in Concrete – Prevention, Diagnosis, Repair*, WILEY-VCH Verlag GmbH and Co. KGaA, Weinheim.

Broomfield, John P. (1997). *Corrosion of Steel in Concrete – Understanding, investigation and repair*, 1st Edition, E&FN SPON.

BS EN 206-1:2000 (2000). *Concrete. Specification, performance, production and conformity*, British-Adopted European Standard.

BS 8110 (1985). *Structural Use of Concrete – Part 1: Code of Practice for Design and Construction*, British Standard Institution.

Clear, K. C. (1976). "Time to Corrosion of Reinforcing Steel in Concrete Slabs", *Volume 3, Performance After 830 Daily Salt Applications*, Federal Highway Administration Offices of Research & Development, FHMA-RD-76-70, Los Angeles, CA.

Cheung, M. S. and Kyle, B. R. (1996). "Service Life Prediction of Concrete Structures by Reliability Analysis", *Construction and Building Materials*, Vol. 10, No. 1, pp. 45–55, 1996.

Concrete Structures Euro-Design Handbook (2004), Karlsruhe, Josef Eibl, Ed.

Hurst, M. K. (1998). *Prestressed Concrete Design*, 2nd Edition, E & FN SPON.

Jiang, Y. and Sinha, K. C. (1989). "Bridge Service Life Prediction Model Using the Markov Chain", *Transport Res. Rec.* 1989, 1223, 24–30.

Mak, C. K. (2005). "Some Aspects of the PPP Approach to Transport Infrastructure Development in Hong Kong", *Proceeding – Conference on Public Private Partnerships – Opportunities and Challenges*.

Mangat, P. and Elgarf, M. (1991). "The Effect of Reinforcement Corrosion on the Performance of Concrete Structures", *BREU P3091 Report*, University of Aberdeen, Aberdeen, Scotland.

Mori, Y. and Ellingwood, B. R.(1992). "Reliability-based Service-life Assessment of Aging Concrete Structures", *Journal of Structural Engineering*, 1992, 119(5) 1600–1621.

Nawy, Edward G. (2006). *Prestressed Concrete - A Fundamental Approach*, 5th Edition, Pearson Prentice Hall 2006.

Neville, A. M (1995). *Properties of Concrete*, Fourth and Final Edition, Addison Wesley Longman Limited.

Parrott, P. J. (1994). *Design for Avoiding Damage Due to Carbonation-Induced Corrosion. In Durability of Concrete*, ed. V.M. Malhotra. ACI SP-145, American Concrete Institute, Detroit, 1994, pp. 293–298.

Rosowsky, D. V. and Stewart, M. G. (1998). "Structural Safety and Serviceability of Concrete Bridges Subject to Corrosion", *Journal of Infrastructure Stsystems, ASCE*, Vol 4, No. 4, December 1998.

Sarja, A. and Vesikari, E. (1996). *Durability Design of Concrete Structures*, Report of the Technical Committee 130-CSL, RILEM.

Schiessl, P. (1988). *Corrosion of steel in concrete*. Report of the Technical Committee, 60-CSC, RILEM.

Siemes A., Vrouwenvelder, A. and van den Beukel, A., "Durability of Buildings: a Reliability Analysis", *Heron*, 30(3), 3–48, 1985.

Thoft-Christensen, P. and Sorensen, J. D. (1987). "Optimal Strategy for Inspection and Repair of Structural Systems" *Civil Engineering System*, 1987, 4 (June) 94–100.

Tuutti, K. (1982). *Corrosion of Steel in Concrete*, Swedish Cement and Concrete Research Institute, Stockholm, CBI Research 4:82, 304 pp.

Zhang, X. Q. (2006). "Markov-based optimization model for building facilities management." ASCE *Journal of Construction Engineering and Management* (in press).

Life-Cycle Cost and Performance of Civil Infrastructure Systems – Cho, Frangopol & Ang (eds)
© 2007 Taylor & Francis Group, London, ISBN 978-0-415-41356-5

Reliability and optimization as design basis for systems with energy-dissipating devices

Luis Esteva
Institute of Engineering, National University of Mexico, Ciudad Universitaria, México, D.F., Mexico

ABSTRACT: An overview is presented of some previous theoretical research results for reinforced concrete frames provided with hysteretic energy dissipating devices (EDD's). Attention is focused on the methods for the evaluation of the seismic vulnerability of the systems of interest, and some results are presented about the sensitivity of the seismic reliability and performance functions to the contribution of the EDD's to the lateral strength and stiffness of multistory frames. The process of structural damage accumulation resulting from the action of sequences of seismic excitations is taken into account in the assessment of life-cycle system reliability and performance, and in the formulation of reliability and optimization criteria and methods for the establishment of structural design requirements and for the adoption of repair and maintenance strategies. Problems related to the transformation of research results into practically applicable seismic design criteria are briefly discussed. Several illustrative examples are presented.

1 INTRODUCTION

The primary role of energy-dissipating devices in structures subjected to seismic excitations is the reduction of dynamic response amplitudes of those systems and, therefore, of the levels of nonlinear behavior and damage associated with them. In the case of viscous dampers, the energy dissipation process does not produce significant damage in the dampers themselves; however, for hysteretic dampers this process may be tied to the development of repeated cycles of highly nonlinear deformations and, therefore, to low cycle fatigue effects. In the latter case, the reduction of damage accumulation in the main system is attained at the expense of concentrating damage on the energy dissipating devices (EDD's); this may create the need for a maintenance program that includes the replacement of those devices deemed to have reached excessive levels of cumulative damage. In this manner, the complex and highly expensive process of repairing a conventional structural system subjected to high levels of cyclic nonlinear response produced by a high intensity earthquake is transformed into a much simpler and less costly one, consisting in the replacement of some EDD's. The main objective of this paper is the presentation of optimization criteria and methods for the establishment of structural design requirements and for the adoption of repair and maintenance strategies, within the framework of the expected life-cycle of the system of interest.

For purposes of illustration, this presentation examines some research results for reinforced concrete frames provided with hysteretic energy dissipating devices. The former are supposed to be integrated by beams and columns with strength- and stiffness-degrading constitutive functions, while the latter are characterized by a stable hysteretic behavior, but are exposed to brittle failure resulting from a low cycle fatigue condition. Attention is focused on the following concepts:

(a) Identification of parameters to be used to describe the contribution of the EDD's to the mechanical properties of the systems of interest and study of their influence on the seismic dynamic response of those systems.

(b) Seismic vulnerability analysis of complex nonlinear systems.
(c) Influence of the EDD's on the seismic reliability and performance functions of structural systems for given earthquake intensities.
(d) Structural damage accumulation resulting from the action of sequences of seismic excitations.
(e) Seismic design criteria based on reliability and performance targets for systems with EDD's.
(f) Optimization analysis in the establishment of seismic design criteria, as well as repair and maintenance strategies for systems with EDD's.
(g) Transformation of research results into practically applicable seismic design criteria.

2 SEISMIC VULNERABILITY FUNCTIONS

A seismic vulnerability function of a structural system is an algebraic expression relating the expected cost of the damage produced by an earthquake on that system with the intensity of that event. For engineering applications, the intensity is usually measured by the ordinate of the linear displacement or pseudo-acceleration response spectrum for 0.05 percent damping for a characteristic natural period of the system, such as the value of the fundamental period calculated either with the nominal values of the masses and of the mechanical properties of the structural members or with the expected values of those variables. The expected cost of damage should include the contributions corresponding to the conditions of both survival and collapse of the system. Both the intensity and the expected cost of damage can be represented in a normalized manner, the former with respect to the initial construction cost and the latter with respect to the lateral yield displacement or to the deformation capacity of the system estimated by means of a pushover analysis:

$$\bar{\delta}(y) = \bar{\delta}(y\,S)(1 - p_F(y)) + \bar{\delta}_F\,p_F(y) \qquad (1)$$

In this equation, $\bar{\delta}(y)$ is the expected cost of damage when the system is subjected to an earthquake with intensity equal to y, $\bar{\delta}(y|S)$ is the expected cost of damage conditional to the survival of the system, $\bar{\delta}_F$ is the expected cost of collapse in case it occurs and $p_F(y)$ is the probability of collapse under the action of intensity y. In all cases, the expected costs of damage are normalized with respect to the initial construction cost, C_0. The expected costs of damage include both direct and indirect costs. They are estimated from adequate indicators of expected physical damage, calculated in terms of the dynamic response of the system (Esteva et al, 2002a–c).

According to the principles of displacement-based earthquake resistant design, the ultimate capacity (collapse) failure mode of a structural system occurs if the dynamic lateral roof displacement (response demand, D) exceeds the lateral deformation capacity (C) of the system (Moehle, 1992; Collins et al, 1996). The failure probability is calculated as the probability that D is greater than C. The deformation capacity can be handled either as a random variable with mean and dispersion estimated by means of a pushover analysis or as a value corresponding to very high damage levels, as determined from experiments or from observations during actual earthquakes. Because of the large uncertainties associated with the actual deformation capacity of a complex multi-degree-of-freedom system, which may be significantly influenced by variables such as the lateral distortion configuration of the system at the instant of impending collapse, or the strength and stiffness degradation that results from its cyclic response to the ground motion excitation, alternative approaches have been proposed to estimate the probability of ultimate capacity failure under a random ground motion time history with a specified intensity. Some of them are discussed in the following paragraphs.

Esteva and Ismael (2003) define the collapse condition in terms of a secant-stiffness reduction index $I_{SSR} = (K_0 - K)/K_0$, where K_0 is the initial tangent stiffness associated with the base-shear *vs* roof displacement curve resulting from pushover analysis and K is the secant stiffness (base shear divided by lateral roof displacement) when the lateral roof displacement reaches its maximum absolute value during the seismic response of the system. The failure condition is expressed as $I_{SSR} = 1.0$. The reliability function, expressed in terms of the safety index β (Cornell, 1969), is obtained from a sample of pairs of values of I_{SSR} and y. If the sample contains cases both with I_{SSR} smaller than or equal to 1.0, the reliability function can be obtained by means of a maximum likelihood analysis.

The Incremental Dynamic Analysis was developed by Vamvatsikos and Cornell (2002), as a method to estimate both the intensities producing collapse of a complex nonlinear system and the corresponding deformation capacities. It consists in obtaining a sample of curves, each representing the peak roof response amplitudes of the system for a given ground motion time history multiplied by different scaling factors. If the ground motion intensities are represented as ordinates and the response amplitudes are represented as abscissas, it is observed that, as the scaling factor grows, the slope of the curve decreases until it reaches an excessively low value. The point where this condition is reached is assumed to represent both the failure intensity and the maximum deformation capacity. The reliability index β for a given intensity can be readily determined from the results of a sample of curves similar to that described in the last few lines.

Esteva and Díaz-López (2006) use the results of samples of pairs of random values of the intensity, now denoted as Y, and the stiffness reduction index, I_{SSR}, to estimate means and standard deviations of $Z(u)$, the latter defined as the natural logarithm of the random intensity Y that corresponds to $I_{SSR} = u$. The values of Z in the sample that correspond to values of I_{SSR} equal to 1.0 are upper bounds of $Z_F = \ln Y_F$, where Y_F is the minimum value of Y required to produce system failure. In order to get sufficiently accurate estimates of the mean and the standard deviation of Z_F, the mentioned authors present an algorithm for the generation of samples containing large numbers of values of I_{SSR} slightly smaller than 1.0. That algorithm includes two basic steps: a) an approximate probabilistic relation between ISSR and Y is first obtained from a small sample of those variables, and b) a new sample of values of I_{SSR} is determined for a set of simulated ground motion records, with their intensities chosen in such a manner that they have high probabilities of producing damage levels in the required intervals. Conventional minimum-squares regression analysis can then be used to estimate the mean and the standard deviation of Z_F if none of the values of I_{SSR} generated in this manner are equal to 1.0 or if they are neglected; otherwise, they can be included in the estimation, through a maximum likelihood analysis similar to that proposed by Esteva and Ismael (2003).

From the preceding paragraphs, the estimation of the failure probability of a system is transformed into the evaluation of the reliability index, β. For this purpose, for an earthquake with intensity equal to y, a safety margin Z_M can be defined equal to the natural logarithm of the ratio of the system capacity to the amplitude of its response to the given intensity; it can also be defined as the natural logarithm of the ratio of the ratio Y_F/y. In both cases, the first two moments of the probability density function of $Z_M(y)$ can be obtained by Monte Carlo simulation, using the concepts described in the last few paragraphs. However, the magnitude of the computational effort required may be excessive for typical cases in the practice of earthquake engineering and can only be justified for very important systems or for those with infrequently observed configurations of mechanical properties. Otherwise, the means and standard deviations of the global and local response and capacity measures of the system have to be estimated from those obtained with the aid of simplified reference systems (SRS's), modified by suitable sets of factors that will transform them into the statistical properties of the response and capacities that would result using the detailed model of the system. These sets of factors will have to be previously obtained and made available for different types of *generic systems*, including those more frequently found in practical applications (Esteva et al, 2002a–c).

3 DAMAGE ACCUMULATION

It is well known that the long term performance of a structural system built in a seismic environment is sensitive to the process of damage accumulation; this paper deals with the cases where this process is mainly caused by the occurrence of a number of high- and moderate-intensity events during the life of the system. Theoretical estimates of the evolution of this process for a given system, both for a single earthquake or for a sequence of events with different intensities, can

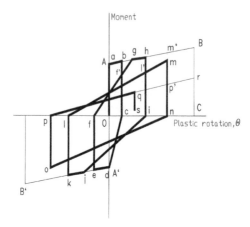

Figure 1. Constitutive function for moment *vs* plastic-rotation at flexural member ends.

be achieved by step-by-step dynamic response analysis for the event of the sequence considered. With the aid of Monte Carlo simulation, this can be done for random values of the intensities of the different events. This can always be done, whether the system is or is not subjected to any post-seismic repair or maintenance actions, which may include, for instance, repair of structural and nonstructural members and preventive replacement of energy-dissipating devices. The latter may not be necessary for viscous damping devices, but it may be advisable for hysteretic ones. Decision criteria related to actions of this type should be based on life-cycle optimization criteria (Esteva et al, 1993, 1998, 1999, 2000; Campos, 2005). The rest of this section and Section 4 are largely taken from Esteva et al (1999).

In continuous frame structures, the process of damage accumulation is usually modeled in terms of the evolution of the moment-curvature relations at the member ends. This implies the assumption that bending failure is much more likely than other failure modes (diagonal tension, axial forces) at all members; this assumption will also be adopted here, as it is consistent with the expected behavior of structures designed in accordance with the principles of modern earthquake engineering. According to this assumption, the mechanical behavior of a structural member or a critical section subjected to alternating-sign deformation cycles will be represented by constitutive force-deformation functions similar to that shown in Figure 1. The model (Esteva et al, 1999, adapted from Wang and Shah, 1987) considers that both local strengths and stiffness values in each loading direction deteriorate gradually, as functions of the damage accumulated in that direction. The model is defined by six parameters: F_y, K_1, K_2, X_F, a and α. The first three are respectively the yield strength and the tangent stiffness before and after yielding; they determine the force-deformation envelope curve. X_F is the deformation associated with the peak value of the load, while a and α serve to transform the low-cycle-fatigue index given by the summation term in the second member of Equation 2 into the damage index D (Equation 3). This index measures the degradation of the internal force corresponding to the maximum deformation amplitude reached during previous loading cycles (See Figure 1).

According to the model proposed here,

$$q = a\sum_{t=1}^{N} \frac{X_t}{X_F} \qquad (2)$$

and

$$D = 1.0 - \exp(-\alpha q) \qquad (3)$$

42

where X_i is the deformation-amplitude in the direction of interest during the i-th cycle and N is the number of cycles. These functions are used here to represent the degrading hysteretic behaviour of plastic hinges at critical sections of flexural members. On the basis of experimental evidence given by Wang and Shah (1987), a was taken equal to unity, and α equal to 0.602. According to Equation 3, the load carrying capacity of the element considered is not completely lost when $q = 1$. This is in better agreement than the original model with both the laboratory results and the behavior of full-scale systems responding to real earthquakes.

A bilinear model with stable hysteresis loops for a large number of cycles will be used to represent the cyclic behavior of energy-dissipating devices. The form and the parameters of the model adopted here are consistent with the laboratory tests performed on assemblages of U-shaped devices developed by Aguirre and Sánchez (1992). An expression of the form given by Equation 4 was fitted to their results:

$$N_F(x) = \exp(ax^{-b} - c) \qquad (4)$$

Here, $N_F(x)$ is the number of cycles to failure under a sequence of cycles with a constant amplitude equal to x(cm); $a = 128$, $b = 0.02$ and $c = 121$.

4 LIFE-CYCLE OPTIMIZATION

Consider a composite structural system made of a conventional frame and a set of energy-dissipating devices. Under the action of a moderate or high intensity earthquake, elements belonging to both subsystems may undergo significant damage. At each location, a damage increment takes place, which is a function of the number of deformation cycles and of the frequency distribution of their amplitudes, as well as of the local residual damage when the earthquake starts. Repair actions on the frame members and preventive-replacement measures on the EDD's are undertaken whenever *empirical evidence or theoretical assessment* lead to consider that current damage levels at individual elements or portions of the system may have reached critical acceptance thresholds. In general, damage on the frame members is apparent in the form of local cracking and/or crushing, while that on the EDD's may be apparent or has to be theoretically inferred on the basis of their estimated responses to the earthquakes experienced.

Let C_i be the initial cost of a system of interest, $T_i, i = 1, \ldots, \infty$, the (random) times of occurrence of earthquakes that may affect it, and $L_i, i = 1, \ldots, \infty$, the losses associated with those earthquakes; they include damage and failure consequences as well as repair and maintenance actions. The following objective function must be minimized:

$$U = C + E\left[\sum_{i=1}^{\infty} L_i e^{-\gamma T_i}\right] \qquad (5)$$

Here, E stands for *expected value* and γ is an adequate discount rate.

For simplicity, the discussion that follows refers to a single-story system with one EDD, as shown in Figure 2. Just after the occurrence of the j-th earthquake, the damage accumulated on that element equals D_{dj}, wile that affecting the structural frame is equal to D_{fj}. After the $(j+1)$-th event, these values become respectively $D_{d(j+1)} = D_{Dj} + \delta_{d(j+1)}$ and $D_{f(j+1)} = D_{fj} + \delta_{f(j+1)}$, where $\delta_{d(j+1)}$ and $\delta_{f(j+1)}$ are the corresponding damage increments. If D_{fj} exceeds a given threshold, designated here as D_{rf}, the frame is repaired in such a manner as to eliminate the damage accumulated, thus restoring its initial strength and stiffness, R_f and K_f. It is assumed that the damage level on the frame can be assessed on the basis of the evidence of physical deterioration, while that on the EDD's is inferred from the estimated value of the low-cycle-fatigue index. This information is used to implement the preventive strategy of replacing the EDD after the occurrence of a number of high-intensity earthquakes, on the basis of a threshold value D_{rd}, to be defined later.

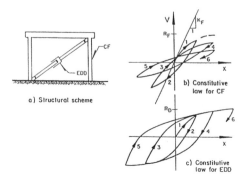

Figure 2.   Single-story frame with hysteretic energy-dissipating elements.

Whether the process of occurrence of earthquake ground motions with different characteristics considers some kind of correlation with previous history or ignores it, the levels of damage accumulated D_{fj} and $D_{dj}, j = 1, \ldots, \infty$, at the end of the j-th earthquake occur as events of a Markov process. The transition probabilities from (D_{fj}, D_{dj}) to $(D_{f(j+1)}, D_{d(j+1)})$ are obtained from the probability density functions of $\delta_{f(j+1)}$ and $\delta_{d(j+1)}$, which depend on D_{fj}, D_{dj} and on the probability density function of Y_{j+1}, the intensity of the $(j+1)$-th event.

In order to determine the conditional probability density functions of $D_{f(j+1)}$ and $D_{d(j+1)}$, given the values corresponding to the end of the j-th earthquake, it is necessary both, to calculate the joint probability density function of the waiting time to the $(j+1)$-th earthquake and its intensity, and to determine the damage states D'_{fi} and D'_{di} of the system's components after carrying out the operations of repairing the conventional frame members and/or replacing the EDD's. The conditional probability functions obtained in this manner are integrated recursively in order to obtain the marginal probability distributions of all D_{fj} and D_{dj}. Details are presented by Esteva and Díaz (1993).

The amount of computational work needed to perform the recursive integrations mentioned in the foregoing paragraph is very large, which makes those operations extremely cumbersome and costly. Monte Carlo simulation offers a feasible alternative, which consists in making detailed simulations of structural response for each earthquake, starting with an initial damage resulting from previous earthquakes, modified by the repair and replacement operations that may have taken place, and computing the final damage on the basis of the detailed structural response.

5   RELIABILITY- AND PERFORMANCE-BASED SEISMIC DESIGN CRITERIA

Hysteretic energy dissipating devices contribute to both the strength and the stiffness of a structural system. As a consequence, accounting for their influence on dynamic response cannot simply be done by the introduction of an additional "equivalent" viscous damping. Their contribution to reduce the peak ductility demands on a conventional frame where they are installed arises mainly from the fact that, contrary to typical conventional frames, their strength and stiffness do not suffer significant degradation when subjected to high-amplitude (and therefore, high ductility) cyclic distortions. Performance-based design criteria should therefore be formulated taking as control variables the peak values of the story ductility demands. For a single-story system, the relevant parameters describing its global mechanical properties can be represented as shown in Figure 3, where δ_{yc} and δ_{yd} are respectively the yield displacements of the conventional frame and of the energy-dissipating systems, δ_m is the peak value of the relative story displacement and δ_y is taken as an equivalent yield displacement for the combined system.

Figures 4 and 5 represent the expected values of the peak story ductility demands on two sets of single-story systems with mechanical properties idealized as shown in Figure 3. The seismic excitation was a set of simulated ground motion records with statistical properties similar to those

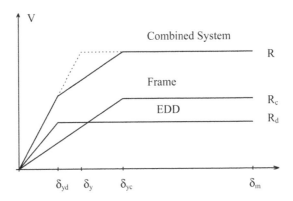

Figure 3.   Base shear *vs* roof displacement for the combined system and for each of its components.

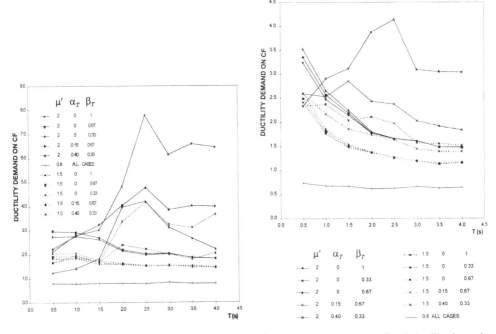

Figure 4.   Expected values of peak ductility demands for conventional single-story system.

Figure 5.   Expected values of peak ductility demands for system with EDD's; $\alpha K_d / K_c = 0.5$, $\delta_{yc} / \delta_{yd} = 2.0$.

recorded at the SCT soft soil site in Mexico during the destructive earthquake of 19 September 1985. Figure 6 shows the expected values of the pseudo acceleration nonlinear response spectra for elasto-plastic systems with viscous damping equal to 0.05 of critical. In all cases, the behavior of the EDD's was represented by a non-degrading hysteretic elasto-plastic constitutive function, while that of the conventional frame was represented by Takeda's hysteretic model (Takeda et al, 1970). Figure 4 corresponds to conventional frames; Figure 5 to systems with EDD's such that the ratios K_d / K_c and $\delta_{yc} / \delta_{yd}$ are respectively equal to 0.5 and 2.0, where K_d and K_d are values of the lateral story stiffness and the other variables were defined above. Each figure contains several curves, corresponding to different assumptions about the target ductility demands of the combined system and about the parameters of Takeda's model.

A comparison of Figs. 4 and 5 shows that, for the two values of the target ductility μ' that were studied, the ductility demands on the CF are largest for $\alpha_T = 0.4$ and $\beta_T = 0.33$. Here, α_T and β_T

45

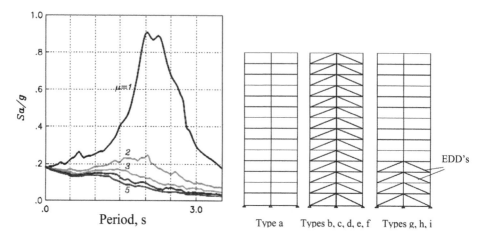

Figure 6. Expected values of elasto-plastic narow-band pseudo-acceleration spectra.

Figure 7. Systems considered in parametric study (Campos, 2005).

are the parameters of Takeda's model, such that $\alpha_T = 0$ means that the stiffness of the unloading branch is not reduced, while $\beta_T = 1.0$ corresponds to the case where the re-loading branch reaches the original force-deflection curve at the yield-point for the first load cycle. For $\mu' = 2$, the use of EDD's reduces in 44 percent the maximum ductility demand, which occurs for a natural period equal to 2.5s. Also, for both values of μ' the use of EDD's causes an increase of the response for systems with $\alpha_T = 0$, regardless of the value of β_T, for natural periods shorter than 1.5s. Thus, the change in dynamic properties resulting from the use of EDD's is not always beneficial to the expected behavior of a structure. More systematic studies are necessary to clarify this point.

According to the last few lines, the seismic performance of a system was expressed in terms of the expected value of its peak ductility demand for a given earthquake intensity. However, according to a more rigorous approach, uncertainties about response amplitudes for a given intensity, as well as about future earthquake intensities, should also be taken into account. A formulation to obtain probabilistic estimates of nonlinear response amplitudes with the aid of simplified reference systems (SRS) has been presented by Esteva et al (2002a). It is based on the determination of expected values and dispersion measures of the random factors transforming the expected values of ductility demands estimated by means of the simplified model into the corresponding probabilistic measures for the response of a detailed system model. Life-cycle indicators of performance can be expressed in terms of expected failure rates or expected damage rates per unit time (Esteva and Ruiz, 1989; Esteva et al, 2000c). Parametric studies about these concepts have only been developed for conventional building frame systems; very little has been done for combined systems including EDD's.

6   ILLUSTRATIVE EXAMPLES

6.1   *Influence of spatial distribution of EDD's on ductility demands and expected damage in multistory frames*

Esteva et al (1998) and Campos (2005) present the results of a study about the nonlinear dynamic responses of several fourteen-story reinforced concrete multistory frames subjected to a set of simulated ground motion time histories as described in Section 5 and the pseudo-acceleration spectra shown in Figure 6. These spectra were also taken as the basis for seismic design of the systems studied, adopting for that purpose the safety factors (load amplification and strength reduction) specified in the 1993 issue of the Technical Norms for Seismic Design of Mexico

Table 1. Cases considered in parametric study.

Type	Stories with EDD's	Period (s)	k_d/k	δ_{yd}/δ_{yc}	μ	Safety factor
a	None	1.46	–		4	1.0
b	All	1.48	0.25	0.50	4	1.0
c	All	1.49	0.25	0.50	5	1.0
d	All	1.48	0.50	0.50	4	1.0
e	All	1.55	0.50	0.50	5	1.0
f	All	1.44	0.50	1.00	5	1.0
g	1–4	1.41	0.50	1.00	5	1.0
h	1–4	1.41	0.50	1.00	5	1.1
i	1–4	1.41	0.50	1.00	5	1.2

City Building Code. The relations between the nominal values of loads and mechanical properties (strength, stiffness) and their corresponding expected values and uncertainty measures were taken equal to those implicitly assumed in the formulation of the Norms; these relations have been reported by Esteva and Ruiz (1989).

The general properties of the frames studied are summarized in Figure 7 and Table 1. One of them (case a) was a conventional frame, while the other eight included diagonal members with EDD's as elements contributing to their lateral strength and stiffness. In five of the latter cases (frames b, c, d, e, f) the ratios $r_k = k_d/k$ and $\lambda = \delta_{yd}/\delta_{yc}$ (where k is the lateral story stiffness and k_d the contribution of the corresponding EDD's; δ_{yc} and δ_{yd} are the lateral story yield displacements of the EDD's) are constant along the building height, while the other three (g, h, i) contained energy dissipators only in the lowest three stories. As shown in Table 1, systems a, b and d were designed assuming a ductility factor of 4; a value of 5 was assumed in all other cases. Systems g, h and i were conceived with the aim of concentrating energy dissipation at the lowest stories, trying to reduce ductility demands in the upper portion of the system. For this purpose, additional load factors as shown in the last column in Table 1 were applied at the upper stories.

Step-by-step dynamic response studies were performed for systems just described under a sample of seismic excitations belonging to the set used to generate the response spectra shown in Figure 6. Uncertainties about their actual gravitational loads and mechanical properties were taken into account by Monte Carlo simulation. The results are summarized in Figures 8 and 9 which show, respectively, expected values of the story damage index $(D_k = (K_0 - K)/K_0)$ and ductility demand (μ). Some of the results confirm our expectations about the beneficial influence of the EDD's on the control of the ductility demands; others openly contradict engineering intuition. Examples of the former are the near coincidence of curves for systems a and c, or b and e; examples of the latter are the large ductility demands on the upper stories of systems g, h and i, as compared with those observed at the lowest stories. Similar apparent inconsistencies have been observed in systematic studies about the dynamic nonlinear response of soft-first-story building frames (Esteva, 1992). They have been ascribed to the dynamic interactions that may occur during the response of nonlinear complex systems.

6.2 Reliability functions of multistory systems with EDD's

The results presented in this section have been taken from unpublished research results (Rangel, 2006) covering nine multistory frames, with heights of five, ten and fifteen stories, respectively. Three of those systems are conventional reinforced concrete frames; the other six are provided with energy dissipating devices designed to resist 0.25 and 0.50 of the lateral forces representing the seismic excitation on the whole composite system. In the following, each system is identified by the number of stories and of the percentage p_d of the base shear and lateral stiffness taken by the conventional frame (100, 75 or 50).

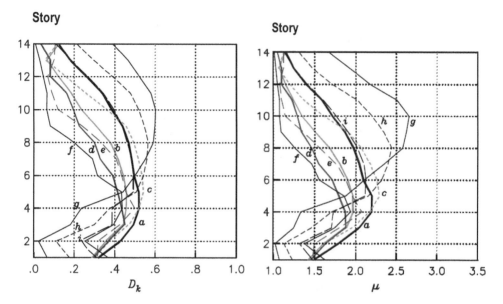

Figure 8. Expected values of stiffness-degradation index for systems in Figure 7.

Figure 9. Expected values of story ductility demands for systems in Figure 7.

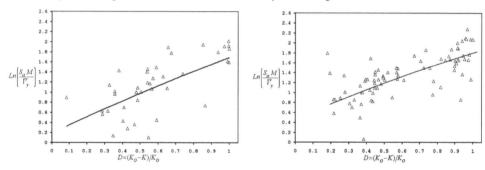

Figure 10. Normalized intensity vs I_{SSR}, $N = 20$, $p_d = 0$.

Figure 11. Normalized intensity vs I_{SSR}, $N = 20$, $p_d = 25$.

Figures 10–12 show values of the normalized intensity $Z = S_a M / \bar{V}_y$ vs the stiffness reduction index I_{SSR} for the three systems with values of p_d equal to 0, 25 and 50 percent. These points were used to estimate the mean and the standard deviation of Z for $I_{SSR} = 1.0$, which were later used to obtain the reliability functions $\beta(Z)$ depicted in Figure 13. The influence of the EDD's is evident, raising the reliability levels for a wide range of values of the normalized intensity. Slightly different results are obtained for other building heights.

6.3   Performance-based rehabilitation of a building, using EDD's

Esteva et al (2005) present a study for the performance-based rehabilitation of a reinforced concrete multistory building assumed to be located at a soft soil site in the Valley of Mexico. It was also assumed that the structure was originally designed for office use, in accordance with the Technical Norms for Seismic Design of the Mexico City Building Code (TNSD-MCBC), but a change in its use made it necessary to multiply by 1.5 the intensity of the design earthquake. A system of EDD's was added in order to make the system comply with the new strength and stiffness requirements derived from the new design intensity. Performance-based design criteria were expressed in terms

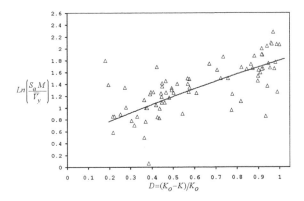

Figure 12. Normalized intensity *vs* I$_{SSR}$, $N = 20$, $p_d = 50$.

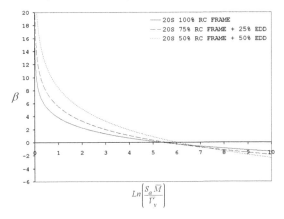

Figure 13. Reliability function in terms of normalized intensity, $N = 20$.

of reliability targets expressed in terms of expected values of the annual expected rates of exceeding the serviceability and ultimate failure limit states. Target values for these rates were established by calibration with the values implicit in the Technical Norms mentioned above for similar types of building structures and intended use.

Figure 14 shows a seismic hazard curve for the soft soil site considered, where the ground motion intensity is expressed in terms of spectral accelerations of linear response for a natural period equal to 1.03s, which corresponds to the system to be rehabilitated. By calibration with the current version of the TNSD-MCBC, it was concluded that for reinforced concrete multistory buildings with the new type of use an annual expected rate approximately equal to 0.0145 is accepted for the occurrence of story distortions in excess of 0.004. For the purpose of this illustration, this failure rate was taken as the reliability target for the serviceability limit performance condition. Under the action of the serviceability-condition design intensity, the peak values of the story distortions Ψ_i ranged between 0.00095 and 0.00819. Because the system did not comply with the codified requirements for the serviceability condition, it was decided to retrofit it, adding hysteretic energy dissipating elements, at least at those stories where the allowable value of the peak distortion was exceeded. A preliminary design was established in order to initiate an iterative process leading to the target failure rate. Denoting by Ψ_i^* and Ψ_{tol} the calculated and the tolerable values of the peak story distortion under the action of the design intensity, the stiffness K_{di} of the EDD's that should be placed at the *i-th* story, can be calculated as follows:

$$K_{di} = \alpha K_{ci} = K_{ci} \frac{\Psi_i^* - \Psi_{tol}}{\Psi_{tol}} \tag{6}$$

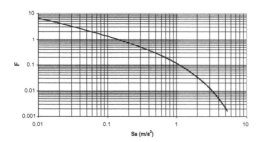

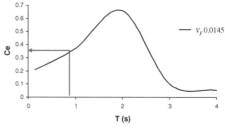

Figure 15.   Reliability spectrum for $v_F = 0.0145$,
$\alpha = 0.26$, $\gamma = 1.0$, $\mu_a = 1.0$.

Figure 14.   Seismic hazard at SCT site, $T_0 = 1.03$s.

The average value of of α resulting from applying this equation was equal to 0.36 for the stories that required the addition of EDD's (Rivera, 2006). Computed along the building height, this average became 0.22, which was initially adopted to characterize the simplified reference system (SRS) used to obtain approximate estimates of the rate of occurrence of story distortions greater than 0.004. This system consisted in the superposition of a single-story frame, representing the conventional sub-system of the detailed model, and a hysteretic element, representing the energy dissipating elements. As a result of adding the EDD's, the fundamental period of the system decreased from 1.03 to 0.89s. The value of α for the SRS was later increased to 0.26, in order to maintain the global distortion of the SRS within the range of allowable values. The value of C_e corresponding to the newly determined values of α and of the fundamental period is equal to 0.35. Figure 15 shows the response spectrum corresponding to an annual rate of being exceeded equal to 0.0145, for systems with $\alpha = 0.26$, $\gamma = 1.0$, $\mu_a = 1.0$. Here, $\gamma = \delta_{yd}/\delta_{yc}$.

The peak story distortion is related as follows to the expected value of the ordinate of the nonlinear response spectrum for the specified intensity:

$$\Psi_i = \rho_i \varepsilon_i \rho \gamma_D \frac{S_d(T, \alpha, \gamma)}{H} \tag{7}$$

In this equation, both Ψ_i and ρ_i are random, ε_i is a deterministic shape parameter derived from a pushover analysis of the combined system and $\bar{S}_d(T, \alpha, \gamma) = \bar{S}_a (T, \alpha, \gamma)(T/2\pi)^2$, where $\bar{S}_a$ is the expected value of the ordinate of the pseudo-acceleration response spectrum for the design intensity. This variable is represented in the vertical axis in Figure 15. For the return interval considered, $\bar{S}_d = 6.58$ cm, which leads to $\Psi = 6.58/3000 = 0.0022$; this value corresponds to the lateral distortion of the SRS for an annual rate of being exceeded equal to 0.0145. The rate of occurrence of story distortions Ψ_i for the detailed model can be calculated by recognizing the uncertainty in Ψ_i in the calculation of the failure rate v_F for any given tolerable value of the story distortion. In Figure 16, a comparison is presented of the values of v_F for different rates of being exceeded, both for the SRS and for the detailed model, including the influence of the uncertainties in Ψ_i mentioned in relation with Equation 7. The ratio of the value of Ψ_i derived from the study of the detailed model and that corresponding to the SRS for the same return interval can be taken as a correction factor C_Ψ, which is a function of the mentioned return interval. It can be observed that the story distortion in the detailed model, for an annual rate of being exceeded equal to 0.0145, is equal to 0.0039, which is lower than the target value of 0.004. Therefore, the retrofitted system satisfies the serviceability limit performance requirements. Details about the revision of the ultimate failure limit state requirements have been presented by Rivera and Ruiz (2006).

6.4   Life-cycle present values of expected costs

Campos (2005) made systematic studies about the values adopted by the utility function U defined in Equation 5. For this purpose, he considered the same structural systems summarized in Table 1

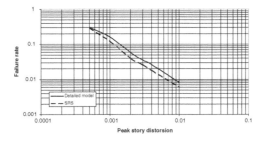

Figure 16. Failure rates for the serviceability performance limit, estimated with the detailed model (step-by-step) and with the aid of the SRS.

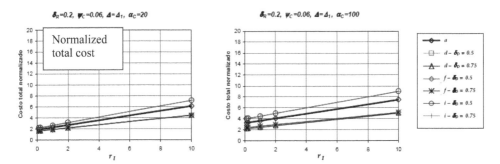

Figure 17. Expected costs of systems, normalized with respect to initial cost of system a, $\psi_C = 0.04$.

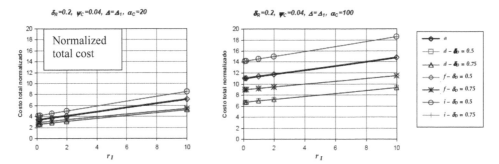

Figure 18. Expected costs of systems, normalized with respect to initial cost of system a, $\psi_C = 0.06$.

and examined the influence of the following variables, among others: Ψ_C = deformation capacity, measured as the lateral distortion corresponding to ultimate failure; r_1 = ratio of indirect to direct repair costs; α_C = expected cost of collapse divided by the initial construction cost; δ_R = threshold damage ratio for repair of a member of the conventional system; δ_D = threshold damage ratio for replacement of an EDD. Some results are presented in Figures 17 and 18. All costs are normalized with respect to the initial construction cost of system a defined in Table 1. The lowest costs correspond to case d, followed by case f. The influence of the threshold values for the replacement of EDD's on the expected costs is negligible.

6.5   *Optimum repair and maintenance policies*

The concepts introduced in Section 4 have been applied by Esteva et al [1999] to the study of the time-dependent process of damage accumulation and reliability evolution in building frames. They

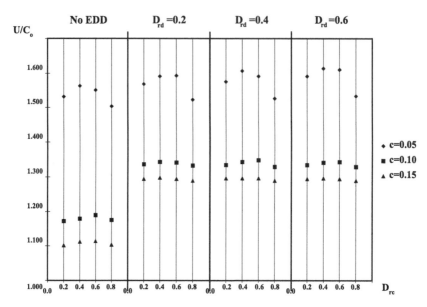

Figure 19.   Life-cycle utility values for 15-story system.

were also employed by the same authors for the study of optimum design criteria and maintenance strategies for structural frames with hysteretic energy-dissipating devices. Because of the complexity of the probability transition matrices involved, extensive use has been made of Monte Carlo simulation. One of the cases studied corresponds to a two-bay, fifteen-story frame system with hysteretic energy dissipating devices. The system was supposed to be built at a site in Mexico City where seismic hazard was represented by a Poisson process characterized by the function relating seismic intensity with annual exceedance rate. The state of damage at the end of each earthquake was measured by the maximum value attained at any story by the index I_{SSR} defined above. Searching for a life-cycle optimum solution, several options were explored regarding the seismic design coefficient c and the threshold values of I_{SSR} adopted as a condition for repair of the main frame and for replacement of the energy dissipating devices (D_{rc} and D_{rd}, respectively). Values of a negative utility function U, calculated as the sum of initial construction cost, C_0, and expected present values of future expenditures, were obtained for each option. The resulting values of U, normalized with respect to the initial construction cost for the main frame designed for gravitational loads only, are depicted in Figure 19. The first section corresponds to the plain conventional frame, while the other three sections contain energy dissipating devices that contribute 75 percent of the lateral strength and stiffness of each story. It can be observed that the negative utility function is sensitive to both the seismic design coefficient and the repair and replacement strategies. The large values that resulted for the initial and long term costs of the systems with EDD's are probably due to the excessively large values assumed for the contribution of those devices to the lateral strength and stiffness of the combined system. This fact may also be responsible for the high values shown by the optimum threshold damage level for repair of the conventional frame. Because no analysis has been made of the sensitivity of these results to the constitutive functions, the damage–response models, the repair and replacement costs, and the consequences of ultimate failure, their value is limited to their role for the purpose of illustrating the application of the proposed life-cycle optimization criteria.

7   CONCLUDING REMARKS

Modern earthquake engineering is based on the concept that optimum structural design criteria must achieve a balance between safety, initial construction costs and life-cycle expected damage. As a consequence, seismic design criteria for structural systems built at sites with significant seismic hazard levels are implicitly accepting the occurrence of moderate structural damage levels under the action of high intensity earthquakes. This generates the problem of damage accumulation during the life of a structure. Energy-dissipating devices provide means to reduce the damage levels produced by each earthquake and to concentrate damage at those devices, thus facilitating post-earthquake maintenance and repair activities. Optimum decisions concerning their use can be made with the aid of available methods and tools for the evaluation of the reliability and expected performance of structural systems in seismic environments. This includes the establishment of optimum earthquake resistant design criteria, as well of optimum repair and maintenance policies. The following are a few of the immediate technical challenges to be faced in the development of these concepts:

(a) Improving our knowledge about the constitutive functions of structural members and assemblages, as well as of energy-dissipating devices, subjected to high-amplitude cyclic excitations.
(b) Developing and calibrating simplified methods, applicable in engineering practice, for the probabilistic estimation of the dynamic response of complex nonlinear systems subjected to high intensity ground motion. Assessing the accuracy of those methods.
(c) Improving criteria and methods for the assessment of system reliability and expected performance functions in terms of earthquake intensity.
(d) Obtaining optimum values of seismic design parameters for some typical families of systems.
(e) Transforming research results into practice-oriented seismic design criteria intended to lead to pre-established reliability and performance levels.

ACKNOWLEDGMENT

The author expresses his most profound acknowledgment to Dr. Orlando Díaz-López, for his multiple contributions to the research work reported in this keynote lecture, as well as for his valuable support in the preparation of this written version.

REFERENCES

Aguirre, M. & Sánchez, R. 1992. A structural seismic damper. *ASCE Journal of Structural Engineering* **118** (5): 158–1171.

Campos, D. 2005. Diseño sísmico óptimo de edificios con disipadores de energía. (Optimum seismic design of buildings with energy dissipators) *Ph. D. Thesis, Graduate Program in Engineering, National University of Mexico*.

Collins, K.R., Wen, Y.K. & Foutch, D.A. 1996. Dual-level seismic design: a reliability-based methodology. *Earthquake Engineering and Structural Dynamics* **25** (12): 1433–1467.

Cornell, C.A. 1969. A probability based structural code. *Journal of the American Concrete Institute* **66** (12) 974–985.

Esteva, L. & Ruiz, S.E. 1989. Seismic failure rates of multistory frames of multistory frames. *ASCE Journal of Structural Engineering* **115** (2): 268–284.

Esteva, L. & Díaz, O. 1993. Optimum decisions related to design and replacement of seismic energy dissipators. *Proc. 6th International Conference on Structural Safety and Reliability, ICOSSAR 93, Innsbruck, Austria* 1: 653–660.

Esteva, L. 1992. Nonlinear seismic response of soft-first-story buildings subjected to narrow-band accelerograms. *Earthquake Spectra* **8** (3) 373–389.

Esteva, L., Díaz, O. & García, J. 1998. Practical seismic design criteria and life-cycle optimization for structures with hysteretic energy-dissipating devices. *Proc. Asia-Pacific Workshop on Seismic Design and Retrofit of Structures, Taipei, Taiwan*.

Esteva, L., Díaz-López, O. & García-Pérez, J. 1999. Life cycle optimization of structures with seismic energy-dissipating devices. *Case Studies in Optimal Design and Maintenance Planning of Civil Infrastructure Systems, American Society of Civil Engineers*. Edited by D. Frangopol.

Esteva, L. & Heredia-Zavoni, E. 2000. Health monitoring and optimum maintenance programs for structures in seismic zones. *Proc. 3rd International Workshop on Structural Control, Paris*.

Esteva, L., Díaz-López, O., García-Pérez, J. Sierra, G. & Ismael, E. 2002a. Simplified reference systems in the establishment of displacement-based seismic design criteria. *Proc. 12th European Conference on Earthquake Engineering:* paper 419.

Esteva, L., Alamilla, J. & Díaz-López, O. 2002b. Failure Models, Significant Variables and Reference Systems in the Reliability-Based Seismic Design of Multistory Buildings. *Proceedings of the 7th US National Conference on Earthquake Engineering, Boston, Mass., USA.*

Esteva, L., Díaz-López, O., García-Pérez, J. Sierra, G. & Ismael, E. 2002c. Life-cycle optimization in the establishment of performance-acceptance parameters for seismic design. *Structural Safety* **24** (2-4): 187–204.

Esteva, L. & Ismael, E. 2003. A Maximum Likelihood Approach to System Reliability with Respect to Seismic Collapse. *Proceedings of the IFIP WG7.5 Working Conference, Banff, Canada.*

Esteva, L., Ruiz, Sonia, E. & Rivera, J.L. 2005. Reliability- and performance-based seismic design of structures with energy-dissipating devices. *Proc. 9th World Seminar on Seismic Isolation, Energy Dissipation and Active Vibration Control of Structures, Kobe, Japan.*

Esteva, L. & Díaz-López, O. 2006. Seismic vulnerability functions for complex systems based on a secant-stiffness reduction index. *Proc. IFIP WG7.5 Working Conference, Kobe, Japan.*

Moehle, J.P. 1992. Displacement-based design of RC structures subjected to earthquakes. *Earthquake Spectra* **8** (3): 403–428.

Rangel, J.G. 2006. Funciones de confiabilidad sísmica de sistemas con elementos disipadores de energía histeréticos (Seismic reliability functions for systems with hysteretic energy dissipators). *M. Sc. Thesis, Graduate Program in Engineering, National University of Mexico* (in process).

Rivera, J.L. & Ruiz, S.E. 2006. Design approach based on UAFR spectra for structures with dissipating elements. *Earthquake Spectra* (accepted for publication).

Takeda, T., Sozen, M.A. & Nielsen, N.N. 1970. Reinforced concrete response to simulated earthquakes. *ASCE Journal of the Structural Division* **96** (12): 2552–2573.

Vamvatsikos, D. & Cornell, C.A. 2002. Incremental dynamic analysis. *Earthquake Engineering and Structural Dynamics* **31** (3) 491–514.

Wang, M.L. & Shah, S.P. 1987. Reinforced concrete hysteresis model based on the damage concept. *Earthquake Engineering and Structural Dynamics* **15**: 993–1003.

Life-Cycle Cost and Performance of Civil Infrastructure Systems – Cho, Frangopol & Ang (eds)
© 2007 Taylor & Francis Group, London, ISBN 978-0-415-41356-5

Bridge management system developed for the local governments in Japan

M. Kaneuji
Civil Engineering Management Division, Kajima Corporation, Tokyo, Japan

N. Yamamoto
Road Management Division, Aomori Prefectural Government, Aomori, Japan

E. Watanabe
Kyoto University & Regional Planning Institute of Osaka, Osaka, Japan

H. Furuta
Department of Informatics, Kansai University, Takatsuki, Osaka, Japan

K. Kobayashi
Graduate School of Civil Engineering, Kyoto University, Kyoto, Japan

ABSTRACT: There are over one-hundred-and-forty-thousand road bridges with a span of 15 meters or longer in Japan, and about 35% of them were built within the high growth period of 1960's and 1970's and the aging of the road bridges will be one of the problems for the road administrators in the near future. Since more than 75% of the total number of bridges are managed by the local governments, authors have challenged to develop a Bridge Management System (BMS) which can be used by many local governments in common.

One of the main purposes for the local governments to implement the BMS is to estimate the Life-Cycle Cost (LCC) to maintain the road bridges in safe and proper condition and reduce it as low as possible. In order to reduce the LCC, it is necessary to examine the deterioration process of individual bridge component, which is referred to as "Micro Management", and to establish an overall maintenance plan for all the bridges, which is referred to as "Macro Management". Since the process of establishing a long-term maintenance plan for all the bridges is very complicated, it is usual to treat Micro Management and Macro Managements differently. The authors have developed a new Bridge Management System, which treats both Micro and Macro Management in a closely related manner by utilizing the concept of "Maintenance Scenario".

The deterioration prediction is one of the key issues in the development of BMS. Since the deterioration of the bridge component is heavily dependent on the environmental condition, it is usual to establish a deterioration curve based on the statistical analysis of the inspection data. But this method cannot be applied to most of the local governments in Japan, because most of them do not have enough inspection data to execute the statistical analysis. Therefore, authors have introduced a new method of establishing deterioration curves applicable to local government which have insufficient inspection data.

The developed BMS has been applied to the actual bridge management in Aomori Prefectural Government very successfully, and authors intend to spread it among the local governments in Japan.

1 INTRODUCTION

The intensive construction of the road bridges started in 1950's after the World War II. During the high-growth period of 1960's and 1970's, a large amount of infrastructures was built to support

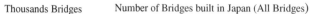

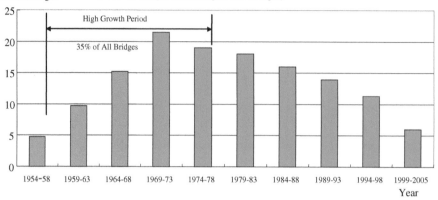

Figure 1.   Number of bridges built in Japan.

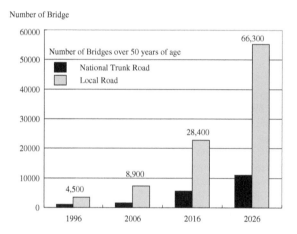

Figure 2.   Number of bridges over 50 years of age in the future.

the economic growth of Japan. Today, there are over one-hundred-and-forty-thousand road bridges with a span of 15 meters or longer, and about 35% of them were built within that period as shown in Figure 1.

According to the Report on the road facilities published by the Ministry of Land, Infrastructure and Transport, the number of bridge older than 50 years is 8,900 in 2006, but it will become 28,400 in 2016 and 66,300 in 2026 as shown in the Figure 2.

Since aging of the road bridges will be one of the problems for the road administrators in the near future, the Ministry of Land and Transportation has announced its mid-term policy to decrease the LCC of the bridge maintenance by extending the average bridge life from 60 years to over 100 years.

Since more than 75% of road bridges are managed by the local governments, authors have challenged to develop a Bridge Management System (BMS) which can be used by many local governments in common.

2   LIFE-CYCLE COST REDUCTION

The main purposes to implement the BMS are the estimation of the Life-Cycle Cost (LCC) to maintain the road bridges in safe and proper condition, and its reduction as small as possible.

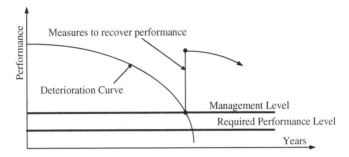

Figure 3.   Management level to secure the required performance.

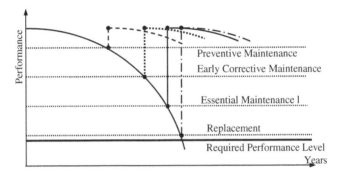

Figure 4.   Different management levels to lower LCC.

Up to the present, the maintenance policy was to keep the bridge condition just above the required level as shown in Figure 3. Unless the structural safety or the traffic condition are degraded and become close to the required level, no maintenance action was executed. Therefore, the preventive maintenance measures have not been applied, even though they are effective in reducing the LCC.

It is widely recognized that the delay of maintenance action until the bridge performance degrades to its lowest acceptable level sometimes makes the maintenance cost expensive. On the other hand, preventive maintenance measures occasionally reduce the LCC effectively. Therefore, in order to reduce the LCC, it is essential to have several choices of the maintenance measures applicable at different performance levels.

The maintenance measures can be classified into four levels as follows:

(A) Preventive maintenance Level: Cleaning, preventive methods such as painting or coating
(B) Early Corrective Maintenance Level: Execution of maintenance measures as soon as deterioration is found
(C) Essential Maintenance Level: Execution of the rehabilitations after the deterioration proceeded.
(D) Replacement Level: Replacement of the bridge elements or components after the deterioration proceeded.

The maintenance level that gives the smallest LCC is dependent on various conditions such as the health condition of the bridge components, type of deterioration, environmental conditions and so on. It is possible to choose a plan with the smallest LCC by comparing the various plans with different maintenance levels as shown in Figure 5.

3   MAINTENANCE SCENARIO

In the process of establishing a long-term management plan for all the bridges, it is inevitable to change some of the individual plans due to the budgetary restriction. The most common restriction

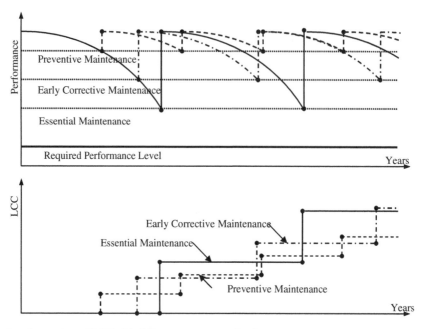

Figure 5. Comparison of LCC with different management levels.

is the uniformity of the budget for a considerable length of time, but the yearly amounts of the LCC to execute the best maintenance plans for all the bridges are not uniform but differ from year to year. Therefore, it is usual to select the bridges to execute the recommended maintenance plans based on the prioritization rule such as a cost/effect analysis, and the maintenance plans for the rest of the bridges are simply postponed.

When the execution of the maintenance plan is postponed, there is a possibility that the originally planned maintenance measures become not applicable because of the progress of deterioration. Therefore, it is necessary to have the second or third best maintenance plans to deal with these situations. A new concept of "Maintenance Scenario", which is the package of the maintenance measures for every bridge component, has been introduced to the developed BMS. The applicable scenarios are selected for all the bridge based on the evaluation of their roles in the road networks and the environmental situations before executing the budget simulations.

3.1 Establishment of maintenance scenarios

Four different management levels have been proposed in the previous section. At each management level, there are usually one or more maintenance measures for each bridge component and each type of deterioration.

It is possible to have many different combinations of maintenance measures for each bridge, but it is more practical to limit the number of combinations from the management point of view. Here, the authors have established four major combinations of maintenance measures with several sub-divisions referred to as Maintenance Scenarios.

(A) Preventive Maintenance Scenario
- Set up management level high and execute preventive maintenance measures to avoid essential maintenance or replacement works that will create traffic restriction.
- Typical management level is 5 or 4 if effective and efficient maintenance measures are available at those condition states.

58

Table 1. Maintenance Scenarios designed for Aomori Prefectural Government.

Category of Scenario	Symbols	Name used by Aomori Prefectural Government
Life Prolongation Scenarios	A1	Strategic Maintenance
	A2	LCC Minimum*
	B1	Early Action High Grade
	B2	Early Action
	C1	Corrective Maintenance (Traffic Safety)
	C2	Corrective Maintenance (Structural Safety)
	CP	Cathodic Protection (Optional Scenario)
Replacement Scenarios	RW	Whole Replacement
	RS	Super-structure Replacement
	RD	Deck Replacement

*LCC Minimum Scenario is the combination of the maintenance measures which give minimum LCC when applied to the newly constructed bridge components.

(B) Early Corrective Maintenance Scenario
 • Take appropriate maintenance actions as soon as any sign of deterioration is observed.
 • Typical management level is 4 or 3 if effective and efficient maintenance measures are available at those condition states.
(C) Essential Maintenance Scenario
 • Delay maintenance actions until the deterioration proceeds to the designated condition state on the assumption that the appropriate maintenance measures will be taken at the designated condition state.
 • Typical management level is 2 if effective and efficient maintenance measures are available at those condition states. The maintenance measures executed at this condition state are sometimes referred to as Rehabilitation.
(D) Replacement Scenario
 • Maintain bridges on the assumption that a part of or the whole bridge will be replaced according to the bridge replacement schedule.
 • There are two kinds of Partial Replacement: Deck Replacement and Superstructure Replacement.

The first three maintenance scenarios, Preventive Maintenance Scenario, Early Corrective Maintenance Scenario and Essential Maintenance Scenario, are referred to as Life Prolongation Scenarios, and each of them has two subdivisions.

For each maintenance scenario, the bridge administrator can choose one maintenance measure against a particular deterioration for every bridge component. If there is one effective maintenance measure available, the same maintenance measure can be selected for every maintenance scenario. For example, if replacement is the only one effective maintenance measure for a particular component, it will be selected as the maintenance measure for every scenario, even for the Preventive Maintenance.

The Table 1 shows the list of Maintenance Scenarios designed for Aomori Prefectural Government.

3.2 *Primary selection of the maintenance scenario*

One of the most important task of the bridge administrator in the process of Micro Management is the primary selection of the maintenance scenario.

In the BMS developed here, once applicable scenarios are selected in the process of Micro Management, they are not changed in the process of Macro Management.

The first step of the primary selection of the maintenance scenarios is the selection of the bridges to be replaced. Severely damaged bridges are the typical examples to apply Replacement Scenario, because the LCC of the replacement scenario is usually less expensive than the repetition of expensive rehabilitations. Also, aged bridge with insufficient strength or insufficient width is another example to apply replacement scenario.

The second step is to select bridges to apply only preventive maintenance scenario because of their unique feature. One of the examples is Jougakura Bridge, which is a steel arch bridge made with weather-proof steel built in the mountain area. The weather-proof steel bridge is regarded as maintenance free, but it is true only when the water is kept away from the weather-proof steel members. If there is constant water leak from the bridge deck, the weather-proof steel member may start to corrode. Therefore, the preventive maintenance scenario has been applied to Jogakura Bridge.

The third step is to select one or several scenarios from preventive maintenance, early corrective maintenance or essential maintenance scenarios for the rest of the bridges, taking various factors into account, such as bridge condition states, environmental conditions and the role of the bridges in the road network.

4 DETERIORATION PREDICTION

In order to calculate the LCC of the bridge component, it is necessary to identify the deterioration curve. The most common way to identify the deterioration curve is the statistical method based on the inspection data. But this is not applicable to most of the local governments in Japan, since they do not have enough inspection data to depend on.

4.1 *Various approaches to establish the deterioration curves*

There are several approaches to establish the deterioration curves.

The first approach is a theoretical and experimental approach. This is based on the theoretical analysis of deterioration phenomena and the experimental data. The process of analyzing the deterioration phenomena is essential and very useful, but the real deterioration process in the field is affected by many factors such as the environmental conditions or the characteristics of the material, thus it is difficult to estimate the deterioration process only from the experimental data. There is not enough theoretical and experimental data to cover every deterioration process.

The second approach is a statistical approach based on the analysis of inspection data. Since the inspection data are supposed to reflect various many conditions which have affected the speed of deterioration, it is desirable to establish deterioration curves based on the existing inspection data. But, the purpose of periodical inspection is to find severe damages or deteriorations that need prompt repair or rehabilitation, it is difficult to know when deteriorations started from the existing inspection data. Therefore, it is difficult to establish deterioration curves based on the existing inspection data.

The third approach is the one based both on the knowledge of experts and on the field data. The first step of this approach is to assume deterioration model curves based on the theoretical and experimental findings on the deterioration phenomena or on the experts' knowledge. The second step is to modify the deterioration model curves by reflecting the inspection data. For the bridge administrators that have not plenty of inspection data, the third approach is practical and useful.

4.2 *Establishing the deterioration model curves*

Since most of the local governments have not many bridge inspection data, the third approach is the most practical way to establish the deterioration curves to start the Bridge Management.

In order to assume deterioration model curves for every bridge component, the survey on the technical data regarding the deterioration curve has been conducted. The length of each condition state has been determined based on these technical data and the experts' knowledge. For example,

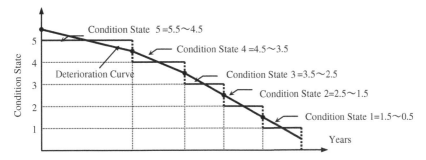

Figure 6.   Deterioration curve.

the length of the Latent Period of concrete members exposed to the salt damage environment is obtained by the Fick's second law of diffusion process, and the length of the following periods are estimated from the existing damaged examples and experts' knowledge.

In Aomori Prefecture, the first periodical inspection has been conducted in three years from 2003 to 2005. These inspection data have been analyzed and reflected to assume the length of each condition states.

Once the length of each condition state is determined, one can predict the condition state after a certain time. But, the deterioration curve with five discrete condition states is not convenient in terms of handling. By re-defining the condition state as to have a certain range, such as the condition state 4 being from 4.5 to 3.5, a deterioration curve with five linear lines is obtained as shown in Figure 6.

4.3   Modification of deterioration model curve by reflecting the inspection data

Even if the deterioration curve is derived from a large amount of inspection data, the condition state of each component defers from the established deterioration curve, due to the local environmental condition of the component.

The discrepancy from the average deterioration curve is usually treated by a probabilistic approach, but a different method was applied in the developed BMS.

When the deterioration speed of a certain bridge component is faster (or slower) than the average deterioration curve, it is quite understandable that there must be some reason for it and the deterioration of that component continues to be faster (or slower) than the average deterioration curve in the future. Therefore, the deterioration curve for that component is modified as to go through the inspected point as shown in the Figure 7. By this modification, the condition state of that element in the future will be much closer to its real deterioration.

5   LCC CALCULATION

Once the deterioration curve is obtained, and the applicable maintenance scenarios are selected, the remaining information necessary to calculate LCC are the data on the maintenance measures and the deterioration curve after the maintenance actions.

Information necessary to calculate LCC are as follows;

(1) Condition state obtained by the inspection
(2) Deterioration curve
(3) Management level indicated by the maintenance scenario
(4) Maintenance measure
(5) Cost of maintenance action
(6) Recovery of condition state after the maintenance action
(7) Deterioration curve after the maintenance action

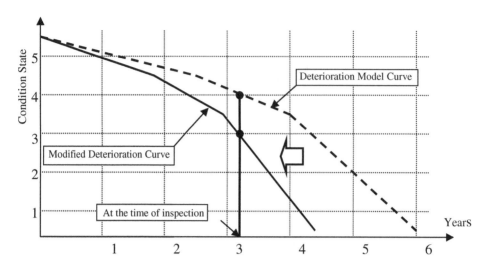

Figure 7. Modification of the deterioration curve reflecting the inspection data.

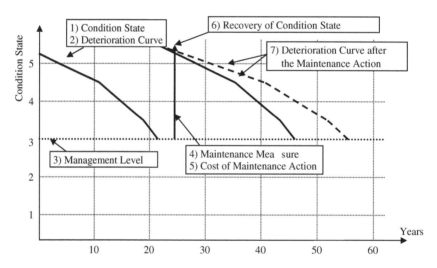

Figure 8. Information necessary for the LCC calculation.

6 BUDGET SIMULATION

After the primary selection of applicable maintenance scenarios and the LCC calculation for every maintenance scenarios are completed, the long-term budget simulation is conducted in Macro Management.

There is always demand from the financial division to reduce the long-term budget and at the same time to equalize the annual budget. Once it is agreed that the function of all the necessary road bridges need to be maintained, the remaining issue is how to establish a feasible long-term budgetary plan. It is accomplished very easily by using the budget simulation function of the developed BMS.

The first step of the budget simulation is to choose the maintenance scenario with the smallest LCC for every bridge and add them all. The second step is to calculate the uniform budget by dividing the summation of LCC by the number of years corresponding to the LCC. The third step is

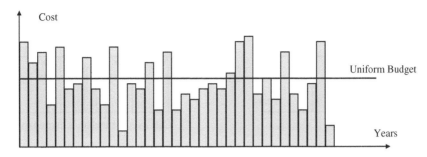

Figure 9. Uniform budget is equal to the average LCC.

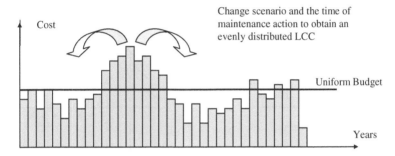

Figure 10. Changing maintenance scenario to obtain a evenly distributed LCC.

to change the maintenance scenario so that the annual amounts of LCC become evenly distributed. It is important to be aware that the change of scenario results in the increase of LCC.

6.1 *Uniform budget*

The maintenance cost of the infrastructures is necessary to keep them in proper condition to be handed over to the next generations of taxpayers and should be paid by the beneficiaries. In order to have tax fairness among generations it is desirable to maintain the infrastructures with uniform annual budgets over the life cycle of the asset.

The uniform budget can be calculated simply by dividing LCC by life cycle years, as shown in Figure 9. The social discount rate can be neglected or regarded as zero for two reasons. The first reason is that the annual budget is uniform through the life cycle period. The second reason is that the maintenance cost for a specific year in the future will be paid with the taxes of that specific year.

When the LCC is not distributed evenly but concentrated in a particular period, the peak of the LCC should be cut off to get a uniform budget plan. The peak cut can be done by changing the maintenance scenarios, as shown in Figure 10.

6.2 *Criterion to evaluate the budget equalization*

In the course of equalization of LCC, it is necessary to establish a criterion to evaluate whether the LCC is evenly distributed so that a uniform budget can cover the LCC. The first thought was to find a combination of maintenance scenario which has the smallest standard deviation of the LCC distribution. But, it was found to be wrong because the LCC distribution with smaller standard deviation has a larger LCC. Therefore, the authors had to look for a different criterion other than standard deviation.

In Figures 11 and 12, two schematic examples of LCC distribution are shown. Both cases have the same average and standard deviation. In Case-A the years with higher cost are scattered evenly,

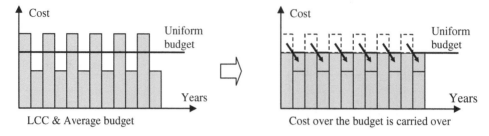

Figure 11.   Case-A: Carried over cost does not increase when LCC is evenly distributed.

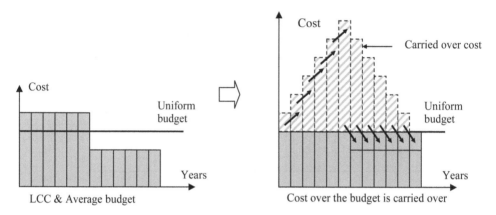

Figure 12.   Case-B: Carried over cost increases when LCC is concentrated in a particular period.

and in Case-B they are concentrated in an early period. If the annual amount of LCC is higher than the uniform budget, the excess amount of LCC is carried over to the next year. In Case-A, the following year's annual amount of LCC is smaller than the uniform budget and the carried over cost is accommodated within the following year's budget. In Case-B, the excess amount of LCC is accumulated since the following year's amount of LCC is also larger than the uniform budget. When the accumulated carried over LCC becomes equal to the uniform budget, the maintenance works scheduled has to be postponed by one year.

Therefore, it was concluded that the allowable amount of carried-over LCC should be the criterion to evaluate the budget equalization.

6.3   Multi level budget configuration

Since a scenario with the smallest LCC was selected as a priority plan, every time the maintenance scenario is changed in the process of budget simulation, the total LCC and the uniform budget increase as shown in Figure 13. If a multi level budget configuration is allowed, there is a possibility to obtain a smaller LCC as shown in Figure 14.

6.4   Examples of budget simulation

The first step of the budget simulation is the summation of the smallest LCC for every bridge as shown in Figure 15. The total LCC for 50 years is 78.7 billion yen. Since A2 Scenario is selected as the priority plan in many cases, and there are many bridge components whose condition states are lower than the management level of the corresponding scenario, a large amount of maintenance action is necessary in the first year.

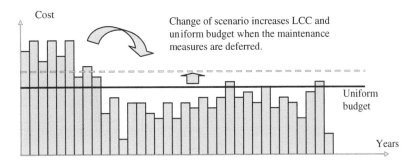

Figure 13.   Change of scenario increases uniform budget.

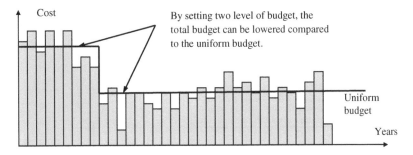

Figure 14.   Budget increase is lowered with multi level budgeting.

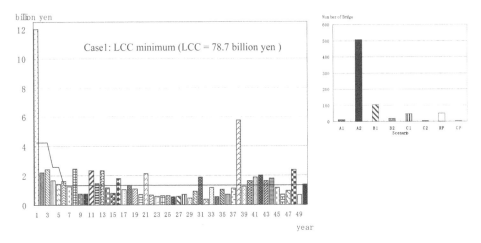

Figure 15.   LCC of the smallest scenarios and the distribution of the corresponding scenarios.

The next step is the budget simulation according to the budget equalization rule. Figure 16 shows the result of budget simulation for the smallest LCC scenarios, allowing the delay of the maintenance actions up to three years. The budget configuration is 4.25 billion yen for the first three years, 2.6 billion yen for the next two years and 1.35 billon yen for the rest with the total LCC of 78.7 billion yen.

The next step is to find a feasible uniform budget for 50 years. Figure 17 shows the result of the budget simulation for a uniform budget. The total budget necessary to maintain all the bridges for 50 years with a uniform budget is 94.7 billion yen, which is 16 billion yen larger than the smallest

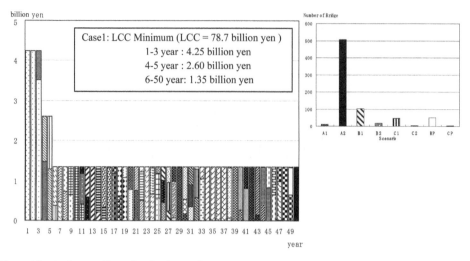

Figure 16. Budget configuration for the smallest LCC scenario.

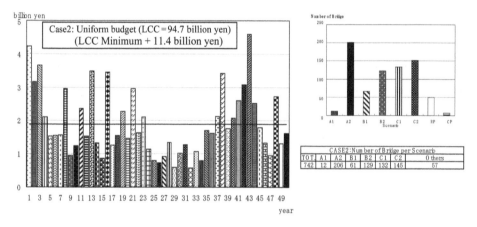

Figure 17. A budgetary plan with uniform budget configuration.

LCC scenario, and the uniform budget is 1.9 billion yen. In order to make the distribution of LCC as uniform as possible, it is necessary to change many scenarios as shown in the Figure 17.

Figure 18 shows one of the budget simulation results with two-level budget configuration. The total budget is 90.1 billion yen for 50 years, which is 11.4 billion yen larger than the smallest LCC, and the budget configuration is 2.2 billion yen for the first five years and 1.76 billion yen for the rest.

After executing many simulations, a budgetary plan with three-level budget configuration shown in Figure 19 was approved for the fiscal year 2006. The total budget for 50 years is 80.7 billion yen, which is only 2.0 billion yen larger than the smallest LCC, and the budget configuration is 3.4 billion yen for the first three years, 2.6 billion yen for the next two years and 1.45 billion yen for the rest.

7 5-YEAR BRIDGE MAINTENANCE PLAN

Once the long-term budgetary plan is established with a combination of maintenance scenarios for every bridge, the five-year maintenance plan is made with the maintenance works listed in an order of urgency as shown in Table 2.

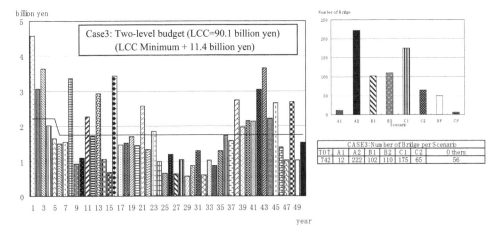

Figure 18. A budgetary plan with two-level budget configuration.

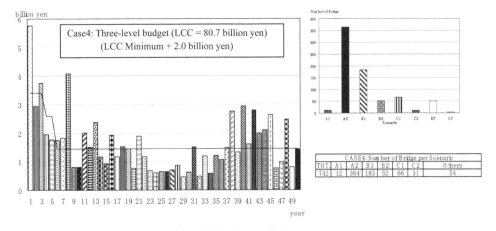

Figure 19. Budgetary plan approved for the fiscal year of 2006.

Table 2.   List of maintenance works for the next five-year bridge maintenance plan.

Number	Name	Scenario	Components	Deterioration	Maintenance measures	Cost	Fiscal year
711010045	Abc	A2	Deck	Fatigue	Steel plate reinforcing	11,680	
112800017	Cda	A2	Steel girder	Corrosion	Re-painting	49,420	
112800017	Cda	A2	Steel lateral	Corrosion	Re-painting	4,100	2006
...	...	...	...	...	...	...	
...	...	...	...	...	...	...	
...	...	...	...	...	...	...	
231310001	Bbc	B2	Concrete girder	Salt damage	Surface repair	120,920	
231310001	Bbc	B2	Con lateral beam	Salt damage	Surface repair	10,010	
720030009	Ddd	B1	Deck end	Frost damage	Surface repair	5,200	2007
...	...	...	...	...	...	...	
...	...	...	...	...	...	...	
...	...	...	...	...	...	...	

67

8 CONCLUSION

The Bridge Management System has been developed and implemented in Aomori Prefectural Government successfully. Authors continue to improve the quality of BMS and endeavor to spread it among the local governments in Japan.

ACKNOWLEDGEMENTS

We do express great appreciation to the members of the Task Committee of Osaka Bridge Management and the Aomori Bridge Management Consortium for their effort in the course of establishment of the Condition Evaluation Standards and the development of the Bridge Management System.

REFERENCES

Kaneuji, M., Asari, H., Takahashi, Y., Ohtani, H., Ukon, H. & Kobayashi, K. 2006: Development of BMS for a large number of bridges, in: Bridge Maintenance, Safety, Management, Life-Cycle Performance and Cost, in Cruz, Frangopol & Neves (eds), *Proc. of the third IABMAS conference*, Porto

Kigure, T., Ishizawa, T., Hosoi, Y., Fujii, H., Iwai, M. & Kaneuji, M. 2006: Development of the inspection support system for bridge asset management, in: Bridge Maintenance, Safety, Management, Life-Cycle Performance and Cost, in Cruz, Frangopol & Neves (eds), *Proc. of the third IABMAS conference,* Porto

Matsumura, E., Senoh, Y., Sato, M., Miyahara, Y., Kaneuji, M. & Sakano, M. 2006: Condition evaluation standards and deterioration prediction for BMS, in: Bridge Maintenance, Safety, Management, Life-Cycle Performance and Cost, in Cruz, Frangopol & Neves (eds), *Proc. of the third IABMAS conference,* Porto

Yamamoto, N., Asari, H., Ishiawa, T., Kaneuji, M. & Watanabe, E. 2006: Implementation of bridge management system in Aomori Prefectural Government, Japan, in: Bridge Maintenance, Safety, Management, Life-Cycle Performance and Cost, in Cruz, Frangopol & Neves (eds), *Proc. of the third IABMAS conference,* Porto

Technical contributions

Life-Cycle Cost and Performance of Civil Infrastructure Systems – Cho, Frangopol & Ang (eds)
© 2007 Taylor & Francis Group, London, ISBN 978-0-415-41356-5

Bridge life-cycle management in the service of the Finnish Road Administration

Matti Airaksinen
Ramboll Finland Ltd, Finland

Marja-Kaarina Söderqvist
The Finnish Road Administration, Finland

ABSTRACT: The bridge management system (BMS) development work in Finland has given a good response, the BMS being a tool for optimal maintenance planning. It includes bridge age behavior modelling and both the network level and project level management systems. The quality improving methods of the inspection system and reliable project level management system have helped to almost double the money used for bridge MR&R in ten years.

Optimal condition targets with budget scenarios steer the maintenance of bridges. The Markov Chain based LCC analysis, which is a combined analysis of performance, life cycle costs and life cycle ecology, is applied to optimization and planning of MR&R activity at both the network level and the project level life cycle management system (LMS). The development goes on. The intention is to find out the ways to improve the quality of the age behaviour models, predictions and life-cycle analysis.

1 INTRODUCTION

The project level bridge management system is a device for the use of road districts for programming of the maintaining operations for bridges. With the help of the system it is possible to draft up repair programs, condition predictions and life cycle analysis for bridges based on budget and condition targets set to the bridge stock. For this purpose deterioration models for bridges and their structural elements have been developed. Deterioration models have been created on the basis of data from general inspections registered in the Bridge Register of the Finnish Road Administration (Finnra), and using the research data from reference bridges. As the data is based on visual inspection the invisible structural elements or deterioration mechanisms are not included in the models. Deterioration models have been created using division by condition classes. The effectiveness models of measures are used to illustrate into which condition category the demanded repair measure will move the structure, when performed correctly. Division between measures has been decreased so that the models describe only basic repair and reconstruction of bridges. Cost models have been created by using cost of different measures as per Bridge Inspection Handbook. Costs mentioned in the Bridge Inspection Handbook are based on data collected in completed projects.

Deterioration and measure models and also cost models are needed for the long and short term analyses of the bridge stock. In the network level bridge management system (HIBRIS) bridges are divided into 14 partial networks according to the bridge type and main construction material, and the environmental conditions. Each partial network is described with the help of deterioration, measure and cost models. With the help of network level bridge models it is possible to review the backwardness in the maintenance and age behavior of the bridges, effectiveness of the measures and the needed funding for the future maintenance.

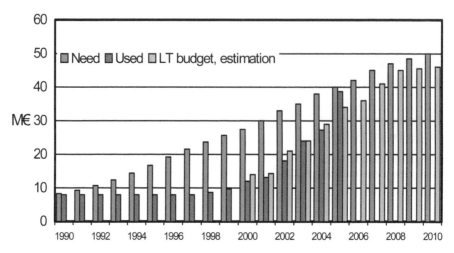

Figure 1. The growing budget needs for bridge MR&R.

1.1 *The profit control and maintenance strategy of Finnra's bridges*

The MR&R of bridges is operated with lower budget than needed in Finland today. Figure 1 shows an estimated need of funding in the near future. In the 90ths the level of yearly funding has been about 10 Million euros. In the beginning of the year 2000 the discussion of bridge MR&R and of the condition of bridges has rapidly reached the press. The common opinion has led to stricter upkeep targets and to growing amount of funding, which now has reached the level of about 40 Million euros per year. From this amount about one Million euros is used for inspections yearly.

1.2 *Bridge upkeep target measure – The Sum of The Damage Points*

An indicator for MR&R preservation and functionality goals of bridges has been developed to better describe the condition of bridges and the need of funding. The development of this Sum of The Damage Points (VPS) for an individual bridge and for the whole bridge stock of the district will be followed and reported in the Bridge Register and the Project Level Bridge Management System. This indicator VPS is a function of the estimated condition, the damage severity class and the repair urgency class of the structural part of the bridge. All these measures are given by the bridge inspector and stored in the database. An example of VPS distribution in the country is shown in Figure 2. The yearly MR&R programmes have a small effect on the VPS decrease.

2 THE BRIDGE MANAGEMENT SYSTEM (BMS)

2.1 *Bridge database*

The whole bridge management system is based on a thorough bridge inspection and condition evaluation. The damages and deterioration detected during the inspections, their severity, exact location and extent are recorded. Also, information on the effect of the damages on bridge bearing capacity, on repair urgency class and the inspector's proposals for repair measure and their costs will be described and recorded. All the information is stored in the bridge database together with bridge structural, administrative and traffic data. Also historical data and information on previous repairs and their realised costs are gathered for further research and bridge age behaviour modelling.

In addition to this Bridge Data module, the Bridge Register is composed of five other important modules like Photos, Reports, Parameters, Feedback and Users. The Photo module completes well the bridge data and is today an indispensable tool of the bridge engineer.

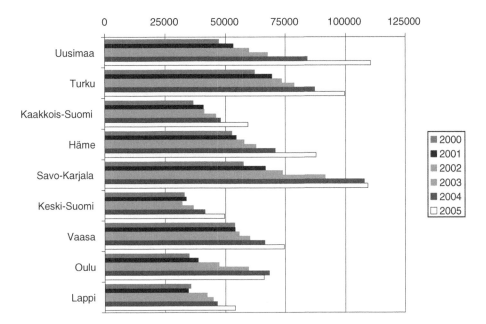

Figure 2.   Sum of the damage points of the bridge stock by road districts.

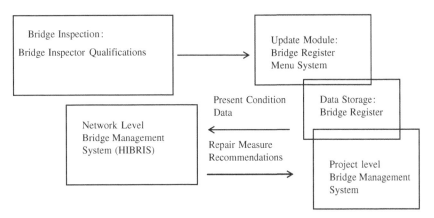

Figure 3.   The connections between the elements of the bridge management system.

There are 68 programmed reports included in the report module. They are classified in categories like: basic data, functionality, inspection, condition, and repair, load carrying capacity and quality reports. These reports as such serve the management of bridges quite well.

2.2   *Network level bridge management system (HIBRIS)*

HIBRIS is a network level management system for analysing maintenance and replacement investments at the Finnish Road Administration. The system is built for analysing paved roads, bridges, gravel roads, and road furnishing and equipment together and/or separately.

The models for network level analysing form an entity with which one section of the road property is examined. The functions of the models are:

- Condition evaluation of the inspected structure with the help of condition variables and condition classification

- Describing of age behaviour
- Describing of measures (influence on condition and costs)
- Describing of effects (in money)

The most central part of the HIBRIS program gives a possibility to do calculations based on linear optimization. The efficiency of these calculations is based on rapidly changeable optimization restrictions and on a possibility to produce and review alternative results. The principal optimization functions are:

- Long term optimization, in which the costs for the society are minimized.
- Short term optimization, in which the difference between the present condition and the future condition is minimized.

2.3 *The Project Level Bridge Management System*

The Project Level BMS, which deals primarily with individual bridges, uses the recommendations and goals from the network level to decide on the repair measures in individual repair projects to create repair and reconstruction programmes. The project level system is the key tool for everyday bridge repair planning in the road districts. The system helps the bridge engineer to plan and schedule the repair projects for individual bridges based on the recommendations and the damage data in the database.

The repair and reconstruction programmes are the central concept of the Project Level BMS. The programmes are produced for a period of six years using repair indexes to find the bridges in need of repair or reconstruction. The length of the period was chosen to correspond to the time span of Finnra's operation and economic plan.

The Project Level BMS is based on:

- Repair index (KTI)
- Rehabilitation and reconstruction index (UTI)
- Deterioration models based on Markov Chain method
- Life cycle analysis with Life Cycle Action Profiles (LCAP)
- Cost analysis

2.4 *The Life Cycle Analyses module*

The Life Cycle Analyses module was programmed for the Project Level BMS use. Condition predictions and efficiency analyses are needed when combining repair and reconstruction needs in an annual work programme. The life cycle (LC) analysis gives the bridge engineer the possibility to compare the effects of different management strategies on the remaining economical and functional life of the bridge. The purpose of this analysis is to find the most feasible and economically most effective maintenance strategy to manage the structures.

Based on the network level LCC and risk analyses, decision trees are built for automatic life cycle planning of structures. Alternative strategies over a defined time frame are compared by LCC analyses together with necessary risk analyses and the most appropriate strategies are selected. Similar decision trees both at the network level and the project level planning enable consistent and comparative results. The results seem to be promising but development and testing are still continuing.

3 DEVELOPMENT OF MODELLING

3.1 *General*

The analysis application of HIBRIS was completed in 2004. The partial network division and condition variables were verified at the same time. The first versions of deterioration, measure and cost models of bridges at the network level have been in test use. The Project level BMS has been in use since 1998 but it has lacked condition predictions and life cycle analyses, which are essential for planning of measures. This year new models were developed for analysing age behaviour, for needed measures and for needed costs both at network and project level. A division by structural elements was done for the project level system; the decision tree leading to making decisions by structural elements, deterioration models of structural elements, effectiveness models of repair activities, and estimation of repair costs. The system was developed so that it covers best possible the most common repairs and their costs. Separate models have been created for concrete, steel, wood and stone structures. The deterioration models developed in the project are based on data collected on bridges and they deviate clearly from the former deterioration models based on Deplhi questionnaires or theoretical calculation methods. Creation of models was done in the following order:

- structural division
- decision trees
- deterioration models
- effectiveness models for activities
- costs

The actual programming work and testing of the models in the BMS program will be the following development project.

3.2 Developing of project level models

3.2.1 Division per structural elements
Division of structural elements was done so that it is possible to make models for each structural element on the basis of the data in the Bridge Register or on the basis of research data. On the other hand models should be simple and general enough so that they are manageable in the system. The final structural division of the decision tree is as follows:

- Concrete structures (5 structural elements)
- Steel structures (4 structural elements)
- Wooden structures (3 structural elements)
- Stone structures (2 structural elements)

14 structural elements in total.

The number of structural elements was compressed to the smallest possible because the system will be held as simple as possible. This is why homogenous structural elements with similar common ways of repairing and reasons leading to repairs (amongst others reinforced concrete beams and prestressed beams) were combined.

3.2.2 Decision trees
The system uses decision tree for finding a repair method for different structural elements on the basis of the condition estimate, circumstances and previous repairs. The data needed by the decision tree must be such that required starting data can be found in the Bridge Register. Decision trees have been created based on the previously mentioned structural element division. With the models it is possible to get cost effective solutions which have the right effect for structural elements belonging in different condition classes.

3.2.3 Deterioration models
The deterioration models were created on the basis of data from the Bridge Register. The models were adjusted to the Markov coefficients so that deterioration according to them is slightly quicker than on the auxiliary diagram formed on the basis of collected data. Deterioration models are based on condition of different structural elements in the Bridge Register whose classification criteria differ between different structural elements. Condition estimate describes the general condition of the structural element in question on scale 0–4.

Edge beam, no anti-icing
Transition matrix

Condition	0	1	2	3	4
0	0,9	0,1	0	0	0
1	0	0,986	0,014	0	0
2	0	0	0,987	0,013	0
3	0	0	0	0,99	0,01
4	0	0	0	0	1

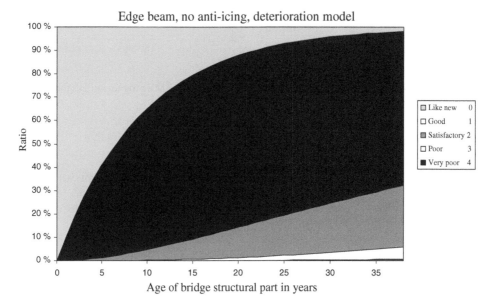

Figure 4. Transition matrix adjusted to condition estimate distribution of edge beams on roads without anti-icing salt, and condition estimate distribution by deterioration model.

3.2.4 *Effectiveness models of measures*

The effectiveness models of repair measures are used for describing into which condition estimate category the chosen repair measure moves the structural element, when performed correctly. The main principle is that only the changing of a structural element moves the structure in category 0 (equivalent to new), otherwise category 1 (good). Effectiveness models for measures are presented in the decision trees.

Some of the repair measures, like coating, are such that they change the durability of the structure compared to the original structure. In these cases deterioration in structure after the measure is similar to the deterioration model according to effectiveness model after the measure.

3.2.5 *Cost models*

Cost models have been formed using the costs for different measure as mentioned in the Bridge Inspection Handbook. Costs in the Bridge Inspection Handbook are based on costs collected from completed projects. Costs are average costs and they show the magnitude of costs. Cost models should be updated in the future with index increases and on the basis of feedback from completed projects.

Cost models were changed so that for each measure mentioned in the decision tree it is possible to give a cost using the same measure unit. E.g. regarding the costs for edge beam and railings attention has been paid to the fact that repair work is done for both sides of the bridge and the unit is the total length of the bridge in meters.

3.3   *Developing of network level models*

The network level BMS HIBRIS is meant for budgeting of the maintenance and expanding investments of the structures of road network (paved roads, bridges and gravel roads), and to analyzing the effects related to these. In HIBRIS system the bridges will be handled by bridge type. This is why bridges have been divided into 14 partial networks on the basis of bridge type and environmental impact as for example:

- Prestressed concrete bridges, roads with anti-icing salt
- Prestressed concrete bridges, roads without anti-icing salt
- Steel bridges, roads with anti-icing salt etc.

3.3.1   *Deterioration models for different bridges*

When making deterioration models for partial networks of individual bridges data was searched from the Bridge Register using condition distributions of the bridges from 1960's to 1990's . On the basis of the information it was discovered that the distribution follows very well the distribution according to the Markov chain theory. It was decided that the deterioration models would be slightly quicker models than could be assumed on the basis of the information received from the Bridge Register. The cause for this was that at least bridges built in the 1960's have been repaired in such a way that the influence on the model was not investigated.

The division of measures was changed towards a more practical direction. The possible measures are:

- maintenance and repair
- rehabilitation
- reconstruction

The division of measures was regarded sufficient because thus it is possible to optimize approximately 80% of the funding for the maintenance of bridges.

Effectiveness and cost models for rehabilitation were tested with different ordinary cases where measures were directed to different elements. Typical bridges in need for rehabilitation were searched for each condition state. On the basis of the project level models the effects of repair measures on improvement of the condition state were investigated. Almost in all cases the bridges were returned to category "good" after rehabilitation. In category "extremely bad" rehabilitation does not always return the structure to category "good". Replacing returned the bridge always in category "equivalent to new".

The effect of the condition of the bridge to the road user costs could not be developed for a network level analysing. The problem was that developing of a model for describing the road user cost caused by deterioration of the bridge did not succeed. Neither was it possible to create a reliable model at network level for the effect of deterioration on the bearing capacity of the bridge.

4   FUTURE DEVELOPMENT NEEDS

Durability properties of Finnra's bridges have been investigated in connection with special inspections, general inspections and rehabilitation. The data gathered has been filed in reports and quality manuals. The data is scattered generally in the archives of road districts and all data is not available in the Bridge Register.

By collecting the existing data and analysing it carefully it is possible to further develop the reliability of project and network level systems, and inspections. Predicting of the development of invisible deterioration must be considered in the future. Furthermore it is important to investigate methods for calculating the road user costs and the effects of the condition of the bridge on its bearing capacity.

5 CONCLUSIONS

A short description of the development of the Finnra BMS family was given. The bridge stock can be managed with the new models. The foci of the work were developing of division of structural elements, decision trees and deterioration models. These form the framework of the system and their functionality is the basis for usability of the system. In the development of the system the target has been clear and simple solutions. Thus the system can be taken into use quickly and it can be accomplished when needed. The system was made logic and it complies rather well with Finnra's policy for MR&R. The project level system can even be used for network level analyses if needed.

For network level analyses the partial network division was changed, new deterioration models were created, division of measures was changed and cost models were reviewed.

The division of measures examines merely rehabilitation and reconstruction of bridges. Repair and protections will be included in the annual maintenance costs. Rehabilitation was made possible as measure even when the bridge is in category satisfactory. With the help of network level models it is possible to form a general opinion on the condition of the bridge stock, the development of the condition, funding need for MR&R and the efficiency of the measures.

REFERENCES

Finnra, 2006. *Siltojen verkko- ja ohjelmointitason hallinnan kehittäminen (Development of the Bridge Management System in network- and project level)*, TIEH 32101003-v-06 Helsinki: Edita. (In Finnish)

Finnra, 2006. *Sillantarkastuskäsikirja (Bridge Inspection Manual, The Directives for Bridge Inspection Procedures.* English 1st edition 1989) 7th renewed edition in Finnish, ISBN 951-803-704-1, TIEH 2000020-v-06 Helsinki: Edita.

Söderqvist, M-K. 1999. *Analysis of BMS Reference Bridges in Finland.* TRB International Bridge Management Conference, Denver Colorado, April 26–28 1999.

Söderqvist, M-K. & Vesikari, E. 2003. *Generic Technical Handbook for a Predictive Life Cycle Management System of Concrete Structures (LMS).* LIFECON Deliverable 1.1, EU Project G1RD-CT-2000-00378. Brussels.

Vesikari, E. 1992. *Rakenneosaryhmien rappeutumismallit siltojen hallintajärjestelmässä (Deterioration Models of the Bridge Structural Part Groups in the BMS).* Research Report RAM805/92. The Technical Research Center of Finland VTT. Espoo (In Finnish).

Vesikari, E. 1998a. *Betonirakenteiden käyttöiän arviointi tietokonesimuloinnilla (Estimation of Service Life of Concrete Structures Using Computer Simulation).* Research Report RTE30275/98. The Technical Research Center of Finland VTT. Espoo (In Finnish).

Vesikari, E. 1998b. *Tarkkailusillaston tutkimusten tulosten analysointi (Analysis of Investigation Results of the Reference Bridge Group).* Research Report RTE30516/98. The Technical Research Center of Finland VTT. Espoo (In Finnish).

Vesikari, E. 1999. *Hanketason siltojenhallintajärjestelmän ikäkäyttäytymismallien kehittäminen (Development of Age Behaviour Models for the Project Level BMS).* Research Report RTE8/99. The Technical Research Center of Finland VTT. Espoo (In Finnish).

Vesikari, E. 2000. Siltojenhallintajärjestelmän hanketason elinkaarianalyysit (The Life Cycle Analyses of the Project Level Bridge Management System). Research Report RTE31235/00. The Technical Research Center of Finland VTT. Espoo, 2000. (In Finnish)

Life-Cycle Cost and Performance of Civil Infrastructure Systems – Cho, Frangopol & Ang (eds)
© 2007 Taylor & Francis Group, London, ISBN 978-0-415-41356-5

Criteria for a bridge management system based on inspection, monitoring and maintenance practices

Ferhat Akgül
Department of Engineering Sciences, Middle East Technical University, Ankara, Turkey

ABSTRACT: Numerous bridge management systems have been developed around the world for different bridge administrations and agencies. These systems are mostly built on common tasks and models aiming at cost optimal maintenance of a bridge stock. The main task is the prioritization and ranking of bridges in a network. However, the level of inspections, the procedures used for the evaluation of load carrying capacity and various other criteria for bridge monitoring and safety change from one administration to another. In Turkey, the bridges are managed by the General Directorate of Highways which maintains a database inventory of bridges in which numerous parameters defining the characteristics of these bridges are stored. In addition to the inventory, a visual bridge inspection manual serves as the official guide for the field inspections. It presents the condition states and their descriptions that must be assigned to bridge elements during an inspection. A comprehensive project is also underway at the Directorate for the development of a road management system based on a geographic information system. This paper presents the process of establishing the tasks and requirements for a bridge management system that needs to be tailored to the specific requirements of the Directorate. Currently, the decisions on maintenance, repair and strengthening and prioritization are made based on engineering judgment and the needs of the Regional Directorates. An initial detailed inspection program needs to be developed and implemented to provide the necessary data for the management system. Evaluation of the costs of different maintenance strategies must be determined. Deterioration, cost and optimization models are the main elements of the analysis part of any bridge management system. These models must be carefully selected to fit the needs of the administration at the management level.

1 INTRODUCTION

An OECD report in 1981 defined a bridge management system as a tool for assisting highway and bridge agencies in their choice of optimum improvements to the bridge network that are consistent with an agency's policies, long-term objectives, and budgetary constraints (OECD 1981). The optimum improvement refers to minimization of maintenance and repair costs while maintaining an adequate safety level for the bridge network. The objectives of a bridge management system still remains the same today, however the methods, tools, technological advances, and computational capabilities for developing such systems have evolved tremendously since that time.

In the last two decades, many bridge management systems have been developed around the world by various bridge administrations such as Pontis, BRIDGIT, and Danbro. The types of systems change from as simple as a single computer database to more sophisticated management tools having cost optimization capabilities and advanced functions. The use of reliability index as a safety measure for the life cycle management of bridges in new or existing BMSs has also been studied in the last decade (Frangopol *et al.* 2001, Akgül and Frangopol 2003) and the current theoretical research on time-variant condition, safety and cost of bridges has advanced considerably in recent years (Frangopol and Neves 2005).

The General Directorate of Highways in Turkey, having 5,486 bridges on government and provincial roads (i.e., excluding the highway bridges), is in need of a bridge management system that needs to be tailored to the specific requirements of the administration. A project has recently been initiated to develop such a system. An overview of the bridge management systems around the world is necessary prior to choosing the methods and approaches for such a tool.

In Europe, the BRIME (2001) project attempted to develop a framework for the management of bridges on the European road network. The project results present information that may be considered when developing the bridge management system for the General Directorate of Highways in Turkey. There is a considerable amount of overweight truck traffic over the rural and urban bridges in Turkey. In spite of the truck weighing stations, the official permit application process, regulations and penalties, the enforcement of the regulations for the overweight truck activity over the roads and bridges so far has been limited. The developed bridge management system is expected to form the first step toward the establishment of route checking procedures and guidelines based on the structural safety levels of the bridges along the given routes.

The goals of the project is to meet the needs and requirements of the General Directorate of Highways for management of its bridges considering the limited maintenance budget and resources, to establish the necessary tools for a more developed overweight truck monitoring process following the development of the BMS, and to implement, while performing these tasks, the best practices and methods used around the world in the newly developed system.

2 EXISTING BRIDGE CONDITIONS AND CURRENT PRACTICES

2.1 Existing bridge stock and its characteristics

Existing bridge stock which the General Directorate of Highways in Turkey currently maintains within its 17 Regional Districts includes 5,486 bridges on government (rural) and provincial (urban) roads. The highway bridges are excluded. As shown in Figure 1, 80% of the bridges were built in periods 1950–1970 and 1985–2005. This corresponds to 4,413 bridges. The remaining 20% were built either before 1950 or in period 1970–1985. This shows that as of 2005, almost 45% (44.3) of the existing bridges are older than 30 years and 59% are older than 20 years. In 1985–2005 period, 2246 bridges were built in Turkey. Assuming that half of these bridges were built in 1985–1995 period, the percentage of bridges less than 25 years old as of 1995 was approximately 44%, those between 25 to 45 years old constituted 50%, and those older than 45 years constituted only 6% of all the bridges. For Germany, these numbers as of 1997 were 41% (<20 years), 40% (20–40 years), and 19% (>40 years) (BRIME, 1999). The percentages for France, Norway and Slovenia were also similar. This indicates that the age of the bridge stock in Turkey is new relative to these European countries. An exception was UK, where the percentages for the above mentioned ages were 30%, 65%, and 5%, respectively. The ages of the bridges in UK closely resemble those in Turkey except for the fact that more bridges are in 20–40 year age range in UK (65% in UK versus 40% in Turkey). This indicates that the bridges in Turkey are younger than the bridge stocks in these countries especially when bridges older than 40 years are considered. The BRIME statistics are for bridges on highway and trunk roads only. The rural and urban bridges were excluded. However, since the highway bridges is only a small portion of the bridge stock, it is reflective of the overall

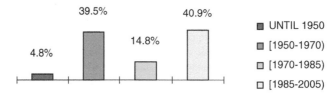

Figure 1.   Percentages of bridges built at different time periods.

population, thus may allow for comparison in general terms. The highway bridge stock is even newer in Turkey since the highway network is relatively quite new. As a result, it is beneficial to have a relatively new bridge stock from a bridge management point of view only if a well designed bridge management system is established at this time and properly implemented in the future to maintain these bridges as they get older.

In Figure 2, percentages of existing bridges in Turkey of different types are shown. As indicted in this chart, an overwhelming 92% of the bridges in Turkey are made of reinforced concrete members. A relatively small number of bridges; i.e., the remaining 8%, are comprised of composite, steel, and stone arch bridges. This fact indicates that the model development for deterioration mechanism in the bridge management system must be performed with special emphasis to concrete bridges.

2.2 *Current practices*

Currently, there is not a bridge management system in use at the General Directorate of Highways in Turkey. However, there exists a basic bridge inventory which contains the majority of the 5,486 existing bridges located on government and provincial roads. The bridge inventory is stored in a Microsoft Access database. The current database contains information only about the physical characteristics of these bridges. The information stored in the computerized database about the conditions of the bridges is highly inadequate. However, the bridges are inspected on an as needed basis and the inspection results are reported to the General Directorate of Highways where the official records are maintained.

In 1999, a study was conducted at the General Directorate of Highways to provide standards and procedures for bridge maintenance inspections. Approximately 200 bridges were investigated. As a result, a Visual Bridge Inspection Manual (Gözle Köprü Muayene El Kitabı, 1999) was prepared. The manual contains a Visual Bridge Inspection Form and presents guidelines and procedures for completing this form. The guidelines include the element condition (or damage) state indexes and the definitions for categorizing the element condition states based on visual inspections. The unique damage states are defined for each bridge element type. The bridge element types included are: pavement, guardrails, expansion joints, slab, supports, beams, piers etc. The damage types include damages such as deformation, cracks, carbonation and spall in concrete, dents and holes, and scour.

Since a central bridge management system does not exist, it was not possible to feed the data obtained from the visual inspection forms into such a computer program. For this reason, the widespread use of the inspection manual has been limited so far. However, the procedures in the manual serve as the official guidelines for performing inspection, monitoring and maintenance actions. Therefore, it would be appropriate during the development of the bridge management system to conform to these guidelines and to utilize these definitions and categorizations.

3 PROPOSED BRIDGE MANAGEMENT SYSTEM

It would be useful to benefit from the guidelines and recommendations established for developing bridge management systems and for improving existing systems worldwide prior to the development

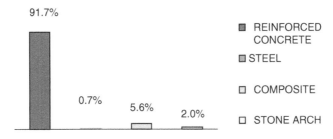

Figure 2. Percentages of bridges of different types.

of a new bridge management system that will be specifically tailored to the requirements of the General Directorate of Highways in Turkey.

The foremost criteria for the bridge management system to be developed are that they should include the fundamental required components of a BMS. The system must at least have the following data, modules, and functions:

- Bridge inventory (the Database)
- Element conditions (through Condition assessment)
- Load carrying capacities (through Structural assessment)
- Deterioration rates
- Costs of maintenance actions
- Optimization of maintenance actions (Optimal maintenance program/strategy)
- Maintenance priorities (Prioritized maintenance program)

These items can be further elaborated for a specific bridge management system tailored for the needs of a bridge administration of a country or a number of countries such as the European Union. For instance, in the final report of the BRIME (2001) project, it is proposed that a framework for a bridge management system for Europe must incorporate the following principal functions:

- Bridge Inventory
- Time variant bridge and element condition
- Load carrying capacity
- Management of operational restrictions and of the overweight routing
- Costs of maintenance strategies
- Deterioration
- Importance of the bridge (indirect costs)
- Optimization with budget constraints
- Maintenance priorities
- Budget monitoring (short and long term)

During the development and implementation phase of these functions, characteristics of the bridges and the local environmental factors must be considered. For instance, the common forms of deterioration of bridges in Turkey and their respective causes must be investigated prior to or during the development phase of the bridge management system. Anatolian peninsula is surrounded by seas along three sides. Bridges in districts near these coastal regions will be subjected to higher levels of corrosion activity due to coastal atmosphere.

In addition to the BRIME study for the European countries, many other countries around the world are also developing bridge management systems of their own or searching better techniques and innovative methods to improve their existing systems. For instance, after using the highly developed bridge management system Pontis for many years, U.S. examined the inspection and management processes and technologies abroad through FHWA's International Technology and Scanning Program. According to Friedland and Everett (2004), a number of issues were identified through contacts with the South Africa and the European countries. Friedland and Everett (2004) listed the following issues that must be improved in order to enhance the existing bridge management system or process:

- Inspection frequency
- Inspector qualifications
- Corridor-based bridge management

The other issues were related to advanced techniques and tools related to waterproofing and concrete permeability. The issues related to inspection planning and asset management must be addressed at the initial development phase of any new bridge management system.

A well-designed bridge or structure management system need to be a multi-task environment capable of supporting both the technical and administrative functions such as maintenance,

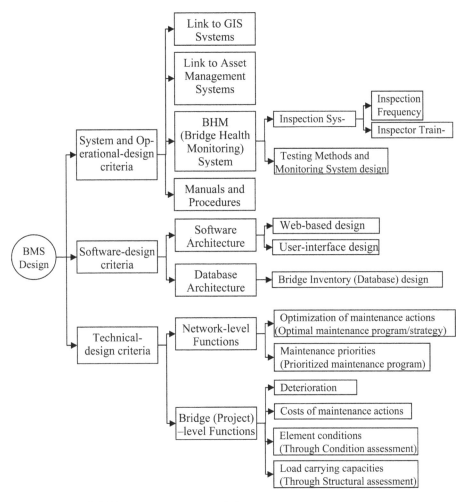

Figure 3.   Phases of a bridge management system design.

management, inspections, budget and quality control. Considering the fact that bridge management systems in the future may indeed ultimately converge toward an overall asset management system for the asset administrations, it is valuable to initially design such systems from the perspective of an Information Management System (IMS). The need for the coordination of the activities mentioned above prompted a few countries in recent years to develop Internet-based Bridge Management Systems (IBMS) such as APT-BMS in Italy (Zandonini *et al.* 2004) and Danbro+ in Denmark (Bjerrum and Jensen 2006). APT-BMS, being a fully web-based application, rests on a software architecture consisting of a database server, a web server (performs data management), and a data analysis application server and has the multi-agency support capability (Zandonini *et al.* 2004).

A proposed strategy showing the phases for the design of the new BMS to be developed for the General Directorate of Highways in Turkey is presented in Figure 3. At the initial phase, all the design criteria pertaining to the system and operational needs, software requirements, and technical aspects must be selected. Numerous selection criteria exists in the following phases of the BMS design also. As an example, there are numerous choices for deterioration modeling. The choices range from theoretical models that are proposed as new methods to models that are tested and are being used in current management systems. The system and operational issues must be well-defined in advance. For instance, it is decided that the BMS for the General Directorate of Highways in

Turkey will eventually be tied into a comprehensive GIS project that is currently in the development phase. Graphical, numerical, and text-based data interchange capability between the BMS and the GIS must be fully utilized in order to avoid the redundancy of data between these systems. It is decided that the software will rest on a web-based architecture and an advanced database system. The user interface will be designed to enable the inspectors at the Regional Districts to enter the inspection results in a straightforward fashion. Bridge-level functions such as element condition and load carrying capacity assessment must be carefully devised considering the advantages and disadvantages of the existing methods and techniques that are being used in BMSs worldwide. BRIME (2001) project emphasized the lack of general methods and straightforward techniques in these areas since the subjects such as the use of reliability index as a safety measure were still in development. However, the recent developments in BMSs indicate the wider use of such performance measures.

Success of any bridge management system in the long term depends mainly on the outcome of continuous inspection, monitoring and maintenance actions assuming that the technical designs of the system (i.e., condition and safety assessment and prioritization techniques) are based on sound methods that are properly calibrated. Therefore, the criteria for performing these operational actions must be well-defined when the BMS is placed in service.

4 CONCLUSIONS

A new study initiated to develop a BMS for the General Directorate of Highways in Turkey is presented. The developed system is expected to enable the General Directorate of Highways to monitor the safety and the maintenance costs of the bridges it owns, to establish the foundation for the development of the overweight truck route checking procedures and to be able to provide justifications, supported by scientific and technical data, presented in the form of technical and administrative reports to higher authorities for unbiased budget requests.

An overall BMS design methodology is discussed including system and operational, software, and technical design aspects. The criteria to be selected for these aspects at the initial design phase will affect the functionality, versatility and ease of use of the developed system in the long term.

REFERENCES

Akgul, F. and Frangopol, D.M. (2003). "Rating and Reliability of Existing Bridges in a Network", Journal of Bridge Engineering 8(6): 383–393.
Bjerrum J. and Jensen F.M. (2006). "Internet-based management of major bridges and tunnels using the Danbro+ system". Bridge Maintenance, Safety, Management, Life-Cycle Performance and Cost; Proceedings of the IABMAS'06 Conference. Porto, Portugal, 2006. Taylor & Francis.
BRIME (1999). "Review of current procedures for assessing load carrying capacity", Deliverable D1, PL97-2220, European Commission 4th Framework Program, p.24.
BRIME (2001). "Final Report", Deliverable D14, European Commission 4th Framework Program.
Frangopol, D.M., Kong J.S. and Gharaibeh E.S. (2001) "Towards Reliability-Based Life-Cycle Management of Highway Bridges", Journal of Computing in Civil Engineering, Vol. 15, No. 1, ASCE.
Frangopol, D.M. and Neves, L.C., (2005). "Condition, safety and cost profiles for deteriorating structures with emphasis on bridges", Reliability Engineering and System Safety 89: 185–198.
Friedland I.M. and Everett T.D. (2004). "AASHTO/FHWA International technology scan: bridge system preservation and maintenance". Bridge Maintenance, Safety, Management and Cost; Proceedings of the IABMAS'04 Conference. Kyoto, October 18–22, 2004: 69–77. Leiden: Balkema.
Gözle Köprü Muayenesi El Kitabı (1999). Köprüler Daire Başkanlığı, Köprü Bakım Şube Müdürlüğü, Karayolları Genel Müdürlüğü, T.C. Bayındırlık ve İskan Bakanlığı.
OECD (1981). Bridge Maintenance, Paris.
Zandonini, R., Zonta, D. and Bortot, F. 2004. "Bridge Management Systems for Medium-sized Local Agencies: Notes from an Italian Experience". Bridge Maintenance, Safety, Management and Cost; Proceedings of the IABMAS'04 Conference. Kyoto, October 18–22, 2004: 69–77. Leiden: Balkema.

Life-Cycle Cost and Performance of Civil Infrastructure Systems – Cho, Frangopol & Ang (eds)
© 2007 Taylor & Francis Group, London, ISBN 978-0-415-41356-5

Optimal risk-based Life-Cycle Cost design of infrastructures

Alfredo H.-S. Ang
University of california, Irvine, USA

ABSTRACT: The role of quantitative risk assessment (QRA) is highlighted and described in the development of optimal design of infrastructures. Optimal design may be determined on the basis of the minimum expected life-cycle cost, E(*LCC*). However, in order to reduce the effect of the epistemic uncertainty in the estimated life-cycle cost, a risk-averse value of the *LCC* may be specified to minimize the chance of under estimating the actual life-cycle cost; similarly, a risk averse value of the safety index may be specified for design. The systematic procedure for these purposes is illustrated with a hypothetical application in the determination of the design height of the levee system in New Orleans with sufficient assurance of protection against flooding from future hurricane-induced surges. The practical implementation of the procedure is emphasized.

1 INTRODUCTION

Structural engineers have the responsibility to provide the proper technical information to decision-makers and stakeholders in the construction of protective infrastructure systems for mitigating a hazard. For this latter purpose, quantitative risk assessment (QRA) methodology provides the tools needed. The main objective of this paper is to present a brief summary of the fundamental elements of the methodology of quantitative risk assessment including an illustration of its application in formulating decisions for optimal design of infrastructures based on minimum life-cycle cost (*LCC*).

In the case of natural hazards, risk is most meaningful and useful when expressed in terms of potential human sufferings and/or economic losses. Besides the probability of occurrence of a hazard, risk must include the potential adverse consequences that can result from the hazard event. The risk associated with natural hazards are very real, such as from strong earthquakes and associated tsunamis, high hurricanes (or typhoons), tornadoes, floods, and massive landslides. The forces created or induced by such natural hazards are usually extremely high and can cause severe damages and failures of engineered systems. Engineers, however, must still plan and design structures and infrastructures in spite of the extreme forces produced by one or more of these natural hazards. How safe should these infrastructures or facilities be for resisting the forces of natural hazards, of course, depends on the capital investments that the stakeholders, such as a government entity responsible for funding, is willing to invest (for safety and reliability) to prepare and protect against or reduce any impending risk to future hazards. In order to make the proper decisions needed for optimal investments, information on risk and associated risk reduction accruable from additional investment, are clearly pertinent. Again, this information would be most useful and effective if presented in quantitative terms.

2 QUANTITATIVE RISK ASSESSMENT METHODOLOGY

Information on risk is often presented in qualitative terms; for example, as *high, medium,* or *low*. More often than not, information in this form is ambiguous and difficult to interpret particularly for engineering purposes; moreover, it is not possible to perform risk-benefit trade off analysis.

For this latter purposes, risk needs to be in quantitative terms, such as potential number of fatalities and injuries, and/or potential economic losses. Similarly, quantitative risk information is needed to assess the benefit of investment in risk reduction, from which the benefit associated with a reduction in risk can be made transparent and useful.

2.1 *Uncertainty in calculated risk*

In assessing risk, especially relative to natural hazards, significant uncertainties can be expected. The occurrence of a given hazard within a given time window, such as a strong-motion earthquake in a particular region of the world, is unpredictable; moreover, the damaging effects of the earthquake are highly variable and difficult to estimate with precision. Also, the human casualties and sufferings, as well as the financial and economic losses that are possible consequences following the earthquake are also highly variable and invariably difficult to estimate. It is, therefore, easy to recognize that there is considerable uncertainty in the estimated risk associated with natural hazards. Such uncertainties, however, are significant and must be taken into account in any quantitative assessment of risk.

Uncertainties may be classified into two broad types (see e.g., Ang and Tang, 2006) – namely, the *aleatory* type and the *epistemic* type. The aleatory type is associated with the natural randomness or inherent variability of a phenomenon, whereas the epistemic type is based on our insufficient knowledge for predicting the phenomenon and in estimating the associated effects and consequences. In this regard, the aleatory uncertainty can be represented by a calculated risk, whereas the epistemic type would define the range of possible risk measures (representing the uncertainty in the calculated risk). Both the calculated risk and its uncertainty are equally important. It is, therefore, important to clearly differentiate the two types of uncertainty; namely, that the aleatory type is data based whereas the epistemic type is knowledge based. Irrespective of the type of uncertainty, the basic tools for its modeling and the analysis of the respective effects require the same principles of probability and statistics.

2.2 *Probability models in QRA*

Probability models, therefore, are the basic tools for quantitative risk assessment (QRA). However, risk is more than just probability; it must include the potential consequences from the occurrence of an event. In the case of natural hazards, the occurrence of a particular hazard in time and location is invariably unpredictable, and its destructive effects on structures and infrastructures are highly variable; finally, the resulting consequences of the destructive effects generally contain significant uncertainty. Therefore, for quantitative considerations, each of these aspects may be evaluated using probability models as follows: QRA will generally consist of three components which may be defined, respectively, as follows:

(1) *hazard analysis;* i.e., the determination of the probability of occurrence of a given hazard within a given time window;
(2) *vulnerability analysis*; i.e., the estimation of the extent and severity of damage to made-made and protective systems, and
(3) *consequence analysis*; i.e., the estimation of the potential consequences caused by the occurrence of the hazard.

The product of the above three components constitutes the estimated risk, R; that is

$$R = H_z \times V_u \times C_q \tag{1}$$

where: H_z = the result of a probabilistic hazard analysis;
V_u = the result of a vulnerability analysis; may be in terms of the probability or fraction of damage to a city or region;
C_q = the estimated potential consequence resulting from the occurrence of the hazard.

As there are epistemic type of uncertainties in estimating or calculating each of the components in Eq. 1, the calculated risk will also contain uncertainty leading to a range (or distribution) of the possible risk measures; in this case, Eq. 1 involves a convolution integral. This risk distribution can then be used to select the level of "confidence" or risk-averseness in the specification of the risk for decision making.

2.3 *Analysis of hazard*

The determination of the occurrence probability of a natural hazard will obviously depend on the particular hazard. For example, probabilistic models for seismic hazard analysis are well established (e.g., Cornell, 1968; Der Kiureghian and Ang, 1977); such models are now widely employed in practice. Similarly, models for the hazard analysis of tornado strikes have been developed by Wen and Chu (1973); whereas, for wind storms and hurricanes, and riverine floods, the respective occurrence probabilities at a given location over a specified period may be estimated from appropriate local or regional statistical data, modeled by extreme-value distributions (e.g., Gumbel, 1954) if appropriate.

Although the probabilistic model for the analysis of a given hazard will depend on the particular hazard, the results of a hazard analysis can generally be expressed in terms of the mean recurrence period, or average *return period*, of the particular hazard, usually in number of years. In estimating the mean recurrence period (or the return period) of a particular hazard, there is invariably some error in the estimation and thus uncertainty (epistemic type) in the estimated return period.

2.4 *Vulnerability analysis*

Given the occurrence of a particular hazard, there is some chance that structures or infrastructures within the affected zone will be severely damaged or completely destroyed. This probability, of course, will depend on the distribution of the maximum force from the hazard relative to the capacity of the structures for resisting such forces. As the maximum forces and the pertinent structural capacities will both contain variability and uncertainty, each may be represented with a probability model. That is, the maximum forces and structural capacities can be represented with respective random variables and associated probability distributions, on the basis of which the probability of failure or damage, pF, of the structures can be calculated.

The resulting failure probability, pF, may be interpreted as the proportion of structures and infrastructures (buildings, bridges, water tanks, etc) in an area or city that will suffer serious damage or collapse; in essence, the vulnerability of the area of concern.

As there are uncertainties in estimating the respective median (or mean) values of the loading and capacity, as well as in the specification of the respective PDF's, there is therefore epistemic-type uncertainty in the estimated vulnerability.

2.5 *Analysis of consequences*

The adverse consequences caused by the destructive forces of a natural hazard can be very severe, particularly for extreme events such as large magnitude earthquakes, high category hurricanes (or typhoons), or massive landslides and mudflows. These would often involve large numbers of fatalities and injuries, high economic and financial losses, major disruptions of utilities and transportation facilities, and related indirect consequences caused by ripple effects. The estimation of the consequences associated with the occurrence of a given hazard is often difficult and may have to be largely judgmental; i.e., relying on judgments from experts with knowledge gained through experience from similar events. Even then, the estimated consequences would contain significant uncertainties (of epistemic type), which may generally have to be expressed as respective ranges of possible losses.

3   DETERMINATION OF OPTIMAL DESIGN

Optimal design of structures or infrastructures may be based on minimizing the expected life-cycle cost, E(*LCC*). It has been shown by Ang and De Leon (2004) that the same optimal design is obtained by using other percentile values of the *LCC* in the optimization process. However, for risk-informed decisions, higher percentile values (such as the 75% or 90% value) rather than the mean or median value, of the *LCC* may be specified in order to minimize the chance of under estimating the actual life-cycle cost for the optimal design.

3.1 *A numerical illustration*

An (hypothetical) example is described numerically below to illustrate the conceptual process of QRA as outlined above. In order to clarify the steps in the QRA process, the problem is necessarily idealized, although the assumptions are reasonably realistic. For this purpose, suppose that an analysis of the hurricane risk for New Orleans (for a period of 20 years) was performed 15 years (say in 1990) before the occurrence of Katrina, a Category 4 hurricane, in August 2005. In this illustrative example, the numerical values used are hypothetical and may not be accurate (as they are pre-Katrina). Nevertheless, they serve to illustrate the quantitative process of assessing the underlying risks and associated uncertainties[1] for the purpose of providing the essential quantitative information for making risk-informed decisions for mitigating a future hazard.

Assume that upon careful examination of the recorded data on hurricanes in the gulf coast region, the return period of a Category 4 hurricane striking the vicinity of New Orleans is determined to be around 100 years; this means that there is a 1% probability each year, and a 20% probability over a 20-year period, that a Category 4 hurricane can be expected to hit the city of New Orleans and its vicinity.

A Category 4 hurricane, with a maximum sustained wind speed of 125–145 mph is bound to cause massive damages to ordinary dwellings and severe damages to some of the engineered infrastructures. Also, as the elevation of the city of New Orleans is 6 to 7 ft. below sea level, the city is protected by the levees and flood walls that kept the water of the surrounding lakes (such as Ponchartrain) and the Mississippi River from inundating the city. It is widely known and reported that the levees were designed and constructed with an average height of around 8 ft for protection against hurricanes of Category 2 or 3. Suppose that the actual levee heights has a symmetric triangular distribution between 7 and 9 ft and that the surges from the lake induced by a Category 4 hurricane can be modeled with a lognormal random variable with an estimated median height of 10 ft and a c.o.v. of 30%. Therefore, under a Category 4 hurricane there is a high probability that the levees will be breached causing massive inundation of the city. With the above assumptions, this probability can be calculated as: P(levee breached) = 0.78.

Furthermore, the vulnerability of much of the houses in New Orleans and vicinity against the hurricane winds would also be very high. Assume that the distribution of sustained wind speed in a Category 4 hurricane is modeled with a Type I extreme-value distribution with a mean speed of 130 mph and a c.o.v. of 40%, and that the wind speed resistance of houses and other structures is a lognormal random variable with a median of 85 mph and a c.o.v. of 30. On these bases, the vulnerability of the building stock and other structures in the city to the hurricane winds would be (evaluated through Monte Carlo simulation), vulnerability of structures = 0.785.

The consequences of the destructive effects of a Category 4 hurricane to the city of New Orleans, therefore, must include those caused directly by the high winds as well as by the surges from the lakes. Assuming that up to 90% of the population (approximately 600,000) in New Orleans will be evacuated before the storm, the potential fatalities may be assumed to range from 1800 to 3000 (i.e., 3% to 5% of those who did not evacuate) and serious injuries between 5000 and 10000, with respective mean values of 2400 fatalities and 7500 injuries; whereas, the economic loss could range between 75 and 150 billion dollars with a mean loss of $112.5 billion. It may be reasonable to assume (prior to the occurrence of Katrina) that the fatalities and injuries will be caused equally by the extreme wind and by the inundation of the city; whereas the economic loss will largely be caused by the failure of the levee system and subsequent inundation of the city.

---

[1] All the calculations in the example were performed through Monte Carlo simulations using MATLAB software with the accompanying Statistics Toolbox.

Table 1. Risk and *LCC* for different levee heights.

Height of levee, ft	Cost of levee, $billion	P(breaching)	E(Econ. risk)	E(*LCC*), $billion
8	0.10	0.78	$17.60 billion	17.7
9	0.25	0.64	$14.40 billion	14.65
10	0.40	0.52	$11.70 billion	12.10
11	0.60	0.38	$8.55 billion	9.15
12	1.00	0.28	$6.30 billion	7.30
13	2.75	0.21	$4.73 billion	7.48
14	5.00	0.12	$2.70 billion	7.70
15	6.50	0.09	$2.03 billion	8.53

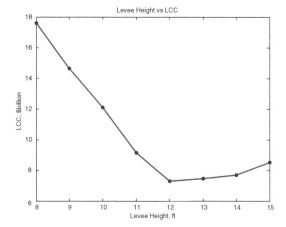

Figure 1. E(*LCC*) vs levee height.

On the basis of the above postulated information, the "best estimate" of the risks to the city of New Orleans can be summarized as follows (based on respective mean values):

Fatality risk $= 0.5[0.20(0.785)(2400)] + 0.5[0.20(0.78)(2400)] = 376$
Risk of serious injuries $= 0.5[0.20(0.785)(7500)] + 0.5[0.20(0.78)(7500)] = 1174$
Risk of economic loss (in dollar) $= 0.20(0.78)(112.5) = 17.55 billion.

There are, of course, epistemic uncertainties in each of the "best estimate" risks indicated above. In light of these epistemic type uncertainties, the risks are random variables; the respective distributions are illustrated in Ang (2006).

Optimal Design Height of Levee – Clearly, New Orleans needs a levee system for protection against inundation of the water from the surrounding lakes and the Mississippi river, as well as for protection of the city against flooding caused by future hurricanes. The cost to improve or retrofit the levee system will depend largely on the design height of the levees. Inversely, of course, the economic risk will decrease with the levee height.

Suppose that the projected cost for improving the levee will vary with the design levee height as shown in the first two columns of Table 1 above.

Suppose that (in 1990) the levee was to be retrofitted to protect against hurricanes of Category 4. Then over a 20-year period, the expected probabilities of breaching for the respective levee heights are shown in Column 3 of Table 1, and the corresponding expected economic risks are given in Column 4. The expected life-cycle cost, E(*LCC*), are therefore those shown in Column 5 of Table 1. Plotting the E(*LCC*) against the respective levee heights yields the results shown in Fig. 1.

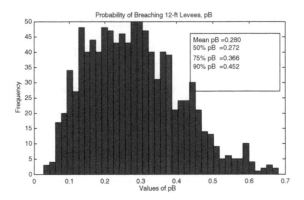

Figure 2.   Histogram of probability of breaching levees, pB.

From Fig. 1, it can be observed that the design height with the minimum E(LCC) would be 12 ft with an E(LCC) of $7.30 billion. However, there are (epistemic) uncertainties in this estimated expected LCC; in particular, several underlying epistemic uncertainties would lead to the uncertainty (and thus a distribution) in the estimated LCC, specifically including the following:

(1) The estimated return period of 100 years for a Category 4 hurricane occurring in New Orleans may actually be between 50 to 150 years. In this case, the annual occurrence probability would range between 0.7% and 2% (in 20 years would be 14% to 40%); the underlying uncertainty may then be represented by a coefficient of variation (c.o.v) of 29%, and may be modeled by a lognormal distribution with a median of 1.0 and a c.o.v. of 0.29, i.e. LN(1.0, 0.29).

(2) Because the specified estimated median surge height of 10 ft in the surrounding lakes is uncertain, the actual median surge could vary between 8 ft and 12 ft. This is equivalent to a c.o.v. of 12% in the median surge height, which may be modeled by a lognormal distribution of LN(1.0, 0.12). Therefore, the probability of breaching the 12 ft levees would also become a random variable and can be described by the histogram shown in Fig. 2 which has a mean value of 0.280, and the following important percentile values:
    50% pB $= 0.272$
    75% pB $= 0.366$
    90% pB $= 0.452$

(3) Finally, the uncertainties in the estimated consequences may be postulated as follows:
    • the expected economic loss, C_E, ranging from $75 billion to $150 billion, assumed to be uniformly distributed within the indicated range; whereas,
    • the estimated cost of construction, C_I, of the 12 ft levees could range from $1.0 billion to $1.50 billion; assumed to be uniformly distributed within this range.

To take account of the above uncertainties, the resulting LCC can be evaluated as

$$LCC = C_I + 0.20 N_H(p_B)(C_E) \tag{2}$$

in which the second term on the right is a convolution integral, where
$C_I =$ the cost of construction of the levee height;
$p_B =$ probability of breaching the levees; the histogram of Fig. 2 contains the uncertainty in the estimation of the median surge height;
$N_H =$ uncertainty in the estimated mean hazard (i.e., return period), prescribed as LN(1.0, 0.29);
$C_E =$ economic loss from inundation of city; assumed to be uniformly distributed between $75 and $150 billion.

In light of the above uncertainties, the LCC would also be a random variable. By Monte Carlo simulation (with 1000 repetitions), we generate the corresponding histogram as shown in Fig. 3 with a mean value of $7.79 billion.

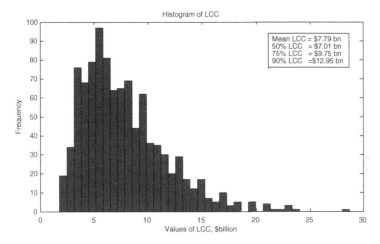

Figure 3.   Histogram of life-cycle cost, *LCC*.

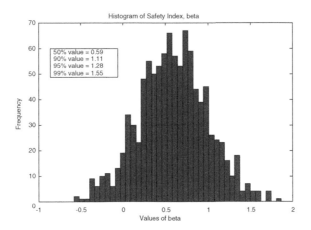

Figure 4.   Histogram of safety index, beta.

Finally, it may be of interest also to observe that the median probability of breaching the 12 ft levee is 0.28 in 20 years, or an annual probability of 0.014. Because of the epistemic uncertainties underlying the calculated probability of breaching, the distribution (or histogram) of the corresponding annual probability can be similarly obtained. From the histogram of the annual probability of breaching, the corresponding histogram of the safety index can also be generated as shown in Fig. 4 in which the 90% value of the safety index is 1.28.

By selecting the 90% value of the safety index for the structural design of the levee, there is reasonably high confidence (so to speak) that the retrofitted 12 ft levee can withstand the forces of a Category 4 hurricane. Observe that the corresponding median (50%) value of the safety index would only be 0.59.

4   INFORMATION AND ADVICE FOR DECISION MAKERS

Technical information obtained or generated from a QRA should be presented to the concerned stakeholders, in terms of the quantitative risk measures obtained as illustrated above. It is essential that this information be presented to the decision makers who are responsible for allocating resources for minimizing the risks to future hazards. In the case of a natural hazard, the most important risk measures would include the fatality and injury risks, and the risk of economic losses.

Information and advice presented in quantitative terms, based on the expertise of engineers, should generally be more convincing to decision makers. These may be in terms of the "best estimate" values of the pertinent risks or the risk-averse (i.e., conservative) values; the latter would serve to reduce the uncertainties underlying the respective estimated risks, or increase the level of confidence in the decision. As with other technical information developed for engineering purposes which are invariably in quantitative terms, risk measures should and can also be developed in the same terms; society would generally expect such information (i.e., supported by quantitative analyses) from the expertise of the engineering community.

5 SUMMARY AND CONCLUDING REMARKS

The fundamentals for the systematic and quantitative assessment of risk, with particular emphasis for hazard mitigation, are summarized. Besides the assessment of the "best estimate" measure of a pertinent risk, the assessment of the uncertainty underlying the calculated risk is equally important. These are illustrated with a quantitative assessment of the risks (for a 20-year period) associated with the occurrence of a Category 4 hurricane in New Orleans on the assumption that the assessment was performed in 1985 (20 years prior to the occurrence of Katrina in 2005).

QRA is illustrated also for developing optimal design of infrastructures based on minimum expected life-cycle cost; the determination of the optimal height of the levee system around New Orleans is used in this illustration. Because of epistemic uncertainties, the actual LCC for an optimal design would be a random variable with corresponding distribution (or histogram); in this light, a risk averse value of the LCC, may be specified in order to minimize the chance of underestimating the actual life-cycle cost.

The fundamentals of QRA, as summarized and illustrated here, show that QRA is a valuable tool available for engineers to generate quantitative technical information on risk and its associated uncertainty. A conservative (or risk averse) measure of risk may be specified to reduce the effects of the underlying (epistemic) uncertainty. QRA is important also for formulating risk-informed decisions in developing optimal design of infrastructures based on minimum life-cycle cost.

REFERENCES

Ang, A. H-S., "Practical Assessments of Risk and its Uncertainty", Proc. IFIP Workshop, Kobe, Japan, October 2006

Ang, A. H-S., and Tang, W.H., *Probability Concepts in Engineering*, 2nd Edition, John Wiley & Sons, Inc, 2007

Cornell, C.A., "Engineering Seismic Risk Analysis", *Bull. of Seismological Soc. of America*, Vol. 58, Oct. 1968

Der Kiureghian, A., and Ang, A. H-S., "A Fault Rupture Model for Seismic Risk Analysis", *Bull. of Seismological Soc. of America*, Vol. 67, Aug. 1977

Gumbel, E.J., "Statistical Theory of Extreme Values and Some Practical Applications", *Applied Mathematics Series 33*, National Bureau of Standards, Washington, DC, Feb. 1954

Wen, Y.K., and Chu, S.L., "Tornado Risk and Design Wind Speed", Proc. of ASCE, Jour. of Structural Div., Vol. 99, No. ST 12, December 1973

Life-Cycle Cost and Performance of Civil Infrastructure Systems – Cho, Frangopol & Ang (eds)
© 2007 Taylor & Francis Group, London, ISBN 978-0-415-41356-5

Lifetime optimization of reinforced concrete structures in aggressive environments

Lorenzo Azzarello, Fabio Biondini & Alessandra Marchiondelli
Department of Structural Engineering, Politecnico di Milano, Milan, Italy

ABSTRACT: The minimum lifetime cost design of reinforced concrete structures under multiple loading conditions is presented. The proposed formulation is based on a new conceptual approach to optimal design of deteriorating structural systems. The time-variant performance over the structural lifetime is taken into account by introducing a proper modeling of structural damage for both component materials, concrete and steel. The effects of maintenance interventions are included by relating the cost of maintenance to the actual damage level of the whole structure. The proposed procedure is finally applied to the lifetime structural optimization of a reinforced concrete frame. The obtained results show that the optimal dimensions of the cross-sections, as well as the optimal amount and distribution of reinforcement, strongly depend on the prescribed damage scenario.

1 INTRODUCTION

In the classical approach to structural optimization, the time evolution of the structural performance induced by the progressive deterioration of the system properties is not properly considered, since the attention is focused on the initial configuration only, in which the structure is fully intact. This approach is not consistent with the actual nature of the design problem, which should lead to concept structures able to comply with the desired performance not only at the initial time of construction, but also during the whole expected service lifetime by taking into account the effects induced by unavoidable sources of mechanical damage and by eventual maintenance interventions.

Based on such considerations, a new conceptual approach to the optimal design of deteriorating structures aimed to overcome the inconsistencies involved in the classical formulation of the optimization problem has been proposed in previous works (Biondini and Marchiondelli 2004, 2006; Azzarello *et al.* 2006). This approach has been developed for truss and framed structures composed by homogeneous members. The aim of this paper is to extend the proposed formulation to the minimum lifetime cost design of reinforced concrete frames under multiple loading conditions. The attention is focused on the damaging process induced by environmental aggressive agents, like sulfate and chloride, which may lead to deterioration of concrete and corrosion of reinforcement (CEB 1992). The structural damage induced by these agents is modeled by introducing a proper degradation law for both materials, concrete and steel, and the structural analysis is carried out at different time instants in order to assess the time evolution of the system performance. The design constraints are related to both the time-variant stress and displacement state, as well as to the amount of structural damage. The objective function is formulated by accounting for both the initial cost of the structure, given by the sum of the costs of the component materials, and the costs of possible maintenance interventions, that are properly discounted over time and assumed to be proportional to the actual level of structural damage.

The effectiveness of the proposed formulation is shown through the lifetime structural optimization of a reinforced concrete frame. The obtained results show that the optimal dimensions of the cross-sections, as well as the optimal amount and distribution of reinforcement, strongly depend on the prescribed damage scenario.

2 LIFETIME STRUCTURAL PERFORMANCE

2.1 *Lifetime parameters*

A structure is safe when the effects of the applied actions S are no larger than the corresponding resistance R, or $R \geq S$. Since both the demand $S = S(t)$ and the resistance $R = R(t)$ may vary during time, the limit state condition $R(t) = S(t)$ is reached after a *performance lifetime* T_P:

$$T_P = \min\left\{ (t - t_0) \mid R(t) \geq S(t), \ t \geq t_0 \right\}$$ (1)

where t_0 denotes the initial time of construction. From the design point of view, structural safety must be verified over a prescribed *service lifetime* T_S, or $T_P \geq T_S$.

A performance lifetime $T_P \geq T_S$ can be achieved by a lifetime oriented conceptual design process and/or by a proper maintenance program. In this context, the elapsed time between two subsequent interventions, aimed to partially or totally restore the initial performance of the damaged structure, is defined as *design maintenance period* T_M. After a maintenance intervention, structural safety must be verified over a prescribed *design monitoring period* $T_D \geq T_M$ without additional maintenance activities.

For the purpose of the present study, the following assumptions are introduced (Figure 1):

(*i*) A maintenance intervention is performed at the end of each monitoring period, or $T_D = T_M$. In this way, the total number of interventions applied during the service lifetime T_S is $r = [\text{int}(T_S/T_D) - 1]$.

(*ii*) The intervention k carried out at time $t_k = (t_0 + kT_D)$, with $k = 1, 2, \ldots, r$, is designed to totally restore the initial resistance of the structure, or $R(t_k) + \Delta R(t_k) = R_0$. If the design loads are constant over time, the variation of the demand associated with the evolution of damage is also fully recovered by the maintenance intervention, or $S(t_k) + \Delta(t_k) = S_0$.

(*iii*) The time evolution of damage is without memory. Consequently, the time evolution of structural performance of the repaired structure does not depend on the previous maintenance history, or $R(t_k + \Delta t) = R(t_0 + \Delta t)$ and $S(t_k + \Delta t) = S(t_0 + \Delta t)$, with $\Delta t \leq T_D$.

Based on such assumptions, the lifetime structural performance needs to be investigated during the first monitoring period only, or in the time interval $t \in [t_0; (t_0 + T_D)]$. In the following, for the sake of synthesis, the parameter T_D will be called *design period*.

2.2 *Multiscale modeling of structural damage in reinforced concrete structures*

Structural damage can be viewed as a degradation of the mechanical properties which makes the structural system less able to withstand the applied actions. In this study the attention is focused

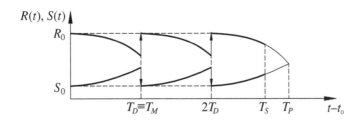

Figure 1. Definition of the lifetime parameters.

on reinforced concrete frames and damage is considered to affect the cross-sectional area $A = A(t)$, the elastic modulus $E = E(t)$, and the material strength $\bar{\sigma} = \bar{\sigma}(t)$ of both concrete and steel:

$$A_c(t) = [1 - \delta_{c,A}(t)]A_{c0} \qquad E_c(t) = [1 - \delta_{c,E}(t)]E_{c0} \qquad \bar{\sigma}_c(t) = [1 - \delta_{\bar{\sigma},c}(t)]\bar{\sigma}_{c0} \tag{2}$$

$$A_s(t) = [1 - \delta_{s,A}(t)]A_{s0} \qquad E_s(t) = [1 - \delta_{s,E}(t)]E_{s0} \qquad \bar{\sigma}_s(t) = [1 - \delta_{\bar{\sigma},s}(t)]\bar{\sigma}_{s0} \tag{3}$$

where δ_A, δ_E, $\delta_{\bar{\sigma}}$, are dimensionless damage indices which provide a direct measure of the damage level within the range [0; 1]. Proper correlation laws may be introduced to define the corresponding variation of other geometrical properties of the cross-section, like the inertia moment, etc.

The time evolution of the damage indices δ_A, δ_E, $\delta_{\bar{\sigma}}$, depends on the physics of the deterioration process, usually related also to the stress state $\sigma = \sigma(t)$ (Figure 2.a). Therefore, a reliable assessment of the time-variant structural performance requires deterioration models suitable to describe the actual damage evolution and its interaction with the structural behavior (Biondini et al. 2004). However, despite the inherent complexity of damage laws, very simple degradation models could be successfully adopted in order to define an effective hierarchical classification of the design alternatives (Biondini and Marchiondelli 2004, 2006*b*).

Without any loss of generality, in this study it is assumed that all material properties undergo the same damage process, or $\delta_A = \delta_E = \delta_{\bar{\sigma}} \equiv \delta$. Moreover, the damage indices $\delta_c = \delta_c(t)$ and $\delta_s = \delta_s(t)$ of concrete and steel, respectively, are correlated to the structural behavior by assuming the following relationships between the rate of damage and the acting stress (Figure 2.b):

$$\frac{d\delta_c(t)}{dt} = \frac{1}{T_{\delta,c}} \left[\frac{\sigma_c(t)}{\bar{\sigma}_{c0}} \right]^{\alpha_c} \qquad \bar{\sigma}_{c,0} = \begin{cases} \bar{\sigma}_{c,0}^+ & \text{if } \sigma_c \geq 0 \\ \bar{\sigma}_{c,0}^- & \text{if } \sigma_c < 0 \end{cases} \tag{4}$$

$$\frac{d\delta_s(t)}{dt} = \frac{1}{T_{\delta,s}} \left[\frac{\sigma_s(t)}{\bar{\sigma}_{s0}} \right]^{\alpha_s} \qquad \bar{\sigma}_{s,0} = \begin{cases} \bar{\sigma}_{s,0}^+ & \text{if } \sigma_s \geq 0 \\ \bar{\sigma}_{s,0}^- & \text{if } \sigma_s < 0 \end{cases} \tag{5}$$

where $\alpha \geq 0$ is a suitable constant, $\bar{\sigma}_0^-$ and $\bar{\sigma}_0^+$ are the minimum and maximum allowable stress at time $t = t_0$, respectively, and T_δ represents the time period which leads to a complete damage under a constant stress level $\sigma(t) = \bar{\sigma}_0$ (Figure 2.c). Moreover, for both materials, the initial condition $\delta(t_{cr}) = 0$ with $t_{cr} = \max\{t|\sigma(t) \leq \sigma_{cr}\}$ is assumed, where $\sigma_{cr} \leq \bar{\sigma}_0$ is a critical stress threshold.

The index δ fully describes the damage evolution in each point of the structure. However, due to its *local* nature, it does not seem handy for design purposes. A more synthetic *global* measure of damage may be derived at different scales from δ by a weighted average over given portions of the structure. With reference to a structural system composed by p structural elements, it is useful to define a global damage index $\hat{\delta} = \hat{\delta}(t)$ as follows:

$$\hat{\delta}(t) = \sum_{i=1}^{p} \hat{\delta}_i(t) = \sum_{i=1}^{p} \int_{V_i} w_i \delta_i(t) dV \qquad \sum_{i=1}^{p} \int_{V_i} w_i dV = 1 \tag{6}$$

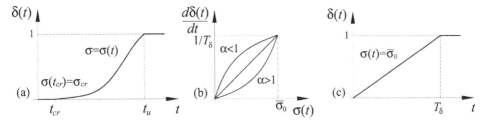

Figure 2. Modeling of structural damage. (a) Time evolution of the damage index $\delta = \delta(t)$. (b) Linear relationship between the rate of damage and the stress level $\sigma = \sigma(t)$. (c) Meaning of the damage parameter T_δ.

where w_i are suitable weight functions, and $\hat{\delta}_i = \hat{\delta}_i(t)$ is computed by the integration over the composite volume $V_i = V_{c,i} + V_{s,i}$ of the reinforced concrete element i.

Finally, since damage is not recovered during time, a synthetic measure of structural damage over the whole design service lifetime T_S may be identified with the damage indices evaluated at the end of the design monitoring period T_D. In particular, the quantities:

$$\tilde{\delta}_i = \hat{\delta}_i(T_D) \qquad\qquad \tilde{\delta} = \hat{\delta}(T_D) \qquad\qquad (7)$$

are assumed as lifetime damage indices of the element i and of the whole structure, respectively.

3 LIFETIME OPTIMALITY CRITERIA

3.1 *Lifetime structural cost*

Several quantities able to represent the structural performance may be chosen as targets for the optimal design. In this study, the adopted design target is the total cost C of the structure over its service life, given by the sum of the initial cost C_0 and maintenance cost C_m:

$$C = C_0 + C_m \qquad\qquad (8)$$

The initial cost C_0 is computed as follows:

$$C_0 = c_c V_{c,0} + c_s V_{s,0} = c_c (V_{c,0} + cV_{s,0}) = c_c V_{c,0}^* \qquad\qquad (9)$$

where $V_{0,c}$, $V_{0,s}$, are the total volumes of concrete and steel, respectively, c_c, c_s, are the corresponding unit costs, $c = c_s/c_c$ is the unit cost ratio, and $V_{c,0}^*$ is the equivalent volume of concrete. The maintenance cost C_m can be evaluated by summing the costs of the individual interventions:

$$C_m = \sum_{k=1}^{r} \frac{C_m^k}{(1+v)^{(t_k - t_0)}} \qquad\qquad (10)$$

where the cost C_m^k of each intervention $k = 1, \ldots, r$ has been referred to the initial time t_0 by taking a proper discount rate v into account (Kong and Frangopol 2003). In the assumed maintenance scenario, all the interventions have the same cost $C_m^k = C_m^1$ and are applied at time $t_k = (t_0 + kT_D)$:

$$C_m = C_m^1 \sum_{k=1}^{r} \frac{1}{(1+v)^{kT_D}} = C_m^1 q \qquad\qquad (11)$$

where the factor $q = q(T_S, T_D, v) \leq r$ depends on the prescribed parameters T_S, T_D, and v only. Since each intervention is aimed to totally restore the initial performance of the structure, the single intervention cost C_m^1 is related to the level of structural damage achieved at the end of the design period T_D. In this study, the following linear relationship is assumed:

$$C_m^1 = C_0 \tilde{\delta} \qquad\qquad (12)$$

Based on this approach, the total lifetime structural cost C is finally formulated as follows:

$$C = C_0 (1 + \tilde{\delta} q) \qquad\qquad (13)$$

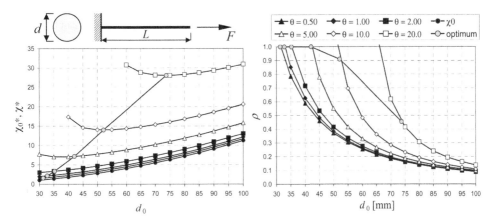

Figure 3. Lifetime performance of a tensioned bar undergoing damage.

Additional cost components, i.e. cost of formwork and so on, may be easily included in this formulation. However, it should be noted that the quantity C is aimed to represent a consistent criterion to compare different designs rather than the actual structural cost in a strict sense.

3.2 The role of maintenance cost

To highlight the actual role played by a prescribed maintenance program on the optimal design, the time-variant performance of the bar shown at the top of Figure 3 is investigated. The bar is made by homogeneous material. By denoting with d_0 the diameter of the undamaged cross-section, the total cost of the bar over the service lifetime T_S is:

$$C = C_0(1 + \tilde{\delta}q) = cA_0L(1 + \tilde{\delta}q) = c\frac{\pi d_0^2 L}{4}(1 + \tilde{\delta}q) \qquad (14)$$

with $\tilde{\delta} = \tilde{\delta}(d_0)$. The diameter d_0 must be chosen in such a way that the acting stress $\sigma = \sigma(t)$ is no larger than the admissible stress $\bar{\sigma} = \bar{\sigma}(t)$ over the prescribed design period T_D:

$$\sigma(t) = \frac{F}{A(t)} = \frac{F}{A_0[1 - \delta(t)]} = \frac{4F}{\pi d_0^2[1 - \delta(t)]} \leq \bar{\sigma}(t) = \bar{\sigma}_0[1 - \delta(t)] \qquad \forall t \in [0; T_D] \qquad (15)$$

with $\delta(t) = \delta(d_0, t)$. In case damage is not considered ($\delta = \tilde{\delta} = 0$), the minimum cost solution d_0^* is simply given by the minimum diameter d_0 which satisfies the stress constraint:

$$d_0^* = d_{0,\min} = \sqrt{\frac{4F}{\pi \bar{\sigma}_0}} \qquad C^* = C(d_0^*) = C_0(d_0^*) = c\frac{FL}{\bar{\sigma}_0} \qquad (16)$$

On the contrary, when damage is properly included in the design problem, the minimum cost solution d_0^* is, in general, no more associated with the diameter $d_{0,\min}$. In fact, higher d_0^* – values may be required to achieve a balance between the maintenance cost and the amount of damage. These aspects are highlighted in Figure 3, where both the cost and structural performance of the bar versus its diameter d_0 are shown for different values of the damage rate $\theta = T_S/T_\delta$, with $\sigma_{cr} = 0$, $\alpha = 1$, and with reference to the following case study: $F = 70\,\mathrm{kN}$, $\bar{\sigma}_0 = 100\,\mathrm{MPa}$, $T_S = 100\,\mathrm{years}$, $T_D = 10\,\mathrm{years}$, and $v = 0$ ($q = r$). In particular, the diagrams shown in Figure 3 refer to the following quantities:

$$\chi_0^* = C_0/C^* \qquad \chi_m^* = C_m/C^* \qquad \chi^* = \chi_0^* + \chi_m^* \qquad \rho = \sigma(T_D)/\bar{\sigma}(T_D) \qquad (17)$$

where C^* denotes the optimal cost without damage ($\theta = 0$). The following remarks can be made:

– The minimum feasible diameter without damage is $d_{0,\min} = 29.9\,\mathrm{mm}$. Its value increases with θ.
– The initial cost χ_0^* increases and the maintenance cost χ_m^* decreases when d_0 increases. For a given value of d_0, the maintenance cost χ_m^* increases with θ.
– The total cost χ^* has a minimum for $d_0^* \geq d_{0,\min}$, and the optimal diameter d_0^* increases with θ.
– The stress ratio ρ decreases with d_0 and increases with θ. The optimal solution d_0^* at the end of the design period T_D may be not fully stressed ($\rho^* \leq 1$).

Similar results are obtained by varying the parameter T_D (Azzarello 2005). Therefore, it has been proved that for homogeneous structures the optimal design solution strongly depends on the time-variant structural performance over the whole service life. The application presented in the following will show that this conclusion can also be extended to non homogeneous structures.

4   LIFETIME OPTIMIZATION OF REINFORCED CONCRETE STRUCTURES

4.1   *Formulation of the optimization problem*

The purpose of a one-target lifetime design process is to find a vector of design variables $\mathbf{x} \in \mathfrak{R}^n$ which optimizes the value of an objective function $f(\mathbf{x})$, according to both side constraints with bounds $\mathbf{x}^-$ and $\mathbf{x}^+$, and inequality time-variant behavioral constraints $\mathbf{g}(\mathbf{x}, t) \leq \mathbf{0}$:

$$\min_{\mathbf{x} \in D} \ f(\mathbf{x}) \qquad D = \left\{ \mathbf{x} \mid \mathbf{x}^- \leq \mathbf{x} \leq \mathbf{x}^+, \ \mathbf{g}(\mathbf{x}, t) \leq \mathbf{0} \right\} \tag{18}$$

Several quantities able to represent the lifetime structural quality may be chosen as targets for the optimal design. Based on the previously introduced cost concepts, the objective function $f(\mathbf{x})$ to be minimized is related to the total lifetime cost C of the reinforced concrete structure as follows:

$$f(\mathbf{x}) \equiv C(\mathbf{x})/c_c = V_{c,0}^*(\mathbf{x})[1 + \tilde{\delta}(\mathbf{x})q] \tag{19}$$

In a lifetime optimum design the behavioral constraints must account for the time-variant structural performance. Focusing the attention on the Serviceability Limit State (SLS), behavioral constraints are related to the structural response at each time instant $t \in [t_0; t_0 + T_D]$ and for each loading condition ℓ with reference to the stress in both concrete fibers $\sigma_{c,i,\ell} = \sigma_{c,i,\ell}(t)$ and steel bars $\sigma_{s,i,\ell} = \sigma_{s,i,\ell}(t)$ of each element i, as well as to the displacement $u_{j,\ell} = u_{j,\ell}(t)$ of each nodal point j:

$$\begin{cases} \bar{\sigma}_{c,i,\ell}^-(\mathbf{x}, t) \leq \sigma_{c,i,\ell}(\mathbf{x}, t) \leq \bar{\sigma}_{c,i,\ell}^+(\mathbf{x}, t) \\ \bar{\sigma}_{s,i,\ell}^-(\mathbf{x}, t) \leq \sigma_{s,i,\ell}(\mathbf{x}, t) \leq \bar{\sigma}_{s,i,\ell}^+(\mathbf{x}, t) \end{cases} \qquad \bar{u}_{j,\ell}^- \leq u_{j,\ell}(\mathbf{x}, t) \leq \bar{u}_{j,\ell}^+ \tag{20}$$

where $\bar{\sigma}_{i,\ell}^- = \bar{\sigma}_{i,\ell}^-(t)$ and $\bar{\sigma}_{i,\ell}^+ = \bar{\sigma}_{i,\ell}^+(t)$ are the minimum and maximum allowable stress, respectively, $\bar{u}_{j,\ell}^-$ and $\bar{u}_{j,\ell}^+$ are prescribed displacement bounds. Clearly, additional constraints related to the Ultimate Limit States (ULS) may be easily introduced, even though they are not considered here. Bounds on both local and global damage may also be considered (Azzarello *et al.* 2006).

4.2   *Lifetime optimization of a reinforced concrete frame*

The previous formulation is applied to the lifetime optimization of the reinforced concrete frame shown in Figure 4.a. A rectangular cross-section is assumed for both beam and columns. With reference to Figures 4.a and 4.b, the optimization problem is defined by $n = 9$ design variables $\mathbf{x} = [b h_1 h_2 | A_1 A_2 A_2' A_3 A_3' | d]^T$, for which the following side constraints are assumed: $b \geq 300\,\mathrm{mm}$; $1 \leq h_j/b \leq 2$, $i = 1, 2$; $A_i \geq 2\,\varnothing\,12$ and $A_j' \geq 2\,\varnothing\,12$, $i = 1, 2, 3$; $0.10 \leq d/L \leq 0.50$.

The three alternative loading conditions shown in Figure 4.c are considered. Since the SLS is investigated, the time-variant structural response in terms of nodal displacements and internal

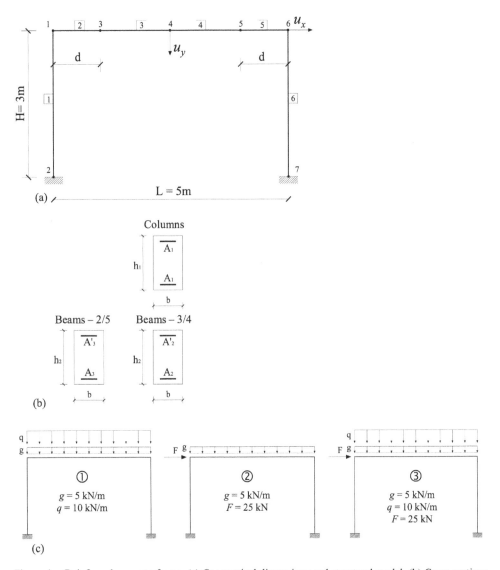

Figure 4. Reinforced concrete frame. (a) Geometrical dimensions and structural model. (b) Cross-sections of beam and columns. (c) Loading conditions.

stress resultants are evaluated by assuming a linear elastic behavior. Based on the general theory for concrete design, the corresponding stress in the materials are computed at the cross-sectional level by assuming for concrete a linear elastic behavior in compression with $E_c = 30$ GPa and no strength in tension ($\bar{\sigma}_c^+ = 0$), and for steel a linear elastic behavior in both tension and compression with $E_s = 15E_c$. The global lifetime damage index $\tilde{\delta}$ is computed by adopting for both concrete and steel the same weight function in each member.

The same behavioral constraints are considered for each loading condition. The stress in the materials is verified in each member by assuming the initial allowable stress $-\bar{\sigma}_{c,0}^- = 15$ MPa for concrete, and $\bar{\sigma}_{s,0}^+ = -\bar{\sigma}_{s,0}^- = 180$ MPa for steel. The displacement constraints $u_x \leq 20$ mm and $u_y \leq 10$ mm (Figure 4.a) over the whole service lifetime are also considered.

Table 1.   Reinforced concrete frame. Damage scenarios ($\theta = T_S/T_\delta = 10$, $T_S = 100$ years).

Damage scenario	Left column	Beam	Right column
A	–	–	–
B	Damage	Damage	Damage
C	–	Damage	–
D	Damage	–	Damage
E	Damage	Damage	–
F	–	Damage	Damage

The lifetime performances are defined by the following parameters: service lifetime $T_S = 100$ years, design period $T_D = 10$ years, unit cost ratio $c = 20$, discount rate $v = 3\%$, and damage rate $\theta = T_S/T_\delta = 10$, with $\sigma_{cr} = 0$ and $\alpha = 1$ for both concrete and steel. Moreover, to investigate the role of the spatial distribution of deterioration, the damage scenarios described in Table 1 are considered.

The optimization problems are solved by using a gradient-based method (Vanderplaats 2001). The direct comparison of the optimal solutions shown in Figure 5 shows that the optimal dimensions of the cross-sections, as well as the optimal amount and distribution of reinforcement strongly depend on the prescribed damage scenario. In particular, it is worth noting that damage leads to a more ponderous design not only for members affected by deterioration. In fact, due to redundancy, damage induces a time-variant redistribution process in which the internal stress resultants tend to progressively move towards the undamaged members. These results prove the fundamental role played by both the time-variant performance and the maintenance planning in the selection of the optimal structural design.

5   CONCLUSIONS

The minimum lifetime cost design of reinforced concrete structures under multiple loading conditions has been presented. The proposed formulation is based on a new conceptual approach to optimal design of deteriorating structural systems. This approach allowed to overcome the inconsistencies involved in the classical formulation of the optimum design problem, where the time evolution of the structural performance induced by the progressive deterioration of the system properties is not adequately considered. In fact, in the proposed formulation, the structural damage is accounted for by means of a proper material degradation law of the mechanical properties for both component materials, concrete and steel, and the design constraints of the optimization problem are related to the corresponding time-variant structural performance over the whole expected service life of the construction. In addition, the objective function is formulated by accounting for the initial cost of the structure, given by the sum of the costs of the component materials, as well as the costs of the possible maintenance interventions, that are properly discounted over time and assumed to be proportional to the actual level of structural damage. The effects of maintenance interventions are included by relating the cost of maintenance to the actual damage level of the whole structure.

The proposed procedure has been applied to the lifetime structural optimization of a reinforced concrete frame under prescribed design scenarios. The results highlighted that the optimal dimensions of the cross-sections, as well as the optimal amount and distribution of reinforcement, strongly depend on the prescribed damage scenario. In particular, the obtained solutions are characterized by a more ponderous design not only for members affected by deterioration. In fact, due to redundancy, damage induces a time-variant redistribution process in which the internal stress resultants tend to progressively move towards the undamaged members. These results proved the fundamental role played by both the time-variant performance and the maintenance planning in the selection of the optimal structural design.

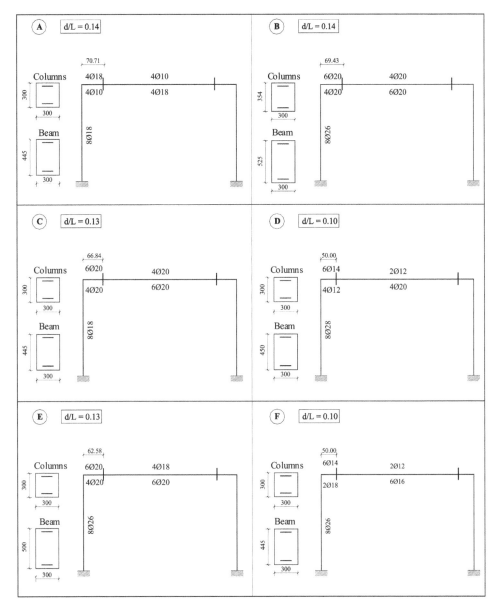

Figure 5. Reinforced concrete frame. Optimal design solutions associated with the damage scenarios described in Table 1 ($\theta = T_S/T_\delta = 10$, $T_S = 100$ years, $T_D = 10$ years, $c = 20$, $v = 3\%$).

REFERENCES

Azzarello, L., 2005. Ottimizzazione di Sistemi Strutturali con Prestazioni Variabili nel Tempo. Degree Thesis, Politecnico di Milano (In Italian).

Azzarello, L., Biondini, F., Marchiondelli, A., 2006. Optimal Design of Deteriorating Structural Systems. *3rd International Conference on Bridge Maintenance And Safety (IABMAS'06)*, Porto, July, 16–19.

Biondini, F., Bontempi, F., Frangopol, D.M., Malerba, P.G., 2004. Cellular Automata Approach to Durability Analysis of Concrete Structures in Aggressive Environments. *ASCE Journal of Structural Engineering*, **130**(11), 1724–1737.

Biondini, F., Marchiondelli, A., 2004. Evolutionary Design of Durable Structures. *2nd International Conference on Bridge Maintenance And Safety (IABMAS'04)*, Kyoto, October, 19–22.

Biondini, F., Marchiondelli, A., 2006*a*. Lifetime Structural Optimization. *European Symposium on Service Life and Serviceability of Concrete Structures (ESCS-2006)*, Helsinki, June 12–14 – Keynote Paper, 1–12.

Biondini, F., Marchiondelli, A., 2006*b*. Evolutionary Design of Structural Systems with Time-variant Performance. *Structure and Infrastructure Engineering* – Tentatively accepted for publication.

Biondini, F., 2004. A Three-dimensional Finite Beam Element for Multiscale Damage Measure and Seismic Analysis of Concrete Structures. *13th World Conf. on Earthquake Engineering*, Vancouver, B.C., Canada, August 1–6, Paper No. 2963.

CEB, 1992. *Durable Concrete Structures – Design Guide*, Thomas Telford.

Kong, J.S., Frangopol, D.M., 2003. Evaluation of Expected Life-Cycle Maintenance Cost of Deteriorating Structures. *ASCE Journal of Structural Engineering*, **129**(5), 682–691.

Vanderplaats, G.N., 2001. *DOT – Design Optimization Tool*, Vanderplaats Research & Development, Colorado Springs, CO, USA.

Life-Cycle Cost tenders for infrastructure projects

Jaap Bakker & Joris Volwerk
*Civil Engineering Division, Ministry of Transport, Public Works & Water Management,
Utrecht, The Netherlands*

ABSTRACT: In the past decade contracts for civil works have changed rapidly. Whereas traditional contracts contain detailed descriptions of the contractors work, more modern contracts aim
at describing the desired result (or functionality) based on the principles of system engineering,
leaving the contractor free to decide how to reach the desired result. Typical forms of modern
contracting are DC (design and construct), DCM (design, construct and maintain) and DCFM
(design, construct, finance and maintain). The last two types of contracts typically aim at a shared
risk between owner and contractor after construction. However, these types of contracts often don't
have a long enough maintenance period to really cover the financial risks in time. DC contracts
(and in practice DCM contracts likewise) do not stimulate the contractor to think about lifecycle
cost aspects. At the Dutch Ministry of Transport, Public Works and Water Management (Rijkswaterstaat) a new approach for tendering DC contracts has been developed: the "LCC tender". Not
the lowest bid, but the lowest lifecycle cost will be used (as one of the criteria) to determine the
economical most advantageous proposal. In this paper different "fit to purpose" approaches are
described for LCC tenders for the construction of new bridges and roads, new "complex structures"
and renovation projects. All three types of tenders are currently being put to practice or ate under
development at Rijkswaterstaat.

1 INTRODUCTION

In the past decade new types of contacts based on design and construct principles have rapidly
gained importance. Based on a functional description of a desired result, a contractor will both
design and the construct. He will make his own choices on the types of elements, construction
forms and materials he uses. Contractors benefit from the freedom in design, leaving more possibilities discernible innovations. Benefits of DC-contracts for authorities are the more explicit
responsibilities for the contractor and an optimal profit from smart and cost effective solutions.

In recent years "Rijkswaterstaat" has adopted the policy to work with functionally specified
contracts (describing a desired result in functional terms) as much as possible. In this kind of
contract specification, it remains to be a challenge to adequately incorporate specifications for
durability. If, for instance, a steel structure is specified to remain durable for 100 years, this is
most often open to interpretation. Does this structure meet demands when periodic conservation
maintenance is needed to achieve this? It appears that durability is not a "fair" or adequate criterion
for comparison. Costs involved with maintaining a steel structure over a lifespan of 100 years
divert considerably from costs for maintaining concrete or timber structures, even though they all,
in essence, fulfill the same function.

Boldly said, it is impossible to compare different designs sufficiently without a decent lifecycle
costs consideration.

This paper documents the introduction of lifecycle costs (LCC) as a tender criterion a.k.a. the
"LCC tender", by describing 3 different examples currently under development or in use within
Rijkswaterstaat.

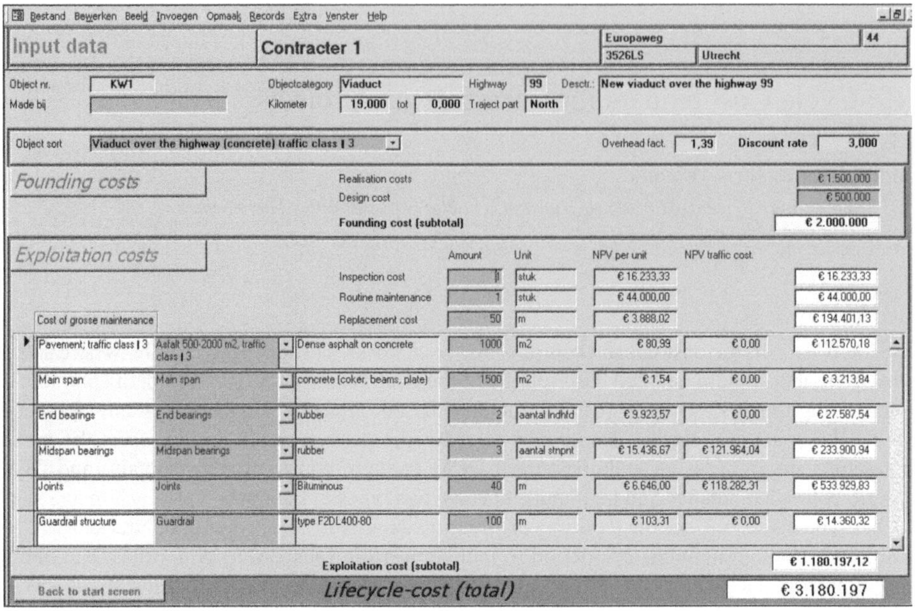

Bestand Bewerken Beeld Invoegen Opmaak Records Extra Venster Help

| Input data | Contracter 1 | Europaweg | 44 |
| | | 3526LS | Utrecht |

| Object nr. | KW1 | | Objectcategory | Viaduct | | Highway | 99 | Descr: | New viaduct over the highway 99 |
| Made bij | | | Kilometer | 19,000 tot | 0,000 | Traject part | North | | |

Object sort Viaduct over the highway (concrete) traffic class I 3   Overhead fact. 1,39   Discount rate 3,000

Founding costs

	Realisation costs	€ 1.500.000
	Design cost	€ 500.000
	Founding cost (subtotal)	**€ 2.000.000**

Exploitation costs

		Amount	Unit	NPV per unit	NPV traffic cost
Inspection cost		1	stuk	€ 16.233,33		€ 16.233,33
Routine maintenance		1	stuk	€ 44.000,00		€ 44.000,00
Cost of grosse maintenance	Replacement cost	50	m	€ 3.888,02		€ 194.401,13

Pavement; traffic class I 3	Asfalt 500-2000 m2, traffic class I 3	Dense asphalt on concrete	1000	m2	€ 80,99	€ 0,00	€ 112.570,18
Main span	Main span	concrete (coker, beams, plate)	1500	m2	€ 1,54	€ 0,00	€ 3.213,84
End bearings	End bearings	rubber	2	aantal lndhfd	€ 9.923,57	€ 0,00	€ 27.587,54
Midspan bearings	Midspan bearings	rubber	3	aantal stnpnt	€ 15.436,67	€ 121.964,04	€ 233.900,94
Joints	Joints	Bituminous	40	m	€ 6.646,00	€ 118.282,31	€ 533.929,83
Guardrail structure	Guardrail	type F2DL400-80	100	m	€ 103,31	€ 0,00	€ 14.360,32

| | **Exploitation cost (subtotal)** | **€ 1.180.197,12** |
| **Back to start screen** | ***Lifecycle-cost (total)*** | **€ 3.180.197** |

Figure 1.   Input sheet for the LCC-application for LCC-tenders for "simple structures".

2   LCC–TENDER FOR "SIMPLE STRUCTURES"

The first example concerns an LCC-tender approach for "simple structures", typified by merely a few structural principals (often monolith) and thus constructed that the required maintenance regime is affected relatively little by variation in local circumstances.

This example concerns the "design and build" of 20 structures (viaducts and sound barriers) invoked by a road widening project. This tender process has recently been finalized.

The method used in this example has been published in more detail [1] and can be summarized as followed. Contractors partaking are required, during the tender process, to name their general choices in structural principals, material and components. On this basis the costs for future maintenance are generated and projected backwards to a Present Value. A specially developed software application is used to generate the lifecycle costs result (see figure 1).

Future maintenance costs for viaducts and sound barriers are gathered based on the standard components the contractor chooses to comprise the structure. The issued list of standard components is the same for each contractor. Attached to each standard component is a standard maintenance prognosis by way of standard costs and standard intervals. As a result a fictional maintenance plan is generated and all cost are projected backwards to a present value. The Present Value is added to the bid price and this result is used to determine the most profitable submission. Costs for traffic measures are partially taken into account and capitalized. The social costs related to traffic congestion are not taken into account at all. Rijkswaterstaat's current maintenance policy states that maintenance to viaducts and other structures in and over Dutch highways will be executed together with road surface works as much as possible. If components have a average maintenance interval greater then 10 years it is safe to assume, insuring an adequate regime for inspection, that they can be maintained or replaced during road surface works. In this case traffic measures and social cost due to traffic congestion don't have to be taken into account. Only if components require a maintenance interval smaller then 10 years or required maintenance can't be executed within the timeframe of the road surface works, costs for traffic measures will be incorporated into the LCC-sum. Social costs are not taken into account because practically all maintenance that can't be combined with road works can be executed during the night.

During the tender a procedure query sessions for each individual contractor where held. This procedure has been allowed by the European Tender Regulation since 2005. During this procedure each individual contractor was given opportunity to present innovative solutions (e.g. materials or components) with substantiated maintenance prognoses. When agreed upon, this solution is added as an additional component/option to the database of that specific contractor.

The contract specifications (standards, norms and regulations) also lay at the basis of the maintenance prognoses. If the quality required in the contract is not met during realization then costs for direct repair as well as costs for additional future maintenance (as a result of poor quality) are claimed.

This tender approach results in a general LCC consideration but explicitly doesn't result in a true insight into the actual maintenance prognosis because factors like the quality of realization aren't taken into account. It does, however, force a contractor to consciously make choices in design, materials and components used at a competitive stage in the process. This, combined with adequate quality control during realization, makes a rough but adequate optimization of lifecycle costs.

The first experience with this tender approach shows that contractors are stimulated to innovate and to find smart design solutions. For example, bridge designs where made without expansion joints, using reinforced asphalt in critical areas as a lifecycle costs effective alternative for traditional design solutions. It appeared that joints are the most substantial cost driver in the lifecycle of a viaduct [2, 3, 7]. The LCC approach also turned out to be beneficial for cost effective sound barrier design.

3   LCC-TENDER FOR "COMPLEX STRUCTURES"

All simple bridge, viaduct and sound barrier structures can be treated reasonably well as a compilation of components suited with a predictable maintenance prognosis. This is not the case when considering a complex structure containing a larger amount of variables and maintenance prognoses are, for the greater part, dependent on usage or variety in construction detail. For instance, maintenance costs for a lock-gate depend greatly on design choices. Maintenance strategies and maintenance costs are very dependent on the possibilities for taking a lock "out of service", and whether a backup lock-gate is available. The maintenance prognosis for a "drive motor" moving the lock-gate to its "open" and "closed" position, is even more unreliable. Different aspects, like "use" intensity, power to strain ratio, the quality of component manufacturing (which will certainly differ per manufacturer), the availability of spare-parts and energy consumption all contribute greatly to final maintenance prognosis. On top of this, refurbishment or replacement is often combined with conservation maintenance to the lock-gate itself. In these cases components can't be considered separately. This requires a different LCC-tender approach to the one described in the previous paragraph.

The following approach is currently under development within Rijkswaterstaat, but has not yet been put to practice. This method is only applicable in case of a tender procedure containing a "competitive dialog" phase. In this procedure, and particularly this phase, the commissioner and each individual contractor come to individual "contract terms" in a set of privately planned "dialog meetings". This procedure has been allowed by the European Tender Regulation under strict conditions since 2005.

This method requires a contractor to develop his own "personal" maintenance prognosis following his particular design and to submit it along with his bid. During the tender procedure the content of this plan is judged by an independent comity of experts on feasibility, foundation and uniformity. The comity delivers a binding verdict for uniformity (in cases of diverting maintenance prognoses for similar components). To facilitate equality as much as possible reference data (standard costs and intervals) is provided for the most frequently used and most obvious components. Contractors primarily use this data, but may deviate from this data if they motivate why it is clearly not applicable in or represent able for their design.

This procedure for LCC-tender is represented in figure 2.

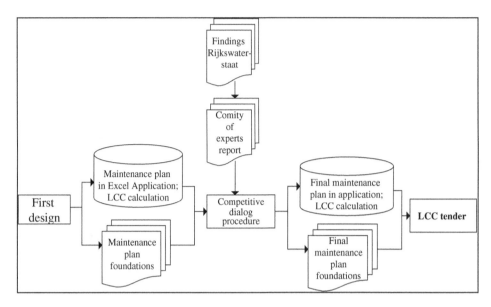

Figure 2.   Flowchart for the LCC-tender procedure for complex structures.

Yearly cost

Regular maintenance (yearly)
Cost: 50000
Uncertainty class: B

Inspection
Cost: 10000
Uncertainty class: B

Energy
Cost: 9000
Uncertainty class: B

[Back to input sheet]

Data at (sub)component-level

Comp_ID	Component	Amount	units	Measure number	Measure	Interval	uncertainty class (interval)	Unit-price	Uncertainty class (cost)
1	component 1.1.a	5	units	1	repairwork	10	B	10.000	B
				2	replacement	20	B	30.000	B
				3					B
2	component 1.1.b	2000	m2	1	replacement	40	C	60000	B
				2					B

Figure 3.   Input sheet for the LCC-application for complex structures.

This approach is also supported by a software application (Excel) that converts the future costs in the maintenance prognoses to present values. Different to the application described in the previous paragraph, this application contains a probabilistic approach. Each maintenance prognosis is given a "qualification of uncertainty". Each "qualification" (A thru D) represents a range of uncertainty (the variety-coefficient) for a given maintenance interval. Unorthodox design solutions (with which little experience concerning actual performance has been gained) will be given a high margin of uncertainty. Other factors related to "high uncertainty" are "sensitivity to flaws during construction" or a historically "known uncertainty in maintenance prognoses". The present value of maintenance prognosis with a high uncertainty is greater then the same one containing little uncertainty.

Similarly to the previous example the present value for future maintenance is added to the bid price to determine the (economically) most profitable bid. An example of the LCC-application "input" and "output" is represented in figures 3 and 4.

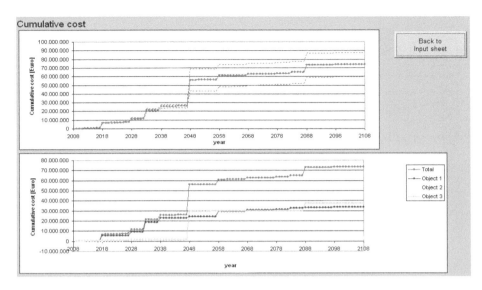

Figure 4. Output graphs for the LCC-application for complex structures.

4 LCC-TENDER FOR MAINTENANCE WORK

A very limited LCC approach is currently being put to practice in the large scale tendering of "overdue maintenance" (involving approx. 1000 structures with a maintenance backlog, mainly viaducts). On top of the implementation of LCC, the work is being commissioned as an "Engineering and Construct" (E&C) contract. This type of commission, in which a contractor is required to both engineer and construct a maintenance solution on the basis of provided functional specifications, is relatively new to the maintenance "branch". This is why this approach is considered a little more explicitly.

The basis for this E&C contract is provided by a specification describing the desired or required functional conditions (hence: "functional specification") for structures to function adequately. These functional conditions are differentiated by the aspects they predominantly address, namely Reliability, Availability, Maintainability and Safety (RAMS analysis framework). In the context of "Systems Engineering" the RAMS analysis framework represents the set of ultimate requirements for each individual structure. These requirements are translated to components by way of specific requirements and codes. The detailing of requirements and codes varies depending on the importance of components and the share they hold in the commission.

In preparation of an E&C maintenance contract a so called "work inventory" is executed. This entails an inspection of relevant structures in which all deviations from conditions described in the "RAMS analysis framework" are registered and described as "problems" (describing what particular functional requirement or code is not met or at risk, and what is causing this). Contractors are commissioned to generate solutions for these "problems". In addition to meeting all these contract specifications a warranty must be given on each problem not reoccurring within minimum of 7 year and on not initiating new problems in relation to the RAMS analysis framework as a result of the applied solution.

During the "work inventory" the "inspection contractor" also provides a general estimate for maintenance costs. This estimate is reliable within 30% and is communicated to all contractors partaking in the tender process as "additional" information, representing the opinion of the "inspection contractor", and as such not having any legal or binding status.

These estimates are utilized to determine a "LCC added-value" for each of the solutions submitted by contractors in the tender.

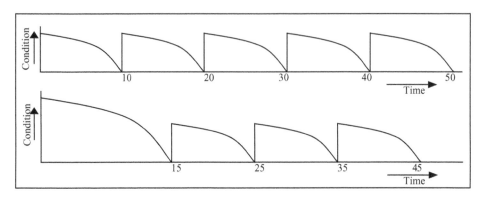

Figure 5. Reference estimate (top) versus the contractor's solution with a longer warranty.

4.1 *Determining the "LCC added-value"*

Arbitrarily it is assumed that each problem that has been solved for a guaranteed period of seven years will on average reoccur after 10 years. This will fit Rijkswaterstaat's infrastructure network maintenance policy, in which roads are subject to an integral maintenance on average every 7–10 years. It is assumed that successive maintenance here after will be combined with road surface maintenance and there for will not cause any additional traffic congestion.

It may well be that a contractor wishes to offer a solution with a warranty exceeding the required minimum of 7 years. Such an offer can be more profitable even though it may cost more because costs for maintenance will arise later in the future. In order to value such differences a specific LCC software application has been developed to facilitate these tender submissions.

This LCC application determines the "added value" (savings in future maintenance) opposed to the estimates provided by the "inspection contractor". It is assumed (arbitrarily) that the solution estimated by the inspection contractor will last 10 years and, as such, that the estimated costs will reoccur every 10 years. To calculate the "added value" for the solution offered by a tendering contractor, it is assumed that the solution offered will applied only once and that this offered solution will always last 3 years longer than the offered warranty period. At the end of this warranty period +3 years the estimate provided by the inspection contractor is reapplied every 10 years to complete the lifecycle costs analysis. The theoretical difference between the offered solution and the reference estimate is represented in figure 5.

The assumption that the lifespan will always reach or be limited to the offered warranty period +3 years is debatable. For instance, if the warranty period is modeled as a 90% reliability interval one may assume a longer warranty period is accompanied by greater range of lifespan extension (as well as uncertainty). In this case the expected lifespan extension would be greater than the (based on a 7 year warranty) warranty period +3 years.

However it is expected that contractors will take greater risks when offering longer warranty periods. By offering a longer warranty period whilst at the same time bidding a higher price to cover the risk of future warranty claims, the offer may still be more profitable than the offer made by a contractor offering low prices and merely 7-year warranties. This can be explained by the fact that the actual return on investments for the contractor is much higher than the 4% discount rate used in the LCC application.

The assumption that the solution offered will be applied only once and will successively be replaced by the estimate provided by the inspection contractor is probably not true to reality. However, in order to fairly compare bids on "LCC added value" (savings on future maintenance costs) an similar reference period (lifespan) must be applied to each offer.

Simulating this way ensures that the offered price for the offered solution influences the bid only, and has no bearing on calculated savings on future maintenance costs ("LCC added value"). This prevents bid manipulation by shifting costs.

4.2 First results

The first LCC-tenders of contracts by way of the method described above have been completed. It appears that contractors are (for the moment) still hesitant to offer longer warranty periods for maintenance commissioned. The LCC-criteria should be related more thoroughly to other requirements and specifications in the contract to avoid negative unforeseen effects. Expansion joints give a good example that illustrates the unforeseen effects of conflicting requirements. Expansion joints can make up 20 to 30% of maintenance costs for viaducts [7]. In addition to this maintenance to joints is always accompanied with congestion or traffic congestion.

In one particular tender a large part of the commission involved joints. To limit or avoid congestion as a result of the commissioned maintenance work, heavy penalties where specified for working outside provided time-boxes (basically night time only). The imposed penalties where so high that it became unfeasible to offer and utilize durable joints. The "LCC added value" was no match for the penalties imposed.

5 CONCLUSIONS

This article describes three ways to incorporate lifecycle costs as a criterion for determining the (economically) most profitable bid in the tender of functionally specified contracts. By considering lifecycle costs in a tender process, contractors are enticed and stimulated to design, develop and offer solutions that are profitable and beneficial on a longer term. This is not solely in the interest of the commissioner, but also offers contractors a better opportunity to distinguish themselves in terms of quality and service. In addition to this innovation is rewarded, as parties' stands to gain from design optimization for maintenance.

LCC-tender experiences thus far within Rijkswaterstaat are encouraging. Contractors are given the opportunity to distinguish themselves by offering a balanced lifecycle design and development of innovations that imply savings on lifecycle cost. By utilizing LCC-tenders better options emerge for commissioners designate and leave more freedom in design and responsibilities for design choices to market parties.

It appears however that this development within Rijkswaterstaat is merely a first step. Market parties are only starting to adjust to these new methods and commissions. On the side of the commissioners there is still a lot of room for improvement. It is of great importance that an adequate, stable and accepted set of reference data is compiled to be used in LCC-tenders. In addition to this experience must be gained with all of the three approaches described above, particularly with regards to effects that contract specifications and requirements related to "(economically) most profitable bid" criteria have on contractors' bid and tendering strategies.

REFERENCES

Bakker, J., Volwerk, J.J., Bosch, R., and Boendemaker 2005. C. Life Cycle Cost integrated in tenders for infrastructure projects; *proceedings 4th International Workshop on Life Cycle Cost Analysis and Design of Civil Infrastructure Systems (LCC4), Florida; 9–11 May 2005*, Cocoa Beach: Balkema.

Klatter, H.E., van Noortwijk, J.M., and Vrisou van Eck, N. 2002. Bridge management in the Netherlands; Prioritisation based on network performance. In J.R. Casas, D.M. Frangopol, and A.S. Nowak, editors, *First International Conference on Bridge Maintenance, Safety and Management (IABMAS), Barcelona, Spain, 14-17 July 2002*. Barcelona: International Center for Numerical Methods in Engineering (CIMNE).

J.M. van Noortwijk, and H.E. Klatter 2004. The use of lifetime distributions in bridge maintenance and replacement modelling. *Computers & Structures*, 82(13–14):1091–1099.

Bakker, J.D., van der Graaf, H.J., and van Noortwijk, J.M. (1999). Model of Lifetime-Extending Maintenance. In M.C. Forde, editor, *Proceedings of the 8th International Conference on Structural Faults and Repair, London, United Kingdom, 1999*. Edinburgh: Engineering Technics Press.

van Noortwijk, J.M., and Frangopol, D.M. 2004. Two probabilistic life-cycle maintenance models for deteriorating civil infrastructures. *Probabilistic Engineering Mechanics*.

Ministerie van Financien 2003. Kabinetstandpunt over waardering van risico's bij publieke uitvoeringsprojecten, http//www.minfin.nl/BZ03-1237.doc, 14 november 2003, Den Haag.

Bakker, J. Volwerk, J.J., and Verlaan, J. 2006. Maintenance management from an economical perspective. In J.R. Casas, D.M. Frangopol, and L.C. Neves, editors, *Third International Conference on Bridge Maintenance, Safety and Management (IABMAS), july 16–19 2006,* Porto: Balkema.

Life-Cycle Cost and Performance of Civil Infrastructure Systems – Cho, Frangopol & Ang (eds)
© 2007 Taylor & Francis Group, London, ISBN 978-0-415-41356-5

Lifetime analysis and structural repair of a cable-stayed bridge

Fabio Biondini
Department of Structural Engineering, Politecnico di Milano, Milan, Italy

Dan M. Frangopol
Department of Civil and Environmental Engineering, ATLSS Research Center, Lehigh University,
Bethlehem, Pennsylvania, USA

Pier Giorgio Malerba
Department of Structural Engineering, Politecnico di Milano, Milan, Italy

ABSTRACT: This paper presents the results of the probabilistic lifetime analysis performed for the structural repair of an existing cable-stayed bridge in Italy. To this aim a novel procedure for lifetime nonlinear analysis of concrete structures subjected to diffusive attacks from external aggressive agents is applied. Based on this procedure, both the deterministic and probabilistic time-variant structural performances are analyzed with respect to proper indicators considering two separate cases: without and with rehabilitation intervention. The effects of each random variable on such indicators are finally quantified and compared by means of suitable time-dependent sensitivity factors. The results demonstrates the applicability and effectiveness of the adopted intervention.

1 INTRODUCTION

The Certosa-Cable stayed bridge in Milan (Italy) was opened in 1989 (Figure 1.a). After about fifteen years of service, the Milan Municipality decided for a general and detailed inspection of the bridge (Figure 1.b). This inspection revealed some damage, especially at the bottom part of the pylons, induced by the interaction with the surrounding environment. Therefore, a structural repair has been carried out in order to protect the structure from future diffusive attacks of external aggressive agents.

The effectiveness the rehabilitation interventions has been investigated (Biondini *et al.* 2006e) by using a novel procedure for durability analysis of concrete structures subjected to diffusive attacks from external aggressive agents, recently proposed in Biondini *et al.* (2004, 2006a). Based on this procedure, both the deterministic and probabilistic time-variant structural performances are analyzed considering two separate cases: without and with rehabilitation intervention. The results of this analysis allow to compare what should happen in the next five decades in case the damaging stressors were left free to act without any intervention and what will be the state of the structure after the repair.

In the following, the main results are summarized. Detailed results are presented in Biondini *et al.* (2006c, 2006d, 2006e). The effects of the uncertainties associated with each random variable on the structural performance are also investigated by means of a time-dependent sensitivity factor based on a linear regression of the data resulting from the simulation.

2 CHARACTERISTICS OF THE BRIDGE

The bridge was designed by Francesco Martinez y Cabrera (2002). The total length of the Certosa bridge is 180 m, with a central span of 90 m and two lateral spans of 45 m, as shown in Figure 2.a.

(a)

(b)

Figure 1.   View of the Certosa bridge (a) at the end of construction, and (b) during repair activities.

The bridge deck is a five cells box girder, 19.80 m wide and 26.7° skewed with respect to the road axis. The transversal section has a depth varying from 1.40 m to 1.80 m and the upper slab has the same small transversal slope as the road (Figures 2.b,c). The upper and lower slabs are 0.22 m and 0.18 m thick, respectively. Along its span the box is stiffened by a set of transversal beams (Figure 2.d). Such beams end with two short cantilevers to which the cables are anchored. The transversal beams over the piers are deeper than the others, to hold the high forces transmitted by the pylons. The deck and the beams are prestressed with cables made up of 14 and 19 strands of 0.6 inches.

The pylons of the cable-stayed bridge are 19.00 m high and have a rectangular cross-section, varying along the height from 1.14×2.00 m at the top to 1.14×2.50 m at the bottom (Figure 2.d). The tapered shape of the frontal view finishes in two corbels which were planned to lift the bridge. The ends of the pylons carry the devices used to anchor the cables. There are in total three seats consisting of steel boxes which make it possible to anchor the cables leaving them crossed but not intersected on the same plane. Each pylon has three pairs of cables, skewed with respect to the horizontal plane as shown in Figure 2.d. The cables are composed, the longest to the shortest, by 2×45, 60, 45 strands of 0.6 inches.

3   INSPECTION, MONITORING AND REPAIR OF THE BRIDGE

The accurate design of the shape of the surfaces and the care spent in the reinforcement detailing, in the determination of an adequate cover thickness and in the choice of effective surface protective materials have contributed to the conservation of the Certosa bridge. Even though the structure had to withstand the severe loading conditions acting on the most important and busy access to Milan,

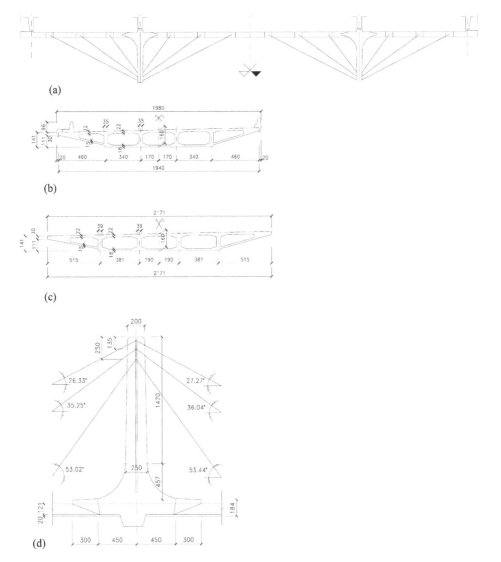

Figure 2. Certosa cable-stayed bridge. (a) Main dimensions of the bridge. (b–c) Main dimensions of the bridge deck: (b) straight cross-section; (c) skewed cross-section. (d) Main dimensions of the pylons.

it was in general in a good state. The deck was intact. Traces of damage were visible at the basis of the four pylons. The results of the first inspections revealed an initial trace of carbonation and local traces of corrosion in correspondence with cast interruptions. Spalling and initial steel corrosion have been detected only at the lower external parts of the pylons, which are directly exposed to traffic emissions.

Based on this inspection, the following repair interventions have been carried out (Biondini *et al.* 2006*c*, 2006*e*). After a strong sandblasting cleaning, the main cracks have been carved up to the basis of the crack and then sutured with tixotropic, anti-shrinkage, polypropylene fiber reinforced high strength mortar. The small cracks and the segregation zones have been repaired with high adhesion cement mortar, added with small polyvinilealcool fibers. The whole skin surfaces have been protected with high adhesion, high elasticity cement mortar, reinforced with double or simple

(a)   (b)

Figure 3.   View of a pylon (a) before and (b) after the rehabilitation intervention.

skin mesh. This protection contributes to contrast the restart of carbonation and it is transpiring outward and waterproof inward. The final surface was painted with a silicate non-pellicular paint. In this way, future diffusive attacks of external aggressive agents are prevented. Figure 3 shows a view of a pylon before and after the rehabilitation interventions.

Special attention has been paid to the definition of the layout of an essential and effective monitoring system based on topographical methods and electronic transducers (Figure 4).

4 DETERMINISTIC PREDICTION OF THE LIFETIME PERFORMANCE

The attention is focused on the cross-section at the base of the pylons, where the most evident traces of deterioration were present. Such cross-section has main nominal dimensions $b = 1.14$ m and $h = 2.50$ m, and is reinforced with 86 bars having nominal diameter $\varnothing = 26$ mm. The time-variant performance of the pylons is evaluated up to 50 years by considering two scenarios. In the first one the structure is left free to undergo damage without any intervention. In the second one the effects of the rehabilitation are taken into account after fifteen years of service by (a) assuming an undamaged state for the restored layer of the concrete cover and (b) placing a diffusive barrier along the boundary of the cross-section in such a way that diffusion of aggressive agents from outside is stopped and future damage can be induced only by the agent already existing inside the structure.

In the investigated scenarios, the cross-section is assumed to be subjected to a diffusive attack from an environmental aggressive agent, which is considered to be located along the whole external perimeter with constant concentration $C(t) = C_0$. The diffusion process is described according to the Fick's laws and is effectively simulated by using a special class of evolutionary algorithms called cellular automata (Biondini *et al.* 2006*a*). With reference to a nominal diffusivity coefficient $D = 10^{-11}$ m²/sec for concrete, the diffusion process associated with the two investigated scenarios, without and with rehabilitation intervention, is described by the maps of concentration shown in Figure 5. The direct comparison of the concentration maps in Figures 5.a and 5.b highlights the

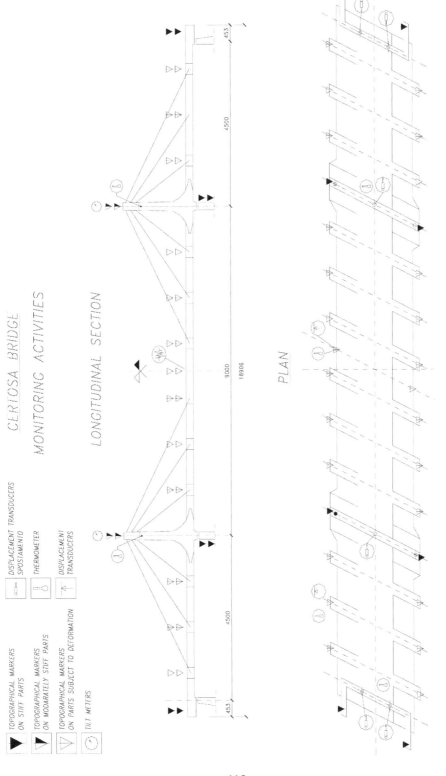

Figure 4. Monitoring system of the Certosa cable-stayed bridge.

115

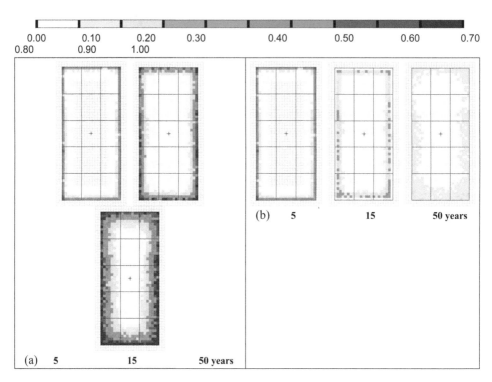

| 0.00 | 0.10 | 0.20 | 0.30 | 0.40 | 0.50 | 0.60 | 0.70 |
| 0.80 | 0.90 | 1.00 | | | | | |

Figure 5. Maps of concentration $C(t)/C_0$ of the aggressive agent after 5, 15, and 50 years from the initial time of diffusion penetration. (a) Damaged structure. (b) Rehabilitated structure.

high effectiveness of the rehabilitation intervention with regards to the limitation of the diffusive attack of the aggressive agent.

Structural damage induced by diffusion is modeled by introducing a degradation law of the effective resistant area for both concrete matrix and steel bars. In this study, damage is coupled to the diffusion process by assuming, for both materials, a linear relationship between the rate of damage and the mass concentration of the aggressive agent. The proportionality coefficients which define such linear relationships are denoted by $q_c = (C_c \Delta t_c)^{-1}$ and $q_s = (C_s \Delta t_s)^{-1}$, where C_c and C_s represent the values of constant concentration which lead to a complete damage of the materials, concrete and steel, after the time periods Δt_c and Δt_s, respectively (Biondini et al. 2006a). For the case under investigation the nominal values $C_c = C_s = C_0$, $\Delta t_c = 25$ years and $\Delta t_s = 50$ years, are adopted.

Several parameters could be adopted as suitable measures of structural performance. As an example, the mechanical damage induced by diffusion can be evaluated from the diagram in Figure 6, which shows the time evolution of the dimensionless bending moment at yielding $m_y = M/(|f_c|A_{c0}h)$ under the axial force $n = N/(|f_c|A_{c0}) = -0.201$, where $N = -20$ MN. The nominal values of the main parameters which define the non linear constitutive laws of the materials are also listed in Figure 6. Such diagram shows that the adopted rehabilitation intervention leads not only to an increase of structural performance, but, mainly, to a relevant decrease of the rate of deterioration and, consequently, to a significant reduction of the costs related to future inspections and maintenance activities.

Such a comparison makes the beneficial effects of rehabilitation very clear from a qualitative point of view. However, the previous deterministic results cannot be used for reliable quantitative predictions because of the unavoidable sources of uncertainty involved in the problem. For this reason, a probabilistic assessment of the service life is also carried out, and the comparison between the two investigated scenarios is made by taking several uncertainties into account.

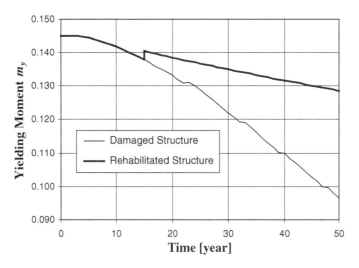

Concrete (Saenz's law):
- compression strength f_c=−35 MPa
- tension strength f_{ct}=0.25| f_c |$^{2/3}$
- initial modulus E_{c0}=9500| f_c |$^{1/3}$
- strain limit in compression ε_{cu}=−0.35%

Steel (elastic-perfectly plastic law):
- yielding strength f_{sy}=500 MPa
- elastic modulus E_s=206 GPa
- strain limit ε_{su}=1.00%

Figure 6. Time evolution of the yielding moment of the cross-section at the base of the pylons.

Table 1. Probability distributions and their parameters.

Random variable ($t = t_0$)	Distribution type	m	σ
Concrete strength, f_c	Lognormal	$f_{c,nom}$	5 MPa
Steel strength, f_{sy}	Lognormal	$f_{sy,nom}$	30 MPa
Coordinates of the nodal points, (y_i, z_i)	Normal	$(y_i, z_i)_{nom}$	5 mm
Coordinates of the steel bars, (y_m, z_m)	Normal	$(y_m, z_m)_{nom}$	5 mm
Diameter of the steel bars, $\varnothing_m$	Normal [*]	$\varnothing_{m,nom}$	$0.10\,\varnothing_{m,nom}$
Diffusion coefficient, D	Normal [*]	D_{nom}	$0.10\,D_{nom}$
Concrete damage rate, q_c	Normal [*]	$q_{c,nom}$	$0.30\,q_{c,nom}$
Steel damage rate, q_s	Normal [*]	$q_{s,nom}$	$0.30\,q_{s,nom}$

[*] Truncated distributions with non negative outcomes are adopted in the simulation process.

5 PROBABILISTIC LIFETIME ASSESSMENT

The probabilistic model for the service life assessment assumes as random variables the material strengths f_c and f_{sy}, the coordinates (y_p, z_p) of the nodal points $p = 1, 2, \ldots$ which define the two-dimensional model of the concrete cross-section, the coordinates (y_m, z_m) and the diameter $\varnothing_m$ of the steel bars $m = 1, 2, \ldots$, the diffusion coefficient D, and the damage rates $q_c = (C_c \Delta t_c)^{-1}$ and $q_s = (C_s \Delta t_s)^{-1}$. These variables are assumed to have the probabilistic distribution with the mean μ and standard deviation σ values listed in Table 1 (Biondini *et al.* 2006b).

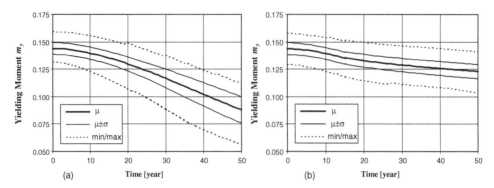

Figure 7.   Time evolution of the yielding bending moment: mean μ (thick line), standard deviation σ from the mean μ (thin lines), minimum and maximum values (dotted lines). (a) Damaged structure. (b) Rehabilitated structure.

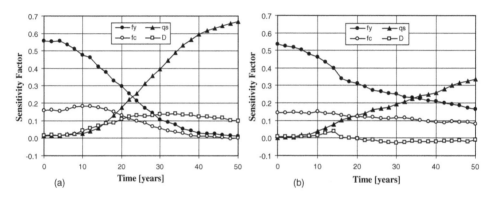

Figure 8.   Sensitivity factors of the yielding moment associated with the uncertainty on material strengths f_c, f_y, concrete diffusivity D, and damage steel rate q_s. (a) Damaged structure. (b) Rehabilitated structure.

A probabilistic measure of the time-variant structural performance is achieved by Monte Carlo simulation. The sampling process is based on the antithetic variables technique and the goodness of the sample size is evaluated by means of a posteriori error estimation. With reference to a sample of 2000 simulations for each one of the two investigated scenarios, Figure 7 shows the time evolution of the statistical parameters (mean value μ, standard deviation σ, minimum and maximum values) of the performance indicators during the first 50 years of service life. The direct comparison of Figures 7.a and 7.b demonstrates the high effectiveness of the adopted rehabilitation intervention. More in detail, Figure 7 highlights the effects of randomness, which lead (1) to reduce the magnitude of the instantaneous increase of performance, and (2) to better emphasize the delayed beneficial effects of rehabilitation, since a noteworthy decrease in the dispersion of the yielding moment is achieved.

Clearly, the relative importance of the effects associated with each one of the random variables listed in Table 1 may significantly vary during time. To investigate this point, a time-variant least squares linear regression is performed on the data samples obtained from the simulation. In this way, the regression coefficients can be assumed as a time-dependent measure of the sensitivity of the yielding bending moment with respect to each random variable (Biondini *et al.* 2006*b*).

The results of the regression analysis are shown in Figure 8 for the random variables which mainly affect the yielding moment: material strengths f_c and f_y, concrete diffusivity D, and damage steel rate q_s. The direct comparison of Figures 8.a and 8.b shows that the rehabilitation intervention strongly influences the time-evolution of the uncertainty effects associated with such variables. In particular, in the undamaged scenario the yielding moment mainly depends on both the steel

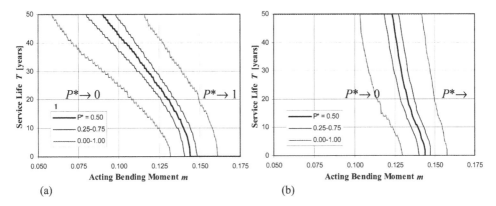

Figure 9. Service life associated to given values of probability of failure P* versus given target levels of the yielding bending moment. (a) Damaged structure. (b) Rehabilitated structure.

strength and concrete strength. For the damaged structure such dependency quickly decreases during time, and after about 25 years the steel damage rate and the concrete diffusivity become the more important parameters for the whole remaining service life (Figure 8.a). For the rehabilitated structure a similar tendency is observed, but the primary role played by the material strength is maintained longer, until about 35 years for the steel strength and during the whole service life for the concrete strength. In addition, at the end of the service life the relative importance of the steel strength and of the steel damage rate remains comparable (Figure 8.b).

Based on the probability functions $P = P(t)$ associated with the failure condition $m < m_y$ and computed from the simulation process, the service life T associated with given failure probability thresholds P^* can be finally evaluated. Figure 9 shows the service life T associated with the five reliability levels $P^* = [0.00, 0.25, 0.50, 0.75, 1.00]$ as a function of the expected values of the performance indicators. These curves allow to assess the remaining service life which can be assured under prescribed reliability levels without maintenance. The direct comparison of Figures 9.a and 9.b shows how the rehabilitation intervention is able to extend the service life of the structure.

6 CONCLUSIONS

This paper presented the main aspects of the activities involved in the structural repair of the Certosa cable stayed bridge in Milan (Italy). An overview of the field inspections and of the analyses carried out for modeling the damage effects, of the design of the repair interventions and of the field activities during the restoration, is given. In particular, in order to outline the importance of an early diagnosis and to investigate the effectiveness of the adopted rehabilitation interventions, a lifetime nonlinear analysis under diffusive attacks from external agents has been developed. In this analysis, both the deterministic and probabilistic time-variant structural performances are analyzed with respect to proper indicators considering two separate cases: without and with rehabilitation intervention. The direct comparison of the results demonstrated that the adopted rehabilitation strategies lead not only to an increase of structural performance over time, but, mainly, to a relevant decrease of the rate of deterioration and, consequently, to a significant reduction of the costs related to future inspections and maintenance actions.

ACKNOWLEDGEMENTS

The Certosa cable-stayed bridge was designed by Francesco Martinez y Cabrera, together with the S.P.E.A. Engineering Technical Office and with the collaboration of Pier Giorgio Malerba for specific structural analysis problems.

REFERENCES

Biondini F, Bontempi F, Frangopol DM, Malerba PG. Cellular Automata Approach to Durability Analysis of Concrete Structures in Aggressive Environments. *ASCE Journal of Structural Engineering*, 130(11), 2004, 1724–1737.

Biondini F, Bontempi F, Frangopol DM, Malerba PG. Probabilistic Service Life Assessment and Maintenance Planning of Concrete Structures. *ASCE Journal of Structural Engineering*, 132(5), 2006a, 810–825.

Biondini F, Bontempi F, Frangopol DM, Malerba PG. Lifetime Nonlinear Analysis of Concrete Structures under Uncertainty. *Bridge Maintenance, Safety, Management, Life-Cycle Performance and Cost* (Edited by P.J.S. Cruz, D.M. Frangopol, and L.C.Neves), Taylor & Francis Group plc, London, 2006b, pp. 149–150, and full 8 page paper on CD-ROM.

Biondini F, Malerba C, Malerba PG, Mantegazza G. Inspection, Monitoring and Structural Repair of the Certosa Cable Stayed Fly Over. *2nd fib Congress*, Naples, June 5–8, 2006c.

Biondini F, Frangopol DM, Malerba PG. Probabilistic Lifetime Analysis of an Existing Cable-stayed Bridge. *European Symposium on Service Life and Serviceability of Concrete Structures* (ESCS-2006), Helsinki, June 12–14, 2006d, 258–263.

Biondini F, Frangopol DM, Malerba PG. Time-variant Performance of the Certosa Cable-Stayed Bridge. *Structural Engineering International*, 16(3), 2006e, 235–244.

Martinez Y, Cabrera F. *Collected Papers – In memory of Francesco Martinez Y Cabrera*, Politecnico di Milano, 2002, 421–426, 427–436.

Life-Cycle Cost and Performance of Civil Infrastructure Systems – Cho, Frangopol & Ang (eds)
© 2007 Taylor & Francis Group, London, ISBN 978-0-415-41356-5

Optimal seismic retrofit and maintenance strategy for steel bridges using Life-Cycle Cost analysis

Hyo-Nam Cho & Hyun-Ho Choi
Hanyang University, An-San, Korea

Kwang-Min Lee
Daelim Industrial Co. Ltd., Seoul, Korea

Kyung-Hoon Park
Korea Institute of Construction Technology, Koyang, Korea

ABSTRACT: This study is to develop a realistic methodology for determination of the optimal seismic retrofit and maintenance strategy of deteriorating bridges. The proposed methodology is based on the concept of minimum LCC which is expressed as the sum of present value of seismic retrofit costs, expected maintenance costs, and expected economic losses with the constraint such as design requirements. The proposed methodology is applied to the LCC-effective optimal seismic retrofit and maintenance strategy of a steel bridge considered as an example bridge in the accompanying proceeding paper. From the numerical investigation, it may be positively expected that the proposed methodology can be effectively utilized as a practical tool for the decision-making of LCC-effective optimal seismic retrofit and maintenance strategy of deteriorating bridges.

1 INTRODUCTION

Bridges are key nodes in any transportation network and essential to regional and/or national economies. Reports from recent strong earthquakes have shown that major earthquakes may cause enormous indirect losses such as degradation of productivity in regional economic activities and difficulties in emergency response and subsequent recovery as well as direct losses such as physical damages and human losses.

These recent significant earthquakes had an enormous impact on awakeneing public concern about potential earthquake hazard mitigation. Since, in Korea, many bridges designed prior to the adoption of seismic design specification still exist without any aseismic retrofit measures, Korea Infrastructure Safety and Technology Corporation (KISTEC) have evaluated seismic performance of existing 2,926 bridges, and have established a seismic retrofit strategy that invest 287 billion won to most venerable 1,757 bridges (KISTEC, 2001). However, the seismic retrofit strategy is mainly based on deterministic approach using current engineering practice and design requirements without consideration of uncertainties in estimation of seismic performance. Thus, it is very doubtful whether the strategy is optimal decision-making with the balance of safety and economy or not.

Thus, in this study, a realistic methodology for determination of the optimal seismic retrofit and maintenance strategy of deteriorating bridges is proposed. The proposed methodology is based on the concept of minimum LCC which is expressed as the sum of present value of seismic retrofit costs, expected maintenance costs, and expected economic losses with the constraint such as design requirements. The proposed methodology is applied to the LCC-effective optimal seismic retrofit and maintenance strategy of a steel bridge considered as an example bridge in the accompanying proceeding paper(LCC5, 2006).

2 FORMULATION OF PROBABILISTIC LIFE-CYCLE COST FUNCTIONS

An optimal decision-making, chosen from multiple alternative seismic retrofit and maintenance strategies, can then be found by minimizing LCC. The formulation for the LCC-effective optimum seismic retrofit and maintenance strategy of deteriorating bridges with design requirements can be represented as follows:

$$\text{Find} \qquad \text{Seismic Retrofit and/or Maintenance Strategy} \qquad (1a)$$

$$\text{Minimize} \qquad E\left[C_T^{PV}\right] = C_{R_i}^{PV}(t_r) + E\left[C_{M_j}^{PV}\right] + E\left[C_{EL_{ij}}^{PV}(t_r)\right] \qquad (1b)$$

$$\text{Subject to} \qquad g_k(\cdot) \le 0 \qquad (1c)$$

where, $E[C_T^{PV}]$ = total expected LCC in present worth; i = an index for seismic retrofit measure; j = an index for maintenance strategy; k = an index for design requirement; $C_{R_i}^{PV}$ = total expected LCC in present worth; t_r = planned seismic retrofit cost in present worth; $E[C_{M_j}^{PV}]$ = time at which seismic retrofit is implemented; $E[C_{EL_{ij}}^{PV}]$ = expected lifetime maintenance cost in present worth; = expected lifetime economic losses in present worth; and $g_k(\cdot)$ = design requirement.

2.1 Seismic retrofit costs

Since uncertainties in estimating the planned seismic retrofit cost may be much less than those of the other cost items, the seismic retrofit cost is considered as a constant in Eq. (2). The present worth of planned seismic retrofit cost can be formulated as the following equation:

$$C_R^{PV}(t_r) = \left(C_{RD} + C_{RC} + C_{RT}\right) \cdot i(t_r) \qquad (2)$$

where, C_{RD} = planning and design cost; C_{RC} = construction cost; C_{RT} = testing cost; $i(t)$ = discount rate function, which can be expressed as $1/(1+q)^t$; and q = discount rate.

2.2 Expected maintenance costs

The maintenance costs can be classified in terms of the costs which are related to bridge performance (e.g., painting cost, corrosion protection, and etc.) and bridge management (e.g., periodic routine maintenance). Since the latter might have a minor influence on the bridge performance, they can be ignored in the LCC formulation. Thus, in this study, the expected maintenance costs which are only related to bridge performances are considered as follows:

$$E\left[C_{M_j}^{PV}\right] = \int_0^L \left[C_{DM} + C_{IDM}\right] \cdot P_{M_j}(t) \cdot i(t) \, dt \qquad (3)$$

Where, j = an index for maintenance strategy; C_{DM}, C_{IDM} = direct and indirect maintenance cost; $P_{M_j}(t)$ = maintenance probability of a bridge at time t; L = life span.

2.3 Expected economic losses

Most of the expected rehabilitation cost functions have been formulated as a function of damage of bridge system. However, except for collapse damage, structural damage is usually an event of damage of structural components rather than that of system. And the countermeasures for rehabilitation of each damaged component may cause different indirect cost due to different traffic closure. Therefore, it may be more reasonable to formulate the problem as a function of component damages as well as system damage. The expected lifetime economic losses are formulated as follows:

$$E\left[C_{EL_{ij}}^{PV}\right] = \int_0^{t_r} \left[E\left[C_{R_{0j}}(t)\right] + E\left[C_{H_{0j}}(t)\right]\right] \cdot i(t) \, dt + \int_{t_r}^{t_L} \left[E\left[C_{R_{ij}}(t)\right] + E\left[C_{H_{ij}}(t)\right]\right] \cdot i(t) \, dt \qquad (4a)$$

122

$$E\left[C_{R_{0j}}(t)\right] = \sum_m \sum_k \left[C_{DR_m}^k + C_{IDR_m}^k\right] \cdot P_{f_m^{0j}}^k(t) + \left[C_{DR_{sys}}^C + C_{IDR_{sys}}^C\right] \cdot P_{f_{Sys}^{0j}}^C(t) \tag{4b}$$

$$E\left[C_{R_{ij}}(t)\right] = \sum_m \sum_k \left[C_{DR_m}^k + C_{IDR_m}^k\right] \cdot P_{f_m^{ij}}^k(t) + \left[C_{DR_{sys}}^C + C_{IDR_{sys}}^C\right] \cdot P_{f_{Sys}^{ij}}^C(t) \tag{4c}$$

where, $0 =$ an index for as-built condition; $i =$ an index for seismic retrofit measure; $j =$ an index for maintenance strategy; $k =$ an index for damage state of components (i.e., slight damage, moderate damage, extensive damage); $m =$ an index for bridge component; $E[C_{R_0}]$, $E[C_{H_0}] =$ expected rehabilitation costs; $E[C_{R_{ij}}]$, $E[C_{H_{ij}}] =$ human loss and property damage cost; $C_{DR_m}^k$, $C_{DR_{sys}}^C =$ direct rehabilitation cost of component m and bridge system, respectively; $C_{IDR_m}^k$, $C_{IDR_{sys}}^C =$ indirect rehabilitation cost of component and bridge system, respectively; $P_{f_m^{0j}}^k(t)$, $P_{f_m^{ij}}^k(t) =$ annual probability of exceeding damage state; $P_{f_{sys}^{0j}}^C(t)$, $P_{f_{sys}^{ij}}^C(t) =$ annual probability of exceeding collapse (system) damage state.

2.4 Indirect costs

2.4.1 Road user cost

To evaluate the rational road user costs, the essential factors such as traffic network, location of bridge, and the information on maintenance and rehabilitations (i.e., work zone condition, detour rate, the change of traffic capacity of traffic network, etc.) must be considered. However, so far, these factors have not considered in the previous road user cost model. To rationally evaluate road user cost, the effect of detour rates in traffic network and the work-zone condition should be considered. For the purpose, a road user cost model proposed by author is introduced in this study. Details on the road user cost model can be founder available references (Lee et al., 2005).

To evaluate the road user cost, the information about the traffic network conditions, such as the average traffic speed on the original and detour routes during normal condition and rehabilitation activity, and the detour rates from original route to detour route, etc, is essential. These data are undoubtedly functions of traffic volume, number of detour routes, number of detour route lanes, and length of detour route, etc., which can be obtained from traffic network analysis. For the purpose, EMME/2 v5.1 (Inro, 1999) is used in this study. And, for the traffic network analysis using EMME/2, ADTVs on Origin-Destination (O-D) and locations of original route and detour routes are necessary as input data. The data can be obtained from Korea transportation data base (http://www.ktdb.go.kr).

Also, the restoration time is very important parameters for reasonable estimation of the road user cost. In this study, the restoration time according to damage state and bridge component is based on expert option to reflect engineering practice in Korea.

2.4.2 Socio-economic losses

Socio-economic losses are the result of multiplier or ripple effect on the economy caused by functional failure of a structure. The indirect socio-economic losses applicable to bridge LCC problems were derived by Cho (2002), based on an extension of the I-O model. However, unfortunately, in Korea, the I-O tables for local areas except for Seoul are not investigated and the I-O table of Seoul is made only from the view point of macro economic analysis rather than micro economic analysis for an area where traffic flowing is influenced. As an alternative way, the suggestion from the reference (Seskin, 1990) may be adopted for the approximate but reasonable assessment of the cost. In the reference, it is reported that socio-economic losses could range approximately from 50 to 150% of road user costs.

3 ALGORITHM FOR LCC ASSESSMENT

Figure 1 shows a conceptual flow for LCC assessment which is used for the LCC-effective optimal seismic retrofit and maintenance strategy of bridges. The algorithm essentially consists of

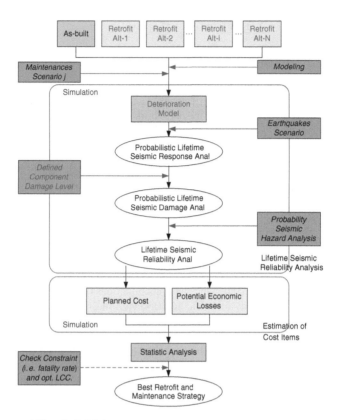

Figure 1. Conceptual Flow for LCC Assessment.

select available seismic/maintenance strategy and modeling for non-linear seismic analysis, lifetime seismic reliability analysis, LCC assessment and selection of best strategy.

A realistic lifetime seismic-reliability based approach is unavoidable to evaluate annual probabilities of exceeding specific damages of deteriorating bridges. For the purpose, a program HYPER-DRAIN2DX-DS which is proposed by author in the accompanying proceeding paper is introduced in this study.

As shown in eq (3), to compute expected maintenance cost, maintenance probability of a bridge is essential. The maintenance probability of a bridge under a maintenance strategy depends on various parameters with uncertainties. Furthermore, to assess annual maintenance probability under sequence of maintenance actions, evaluation of conditional joint probability is necessary. This joint probability problem is accompanied by multiple integral solving steps. This is usually not efficient from the computation point of view. Therefore, in this study, for the computational efficiency, the modified Event Tree Analysis (ETA) proposed by Kong (2001) is introduced. In application of the ETA method, to predict lifetime maintenance probability, information on a serviceable life up to specific condition state is essential. For the purpose, in this study, the time-variant resistance model proposed accompanying proceeding paper is introduced.

4 ILLUSTRATIVE EXAMPLE AND DISCUSSIONS

4.1 *Assumptions*

The proposed methodology is applied to the LCC-effective optimal seismic retrofit and maintenance strategy of a steel bridge considered as an example bridge in the accompanying proceeding paper.

Table 1. Estimation of Seismic Retrofit Cost with respect to Seismic Strategies.

	Planning/ Design Cost	Construction Cost	Testing Cost	Total
Elastomeric Bearing[1]	85.909	6.589	2.448	94.947
Restrainer Cable[2]	23.100	2.100	0.776	25.976

[1]Case-EB: installation 66 Elastomeric Bearings (EBs); [2]Case-REC: installation 66 Restrainer Cables (RECs).

The estimation of LCC of the example bridge will be performed under following assumptions:

(i) It is assumed that the deterioration of the example bridge is already progressed over 25years time periods.
(ii) The life span of the example bridge is set as 50 years. Thus, the time periods for LCC Analysis (LCCA) are from 2005 to 2055 year. And, it is assumed that the seismic retrofit strategy is planned to be implemented in 2005 year.
(iii) Based on statistical analysis from Ministry of Construction and Transportation (2004a), 4.00% discount rate is applied as mean discount rate in this example.
(iv) The example bridge has been constructed as a part of a typical urban expressway which has large ADTV or a typical rural highway which has relatively moderate ADTV (large ADTV have about 2.8 time larger ADTV than moderate ADTV region). Traffic network have two detour rate and original route.

4.2 Data for estimating life-cycle costs

The total expected LCC of a bridge is composed of the present value of seismic retrofit costs, expected direct/indirect maintenance costs, and expected economic losses. First, seismic retrofit cost is estimated based on construction cost (Korea Highway Corp., 2004) and planning/design cost and testing cost (Ministry of Science and Technology, 2004). Table 1 shows the estimation of seismic retrofit cost with respect to seismic strategies.

Second, for the estimation of expected direct/indirect maintenance costs, as presented in accompanying proceeding paper, the time-variant resistance degradations in piers due to the corrosion of steel reinforcement are considered. In this study, the minor repair method is applied as the maintenance countermeasure of piers since it is assumed that the grown crack on cover concrete and corroded steel reinforcement are perfectly recovered to intact state by a maintenance action. The unit maintenance cost is taken as 253,670 won per surface area. And the average amount of virtual defect for condition state C and D is selected as 14.8% and 19.2% of surface area, respectively, based on KISTEC (2000). From the above information, the direct maintenance costs using minor repair method for condition state (CS)-C and CS-D are estimated to be 2.026 and 2.625 Million won.

Third, expected economic losses are consisted of expected direct/indirect rehabilitation costs related to various damage states of bridge components and human losses and property damage costs with or without seismic retrofit and maintenance strategy as formulated in previous section. As shown in Table 2, direct rehabilitation cost and restoration time data Related to damage states are estimated based on Shoji et al. (1997), the CSR's price information (http://www.csr.co.kr), and Experts Opinion.

Data for assessment of indirect rehabilitation costs are based on various available references (Do, 1998; Won, 2000; Lee and Shim, 1997; Kyoung-gi Research Institute, 1999; Seskin, 1990; etc) and available information from web site such as Korea transportation Data Base (http://www.ktdb.go.kr), the construction software research's price information (CSR, http://www. csr.co.kr), the city of Seoul (http://traffic.metro.seoul.kr), Korea Transportation Institute (http://www.koti.re.kr), Korean Statistical Information System (http://wwwsearch.nso.go.kr). Details on Data for assessment of indirect rehabilitation costs can be found in Lee (2006).

Table 2. Direct rehabilitation cost and restoration time data.

Component	DS[1]	Rehabilitation Countermeasure	Ratio of Construction Cost	Restoration Time	Traffic Closure
Deck/Bearing	S	– Replacement bearing[2]	4.3%	5[4]–7[5]–9[6]	–
		– Welding Wedges of EB[3]		–	
	M	– Replacement bearing	11.7%	5–7–9	Type I
	E	– Replacement bearing	130%	5–7–9	Type II
Pier	S	– Minor repair of deck	3.27%	–	–
	M	– Moderate repair of deck	26.2%	12–14–16	Type I
	E	– Replacement pier	130%	45–51–55	Type II
Abutment	S	– Minor repair of abutment	3.27%	–	–
	M	– Moderate repair of abutment	26.2%	17–20–23	Type I
	E	– Replacement abutment	130%	50–65–70	Type II
System Collapse		– Reconstruction	130%	520–540–560	Type II

(1) S: Slight Damage; M: Moderate Damage; E: Extensive Damage   (4) Minimum restoration time (day)
(2) Countermeasure for as-built bridge, retrofitted bridge using REC   (5) Most Probable restoration time (day)
(3) Countermeasure for retrofitted bridge using Ebs   (6) Maximum restoration time (day)
Type I: Operational with some restrictions w.r.t truck and speed limit about 30 km/hr
Type II: No operation during rehabilitation activity

Table 3. Result of Traffic Network Analysis – Average Traffic Speed (km/hr).

	Large ADTV			Moderate ADTV		
	Normal	Type-I	Type-II	Normal	Type-I	Type-II
Original Route	60.45	30.00	–	73.03	30.00	–
Detour Route	56.88	56.09	47.62	71.64	71.12	68.15

Table 4. Result of Estimating Expected Rehabilitation Cost (Million Won).

	Large ADTV			Moderate ADTV		
	Super Structure / Abutment / Pier					
Damage	Slight	Moderate	Extensive	Slight	Moderate	Extensive
Direct	1.43/2.56/2.62	1.43/20.53/20.98	1.43/101.9/104.1	1.43/2.56/2.62	1.43/20.53/20.98	1.43/101.9/104.1
Indirect	-	870/1,741/2,487	5,794/33939/37,251	-	107/214/306	559/3,275/3,595
	Bridge System Collapse					
Direct	1,571.69			1,571.69		
Indirect	447,012.00			43,140.00		

4.3 *Result of traffic analysis and estimating expected rehabilitation cost*

Traffic network analysis is performed using traffic data and EMME/2 program (Inro, 1999). As results, traffic network analysis for original route and detour routes are represented. The reductions of average traffic speed on detour routes due to Type-I traffic closure are small regardless ADTV. However, in case of Type-II traffic closure, the reduction of average traffic speed for large ADTV region is larger than that for moderate ADTV region (traffic congestion effect).

Based on traffic network analysis, the results of estimating expected rehabilitation cost are represented in Table 4. As shown in Table 4, in estimating indirect rehabilitation cost, the socio-economic losses for large and moderate ADTV region are assumed as 1.5 and 0.5 times of the user costs based on Seskin's study (1990). It may be noted that the indirect cost dominates

Table 5. Cases Considered in LCCA.

Case ID	ADTV	Seismic Retrofit Strategy	Case ID	ADTV	Seismic Retrofit Strategy
Case-O-L	Large	Do Nothing	Case-O-M	Moderate	Do Nothing
Case-EB-L		Replacement of steel bearings by EB	Case-EB-M		Replacement of steel bearings by EB
Case-REC-L		Installation of REC	Case-REC-M		Installation of REC

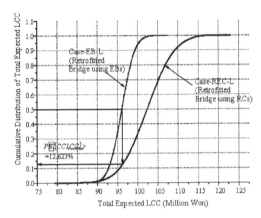

Figure 2. PDF of Total Expected LCC.

the rehabilitation cost if traffic closing is to occur, because, it is about 91~99% of the total rehabilitation cost. Thus, it should be regarded as one of the most important costs in the evaluation of the expected life-cycle cost. It is observed that the indirect rehabilitation costs for large ADTV region (about 2.8 time larger ADTV than moderate ADTV region) are about 10 times expensive then those for moderate ADTV region (also traffic congestion effect).

4.4 Effect of seismic retrofit strategies on total life-cycle cost

The effects of the two seismic retrofit strategies (Replacement of steel bearings by Elastomeric Bearing – EB, Installation of Restrainer Cable – REC) with ordinary maintenance action on total expected LCC of large and moderate ADTV region are investigated using the previously described data and the results of lifetime seismic reliability analysis from accompany proceeding paper.

For As-built bridges (Case-O-L, Case-O-M), the contribution of the expected rehabilitation cost occupies most in total expected LCC. On the other hand, it is observed that the seismic retrofit costs as well as the expected rehabilitation costs for retrofitted bridges contribute most in total expected LCC, since the annual probabilities of exceeding 4 different damages are decreased by using seismic retrofit strategies. It can be found in accompanying Proceeding paper.

For large ADTV region, it may be stated that Case-EB-L (retrofitted bridge using EB) most effective in reducing the expected rehabilitation cost due to pier damages (though Case-EB-L is most economical seismic retrofit strategy, the probability that the total expected LCC of Case-REC-L is less than the mean LCC of Case-EB-L is 12.62%, since the proposed methodology considers uncertainties such as variations of annual probability of exceeding damages, required working days for rehabilitation activity, etc, probabilistic information of total expected LCC can be captured quantitatively).

However, Case-REC-M (the retrofitted bridge using RCs) is most the economical seismic retrofit strategy is selected as the LCC-effective optimal seismic retrofit strategy in case of moderate ADTV

127

Table 6. Result of Estimation of Each Cost Items (I) (Million Won).

Case ID	Seismic Retrofit Cost	Expected Rehabilitation Cost		Human losses and property damage costs	Total Expected LCC
		Direct Reh. Cost	Indirect Reh. cost		
Case-O-L	0	1.223	635.686	0.074	636.983
Case-EB-L	66.463	1.456	28.212	0.007	96.137
Case-REC-L	33.398	0.241	68.788	0.023	102.45
Case-O-M	0	1.223	57.386	0.027	58.635
Case-EB-M	66.463	1.456	2.671	0.003	70.592
Case-REC-M	33.398	0.241	6.832	0.008	40.478

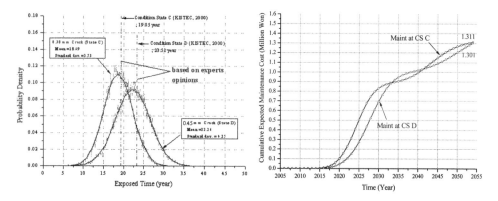

Figure 3. Probability of Maintenance and Cumulative Expected Maintenance Cost.

region (Since the indirect costs for moderate ADTV region are about 0.1 times lower then those for large ADTV region, the contributions of the seismic retrofit costs are to be more important factor then that for the expected rehabilitation costs).

4.5 Effect of seismic retrofit strategies on total life-cycle cost with maintenance strategies

The effects of the two seismic retrofit strategies (Replacement of steel bearings by EB, Installation of REC) without or with two maintenance strategies (maintenance at crack condition state C-Maint-C and crack condition state D- Maint-D) of large and moderate ADTV region on LCC are investigated in this section.

Figure 3 shows probability maintenance and cumulative expected maintenance costs for two maintenance strategies. As shown in the figure, though maintenance cost per each maintenance action of Maint-D (2.625 million won) is more expensive then that of Maint-C (2.026 million), since probability of maintenance for Maint-C is larger then that for Maint-D (the mean application time of Maint-C (18.69 years) is less then that of Maint-C (22.26 years), cumulative expected maintenance cost for Maint-D (1.301 million won) is slightly smaller then that for Maint-C (1.311 million).

Table 7 shows optimal seismic retrofit strategy and maintenance strategies for large ADTV region. Though the expected maintenance cost of Maint-C is more expensive compared with that of Maint-D, the total expected LCC of the retrofitted bridge using REC with Maint-C is more economical in view point of LCC compared with that with Maint-D since the Maint-C provides much more reduced expected rehabilitation cost than the Maint-D. Also, Table 8 shows optimal seismic retrofit strategy and maintenance strategies for moderate ADTV region. Since the indirect costs for moderate ADTV region are about 0.1 times lower then those for large ADTV region, it can easily expected that the contributions of the seismic retrofit and expected maintenance

Table 7. Result of Estimation of Each Cost Items (II): Large ADTV Region (Million Won).

Case ID of Optimal Seismic Retrofit Strategy	Maintenance Strategy	Seismic Retrofit Cost	Expected Maintenance Cost	Expected Rehabilitation Cost		Human losses and property damage costs	Total Expected LCC
				Direct	Indirect		
Case-EB-L	Ordinary	66.463	0.000	1.456	28.212	0.007	96.137
Case-REC-L	Maint.	33.398	0.000	0.241	68.788	0.023	102.450
Case-EB-L	Maint-C	66.463	1.311	1.179	22.724	0.006	91.683
Case-REC-L		33.398	1.311	0.193	55.895	0.018	90.815
Case-EB-L	Maint-D	66.463	1.301	1.223	24.679	0.006	93.672
Case-REC-L		33.398	1.301	0.200	57.198	0.019	92.116

Table 8. Result of Estimation of Each Cost Items (III): Moderate ADTV Region (Million Won).

Case ID of Optimal Seismic Retrofit Strategy	Maintenance Strategy	Seismic Retrofit Cost	Expected Maintenance Cost	Expected Rehabilitation Cost		Human losses and property damage costs	Total Expected LCC
				Direct	Indirect		
Case-EB-M	Ordinary	66.463	0.000	1.456	2.671	0.003	70.592
Case-REC-M	Maint.	33.398	0.000	0.241	6.832	0.008	40.478
Case-EB-M	Maint-C	66.463	1.311	1.179	2.156	0.002	71.112
Case-REC-M		33.398	1.311	0.193	5.558	0.006	40.466
Case-EB-M	Maint-D	66.463	1.301	1.223	2.245	0.002	71.234
Case-REC-M		33.398	1.301	0.200	5.781	0.007	40.687

costs become more important factor then that for the other cost items. Therefore, as expected, the retrofitted bridge using REC which is most the economical seismic retrofit strategy is selected as LCC-effective optimal seismic retrofit strategy. And, similar to large ADTV region, Maint-C is selected as most economical maintenance strategy.

5 CONCLUSION

A realistic methodology for determination of the optimal seismic retrofit and maintenance strategy of deteriorating bridges is proposed in this study. It may be positively stated that the proposed methodology can be effectively utilized as a practical tool for the decision-making of LCC-effective optimal seismic retrofit and maintenance strategy of deteriorating bridges. Also, from the numerical example and discussions, in the case of Korea which may be classified as a moderate seismicicy region, the economical seismic strategy which provides moderate seismic performance and reasonable maintenance action may provide be minimum expected LCC.

REFERENCES

Cho, H. N. (2006), "Seismic Reliability Analysis of Deteriorating Bridges Considering Environmental Stressors", The 5th international workshop on Life-Cycle Cost Analysis and Design of Civil Infrastructure Systems, Seoul, Korea.
Kong, J. (2001), "Lifetime Maintenance Strategy for Deteriorating Structures," Ph.D. Dissertation, University of Colorado.

Korea Infrastructure Safety and Technology Corporation (2001), "Assessment of Seismic Performance and Plan to Strengthening of Existing Road Bridges," Final Report to Korea Road and Transportation Association.

Lee, K. M. (2006), "Life-Cycle Cost Effective Optimal Seismic Retrofit and Maintenance Strategy of Bridge Structures," Ph.D. Dissertation, Hanyang University.

Lee, K. M., Cho, H. N., and Cha, C. J. (2005), "Life-Cycle Cost-Effective Optimum Design of Steel Bridges Considering Environmental Stressors," Journal of Engineering Structures, Vol. 28, Issue 9, July 2006, pp. 1252–1265.

Shoji, G., Fuzino, Y and Abe, M (1997), "Optimal allocation of earthquake-induced damage for elevated highway bridges," Journal of Structural Mechanics and Earthquake Engineering, Japan Society of Civil Engineering, 563(I-39), 79–94.

Life-Cycle Cost and Performance of Civil Infrastructure Systems – Cho, Frangopol & Ang (eds)
© 2007 Taylor & Francis Group, London, ISBN 978-0-415-41356-5

Seismic reliability analysis of deteriorating bridges considering environmental stressors

Hyo-Nam Cho
Hanyang University, An-San, Korea

Kwang-Min Lee
Daelim Industrial Co. Ltd., Seoul, Korea

Hyun-Ho Choi
Korea Infrastructure Safety and Technology Corporation, Kyounggi-Do, Korea

Jong-Kwon Lim
Infra Asset Management Corporation, Seoul, Korea

ABSTRACT: A realistic approach based on lifetime seismic-reliability is unavoidable to perform Life-Cycle Cost (LCC)-effective optimum design, maintenance, and retrofitting of structures against seismic risk. So far, though a number of researchers have proposed the LCC-based seismic design and retrofitting methodologies, most researchers have only focused on the methodological point. Accordingly, in most works, they have not been quantitatively considered critical factors such as the effects of seismic retrofit, maintenance, and environmental stressors on lifetime seismic reliability assessment of deteriorating structures. Thus, in this study, a methodology for lifetime seismic reliability analysis is proposed and a program HPYER-DRAIN2DX-DS is developed to perform the desired lifetime seismic reliability analysis. To demonstrate the applicability of the program, it is applied to an example bridge. From the numerical investigation, it may be positively stated that HYPER-DRAIN2DX-DS can be utilized as a useful numerical tool for LCC-effective optimum seismic design, maintenance, and retrofitting of bridges.

1 INTRODUCTION

A realistic lifetime seismic-reliability based approach is unavoidable to perform Life-Cycle Cost (LCC)-effective optimum design, maintenance, and retrofitting of structures against seismic risk. However, since most developed LCC model has focused on the methodological point, they have not been quantitatively considered critical factors such as the effects of seismic retrofit and maintenance on lifetime seismic reliability assessment of deteriorating structures. Recently, a number of researchers (Shinozuka et al. 2000; Choi, 2002; Lee, 2006; etc.) have conducted comprehensive studies to assess the seismic reliability analysis of structures. However, though, a structure subjected to environmental attack can experience time-variant degradations, unfortunately, these researches have not considered the effects of lifetime maintenance action as well as strength degradation due to environmental stressors on lifetime seismic reliability.

Thus, in this study, a systematic procedure for lifetime seismic reliability assessment considering the effects of seismic retrofit, maintenance strategy and environmental stressors is proposed, and then a program HYPER-DRAIN2DX-DS (latin HYPERcube sampling based DRAIN2DX for lifetime seismic reliability analysis of Deteriorating Structure) is developed to perform the desired lifetime seismic reliability assessment of deteriorating bridges.

2 METHODOLOGY FOR LIFETIME SEISMIC RELIABILITY ANALYSIS OF DETERIORATING BRIDGES

The methodology for lifetime seismic reliability analysis proposed in this study involves the following studies: (i) Probabilistic Lifetime Seismic Response Analysis (PLSRA) considering seismic retrofit and maintenance strategy; (ii) Probabilistic Lifetime Seismic Damage Analysis (PLSDA) using the result of PLSRA; (iii) Probabilistic Seismic Hazard Analysis (PSHA); and (iv) Lifetime Seismic Reliability Analysis (LSRA) using convolution of the seismic hazard with the probabilities of structural damage. Details on lifetime seismic reliability analysis are given in the following sections.

2.1 *Probabilistic Lifetime Seismic Response Analysis*

The objective of Probabilistic Lifetime Seismic Response Analysis (PLSRA) is to compute the main descriptors (i.e., mean, C.O.V, and standard deviation) of structural maximum responses. In the PLSRA, the uncertainties due to the random nature of the seismic loadings and time-variant resistance degradations need to be properly accounted for. Such uncertainties are, generally, taken into consideration by modeling the seismic loadings as stochastic processes and the random time-variant resistance degradations using probabilistic mechanical deterioration processes. In this study, the stochastic processes are modeled using Yeh-Wen model (Yeh and Wen, 1990). Several various environmental stressors are present in bridges (e.g., corrosion, sulfate attack, alkai-silica reaction, and freeze-thaw cycle attack, among others), the most commonly reported environmental stressor is due to corrosion (Enright et al. 1998). Therefore, the time-variant resistance degradations in piers due to the corrosion of steel reinforcement are considered in this study, details on the time-variant resistance model can be found in section 3.

Generally, the PLSRA using Monte Carlo Simulation (MCS) technique requires enormous computing times. Thus, Latin Hypercube Sampling (LHS; Ayyub and Lai, 1989) is chosen as a sampling technique. In addition, an indirect PLSRA is proposed for the approximate but reasonable assessment of the desired efficient PLSRA in this study. Using a number of mean values of time-variant random variables, the relationships between the main descriptors of structural responses can be obtained from regression analyses. Also, it should be noted the relationships between the mean values of time-variant random variables and main descriptors of structural responses can be replaced to the relationship between time and main descriptors of structural responses since the time-variant random variables is function of time.

2.2 *Probabilistic Lifetime Seismic Damage Analysis*

The objective of Probabilistic Lifetime Seismic Damage Analysis (PLSDA) is to compute the probability of exceeding a damage-state at a specific ground motion level. Thus, the PLSDA for a damage state can be performed by computing the conditional probability of exceeding a specific damage-state at various seismic intensities. The conditional probability can be obtained using the functions of regression from PLSDA and Probability Density Functions (PDFs) of time-variant random variables obtained from resistance degradation model, seismic intensity, and definition of damage states of bridge components:

$$f\left(DS_m^k \mid I_{PGA}, t = T\right) = P\left(DS_m^k \mid I_{PGA}, \overline{x_v}\right) \cdot f\left(\overline{x_v} \mid t = T\right) \tag{1-a}$$

$$P\left(DS_m^k \mid I_{PGA}, \overline{x_v}\right) = \int_y P\left[d_m^k < y \mid y = SR_m\right] \cdot f\left(SR_m \mid I_{PGA}, \overline{x_v}\right) dy \tag{1-b}$$

where, m = an index for a bridge component; k = an index for damage state; d_m^k = defined damage state k for component m; I_{PGA} = a seismic intensity; $P(DS_m^k \mid I_{PGA}, \overline{x_v})$ = probability of exceeding a damage state k under given seismic intensity; $f(\overline{x_v} \mid t = T)$ = PDF of time-variant random variable at time T; and $f(SR_m | I_{PGA}, \overline{x_v}) =$ PDF of the seismic response for component m which can be obtained from PLSRA and regression analyses.

In Eq. (1), to obtain the probability of exceeding damage state k under a given seismic intensity, the First Order Reliability Method (FORM) is introduced. Meanwhile, in PLSDA, it is important to define realistic damage states for critical bridge components. For the purpose, based on the HAZUS's definition of damage states (1999) and various data obtained from test and previous earthquakes, the behavior of bridge components, etc., the quantified damage states of the critical components in typical bridges are used in this study.

2.3 *Probabilistic seismic hazard analysis*

There are no available data for all the possible earthquakes that may occur during the lifetime of a structure in Korea. Thus, to account for all the possible earthquakes, a hybrid model that superposes the Poisson-time and Gumbel's type I asymptotic model for non-characteristic earthquakes and renewal-time proposed by Oh and Kang (1992) is introduced in this study.

If assumptions such as a mutually independence of each observed earthquake magnitudes and the temporal behavior of earthquake occurrence with uncertain time are introduced, the occurrence probabilities of exceeding seismic magnitude y at a site during a specified time period can be evaluated as follows:

$$P[Y_r > y] = 1 - \prod_{j=1}^{J} g_j(y) \tag{2-a}$$

$$g_j(y) = Exp\left[-\lambda_j t \cdot Exp\left[-\beta_j\left(y - i_0\right)\right]\right] \tag{2-b}$$

where, j = an index for causative fault regions; λ_j = rate of earthquake occurrence at a causative fault region j; β_j = a reduced scale parameter; and i_0 = standard earthquake magnitude.

As shown in Eq. (2), data related to the rate of earthquake occurrence at a causative fault region and reduced scale parameter can be obtained from Lee's study (Lee, 1989). It should be noted that a seismic magnitude is used as seismic intensity as presented in Eq. (2). However, since PLSRA and PLSDA are formulated as function of seismic accelerations, the earthquake magnitude should

Table 1. Definition of damage state for bridge components.

Damage State	Pier (Hwang et al. 2000)	Abutment (Martin and Yan, 1995)		Superstructure (Mander, 1996; Lee, 2006)		
		Active Action	Passive Action	Fixed Bearing	Expansion Bearing	Deck
Slight	$1.0 < \mu < 1.2$	Half of first yield deformation	Half of first yield deformation	$\delta = 2\,mm$	–	–
Moderate	$1.2 < \mu < 1.76$	First yield deformation	First yield deformation	$\delta = 6\,mm$	–	–
Extensive	$1.76 < \mu < 4.76$	Ultimate deformation	second yield deformation	$\delta = 40\,mm$	Δ_{exp}^{allow}	–
Collapse	$\mu > 4.76$	Twice of ultimate deformation	Ultimate deformation	–	–	Actual unseating width

μ = ductility; δ = deformation of fixed bearing; Δ_{exp}^{allow} = the allowable displacement of expansion bearing which is determined by geometrical configuration of expansion bearing.

be converted into the seismic acceleration. Relationship between the earthquake magnitude and the seismic acceleration can be expressed as follows (Lee, 1984):

$$I_e = I + b_1 \cdot \ln(R/h_f) + b_2 \cdot (R - h_f) \qquad \text{(3-a)}$$

$$I = (\log_{10} a_h - 0.014)/0.3 \qquad \text{(3-b)}$$

where, a_h = seismic acceleration; I = seismic magnitude at a causative fault region (MM scale); I_e = reduced seismic magnitude at a felt area (MM scale); and R = epi-central distance; b_1, b_2 = the regional attenuation coefficients.

Since the focal depth of the 1936 Sanggyesa earthquake is about 10 km (Hayata, 1940) and the 1978 Hongsung earthquake less than 10 km (Lee, 1984), in this study, the focal depth are assumed to be 10 km. In Eqs. (2) and (3), Lee (1989) has determined the regional attenuation coefficients using least square method.

2.4 Probabilistic lifetime seismic reliability analysis

The lifetime damage probability can be obtained using convolution of the seismic hazard with the probabilities of structural damage. Thus, the probability of exceeding a specified damage state k of bridge component m in a year during lifespan can be obtained as

$$f\left(DS_m^k \mid t = T\right) = \left[\int_y P\left(DS_m^k \mid I_{PGA} = y, \overline{x_v}\right) \cdot f_1(y) dy \right] \cdot f\left(\overline{x_v} \mid t = T\right) \qquad \text{(4-a)}$$

$$f_1(y) = -\frac{dP[Y_1 > y]}{dy} \qquad \text{(4-b)}$$

where, $P(DS_m^k \mid I_{PGA} = y, \overline{x_v})$ = probability of exceeding a damage state k under given seismic intensity; $f(\overline{x_v} \mid t = T)$ = PDF of time-variant random variable $\overline{x_v}$ at time T; $f_1(y)$ = PDF of ground acceleration y at the given site in a year which can be determined from PSHA.

2.5 Computational program for lifetime seismic reliability analysis

The program called HYPER-DRAIN2DX-DS (latin HYPERcube sampling-based DRAIN2DX for lifetime seismic reliability analysis of Deteriorating Structure) is developed to perform the desired lifetime seismic reliability assessment of deteriorating bridges. The program developed for this study is a modified version of well-known DRAIN2DX (Prakash et al. 1993), which is extended by incorporating a LHS technique and FORM for the probabilistic lifetime seismic response analysis and probabilistic seismic damage analysis considering the newly developed time-variant resistance model (see section 3). Details on the computational procedure of the HYPER-DRAIN2DX-DS can be found in Lee (2006).

3 TIME-VARIANT RESISTANCE MODEL

In this study, to predict time-variant resistance degradation, a new probabilistic time-variant resistance model is proposed based on previous studies (Liu and Wayers, 1998; Kim and Stewart, 2000; Thoft-Christensen, 1997). Prediction of the lifetime deterioration of RC pier using the proposed model is divided into four steps: Chloride penetration, Corrosion initiation, Initial cracking, and crack evolution. The first two steps are based on chloride penetration into the concrete using Fick's diffusion modeling. The remaining two steps are analyzed by estimating the evolution of corrosion products and the effect of corrosion products on the concrete.

A time to maintenance required crack size or specific condition state defined in 'Guideline for Inspection and Diagnosis of Bridge Structure' (KISTEC, 2003) and lifetime variation of area of steel reinforcement due to corrosion can be expected using the proposed time-variant resistant model. However, in general, a sequence of maintenance actions should be considered to assess annual maintenance probability during life-span. Since previous maintenance probabilities influence remaining maintenance actions, all possible maintenance paths can be computed using conditional probability. The conditional probabilities can be obtained using Event tree method. Details on the time-variant resistance model can be found in Lee (2006).

4 LIFETIME SEISMIC RELIABILITY ANALYSIS OF EXAMPLE BRIDGE

The proposed methodology for the lifetime seismic reliability analysis of deteriorating bridge is applied to a simply supported continuous three span steel bridge (see Fig. 1) with or without retrofit measures. The five retrofit strategies, replacement steel bearings by Elastomeric Bearings (EBs, Case-1), replacement steel bearings by Lead-Rubber Bearings (LRBs, Case-2), installation of Restrainer Cables (RECs, Case-3), replacement steel bearings by EBs and installation of RECs (EBs+RECs, Case-4), and replacement steel bearings by LRBs and installation of RECs (LRBs + RECs, Case-5) are considered to investigate the effectiveness of the various seismic retrofit measures. In addition, the effect of resistance losses in piers due to the corrosion of steel reinforcement is considered and lifetime seismic reliability analyses under costal corrosion environmental condition with or without specific maintenance actions are performed to investigate influences of lifetime seismic reliability.

4.1 Bridge modeling and statistical uncertainties

Since the example bridge consist of elements that may exhibit highly nonlinear behavior (steel bearings, columns, abutments, impact), a 2D nonlinear analytical model of the bridges is developed using DRAIN-2DX (Prakash et al. 1992). The superstructure is modeled using a linear element that represents the stiffness and mass properties of the composite steel girder – reinforced concrete deck.

The columns are modeled using the DRAIN-2DX fiber element. As shown in Fig. 1, the example bridge has EXpansion steel Bearings (EXBs) and Fixed steel Bearings (FB). Experimental tests of steel bearings similar to that in the example bridge in this study were conducted by Mander et al. (1996). The results from these tests were used to develop analytical steel bearing elements. The abutment properties used in this model are based on recommendations by Caltrans (1990) and results from previous experimental studies (Maroney 1994). The model represents the multi-linear

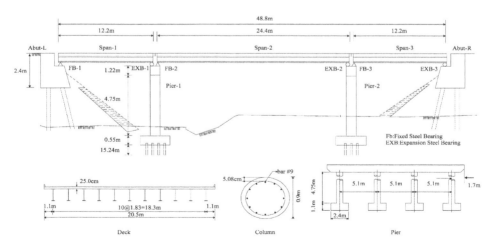

Figure 1.   Bridge profile and typical section of example bridge.

inelastic behavior of the abutments in both active action (tension) and passive action (compression). The contact element approach is used to model impact between decks and deck and the abutments. The pile foundation is modeled using a combination of linear springs in the horizontal and rotational directions. The pile foundation stiffnesses are based on the type and number of piles, as well as the soil properties.

For the analysis of the retrofitted structures, the EBs and LRBs are modeled based on previous study (Kelly, 1997). The dimension of the EBs is 305 mm × 203 mm × 102 mm (L × W × H). The dimension of LRBs is the same as that of the EBs and 63.5 mm is used as the diameter of lead-plug, in this study. The restrainer cables are modeled as tension-only nonlinear elements with a gap. The stiffness for the restrainer cables is based on the cross-sectional area of the cables, A (143 mm^2), modulus of elasticity, E (69,000 Mpa), length of the cable, L (1.5 m), initial slack, ls (12.7 mm), to allow for thermal expansion without producing a force in the restrainers.

The statistical uncertainties recommended by Lee (2006) are utilized as the uncertainties essential for seismic reliability analysis. Uncertain structural properties considered in the seismic reliability analysis are the mass, modal damping ratio, the concrete compressive strength, the reinforcement yield strength.

4.2 *Results of seismic response analysis in intact state of as-built and retrofitted bridge*

Due to space limitation, detailed results on seismic response analysis in intact state of as-built and retrofitted bridge are not presented in this paper, which can be found in Lee (2006).

From the results of PSRA, the trends of probabilistic seismic responses of critical components can be summarized as follows: (i) Since the seat width 250 mm, the example bridge is not in the risk of collapse due to unseating.; (ii) The mean ductility of pier-1 is larger than that of pier-2 since pier-1 responds with deck-2 which is heavier than deck-3. The second reason for the larger ductility in pier-1 is the moving tolerance. Deck-2 on pier-1 has more moving tolerance than deck-3 on pier-2; (iii) The result shows considerable inelastic response in FB-1 since the deformation of FB-1 is primary due to the impact force generated on deck-2. Deck-2 impacts into deck-1, transferring large force to the FB-1 at Abut-L; (vi) Finally, the forces from the fixed steel bearing on abutment dominate the behavior of the abutments in active action (puling action). However, in passive action (pushing action), the pounding force govern this behavior. Therefore, Abut-L with FB-1 has larger force and displacements in both actions. However, Abut-R with EXB-3 has smaller force in pulling action, but it has larger force from pounding.

The results of seismic reliability analysis for venerable components of as-built bridge (Case-6) and retrofitted bridges (Case 1~5) are investigated. Fig. 2 shows the mean annual probabilities of exceeding 4 different damage states. As shown in the figure, for Case-6 (As-built bridge), it is observed that FB-1 has largest annual probability of exceeding slight and moderate damage, while, Pier-1 has largest annual probability exceeding extensive and collapse damage. But, the maximum annual probabilities of exceeding 4 different damages of piers are decreased by using seismic retrofit strategies as shown in the figures. However, the negative effects are partially observed.

For the slight damage, it is observed that the EB-1 of Case-1 (replacement steel bearings by EBs) and Case-4 (replacement steel bearings by EBs and installation RCs) are easily damaged. However, Case-1 is most effective in reducing each damages of pier-1. For example, the annual probability of exceeding slight damage of pier-1 is estimated to be 1.81e-6 (effectively reduced by about 1.016% compared with that from Case-6). Case-3 (installation RCs) is least effective in reducing the annual probability of exceeding damages of Abut-L in active action due to increased interaction transfers the inertia forces of deck to the abutment in active action as shown in Fig. 2. For example, the annual probability of exceeding moderate damage of Abut-L in active action is estimated to be 8.16e-5 (increased by about 7.08 times compared with that from Case-6).

In addition, Fig. 2 (d) shows that pier-1 of as-built bridge has largest annual probability of exceeding collapse damage. However, it is observed that Abut-L in active actions of Cases 2~6 have largest probability of exceeding collapse damage. For Case-1, it is investigated that unseating damage between Deck-1 and 2 have larger probability of collapse damage than other components.

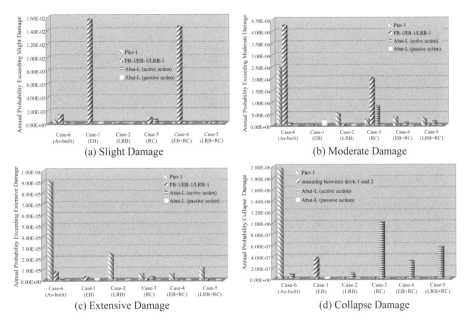

Figure 2. Annual probabilities of exceeding collapse damage.

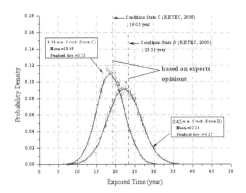

Figure 3. Times to growth of crack width.

4.3 Effect of corrosion environmental condition and maintenance action on lifetime reliability analysis

The probabilistic corrosion analysis of the example bridge in costal corrosion environmental conditions with or without specific maintenance action is performed using time-variant resistance model to investigate effects of corrosion environment and maintenance actions.

The random variables used for the probabilistic corrosion analysis and their associated values and distributions are taken from available references (Thoft-Christensen et al. 1997; and Korea Infrastructure Safety and Technology Corporation-KISETC, 2003).

Fig. 3 shows service life up to crack Condition State (CS)-C (crack size = 0.35 mm) and CS-D (crack size = 0.45 mm) defined in KISTEC (2003). The mean service life up to crack CS-C and CS-D are evaluated to be 18.69 years and 22.26 years, respectively.

Thus, it can be stated that the expected service life up to CS-C and D from the proposed probabilistic corrosion model are similar to those from KISTEC (2000).

137

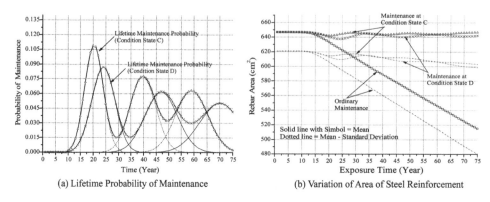

(a) Lifetime Probability of Maintenance	(b) Variation of Area of Steel Reinforcement

Figure 4.　Lifetime probability of maintenance and variation of area of steel reinforcement.

Fig. 4 shows lifetime probability of maintenance and corresponding variation of area of steel reinforcement with ordinary maintenance, maintenance at CS-C and CS-D. It was assumed that a crack in a pier is perfectly repaired using minor repair method. As shown in Fig. 4, it is observed that the mean cross section area of corroded rebar is varied according to the variation of probability of maintenance (the reduction of cross section area of rebar started to decrease where cumulative distribution is greater than 0.5). And, from the figure, it can be easily found that maintenance actions can effectively prevent form degradation of structural performance.

A regression analyses are also performed to obtain the relationship between time-variant random vector (i.e., reduction of cross section area of rebar in the piers) and values for the main descriptors (mean and C.O.V) of structural responses. If the area of steel reinforcement in piers is corroded up to 79% of the intact area (see Fig. 4 (b), it is observed that the responses of Pier-1, FB-1, Abut-L in active and passive action are increased by about 5.898, 0.100, 0.604, and 6.399%, respectively, compared with intact state. Thus, it can be stated that the variations of structural response of pier and Abut-L in passive action for the deterioration due to corrosion are much more sensitive then those of other components. And, a linear relationship is investigated between the mean of the seismic responses and the area of steel reinforcement. Meanwhile, it is investigated that the C.O.Vs of the seismic responses are almost constant regardless the variation of the area of steel reinforcement. Thus, in this study, linear regression functions and constant values are used for the desired LSRA, respectively.

To demonstrate the effects of maintenance actions on annual probabilities of exceeding the 4 different damages, the results of LSRA of as-built bridge with or without specific maintenance actions (maintenance at CS-C and CS-D) are comparatively discussed. Fig. 5 illustrates lifetime annual probabilities of exceeding 4 different damages of critical components (i.e., FB-1 and Pier-1) with or without specific maintenance actions for costal corrosion environmental condition. From the figures, it should be noted that the increments of the mean annual probabilities of exceeding each damages of pier-1 are larger then those of FB-1 since the behavior of pier-1 for corrosion of steel reinforcements is more sensitive then that of FB-1 as discussed above. From the result, it is found that the effects of maintenance actions on annual probabilities of exceeding damages of pier are predominantly observed compared with FB-1. The mean annual probabilities of exceeding slight damage of FB-1 with maintenance at CS-C and CS-D are slight reduced by about 4.99%~5.98% compared with those for ordinary maintenance cases. However, the mean annual probabilities of exceeding extensive damage of pier-1 with maintenance at CS C and CS D are significantly reduced by about 30.18%~31.34% compared with those for ordinary maintenance cases. These results may also be attributed that the behavior of pier-1 for corrosion of steel reinforcements is more sensitive then that of FB-1.

From the results, it may be stated that the effects of maintenance actions on the annual probabilities of exceeding damages are highly related to relationship between structural behaviors. And, it can also be stated that maintenance actions can effectively prevent form increment of the lifetime probability of exceeding damages of components as time passes.

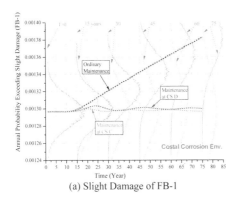

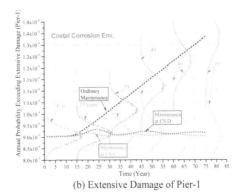

(a) Slight Damage of FB-1   (b) Extensive Damage of Pier-1

Figure 5.   Lifetime annual probabilities of exceeding damage state of FB-1 and pier.

5   CONCLUSIONS

This paper presents the methodology of a realistic lifetime seismic reliability analysis considering the effects of seismic retrofit, maintenance strategy, and environmental stressors. From the illustrative example and discussions, some important conclusions and remarks would be pointed out as follows:

(a) From numerical investigation, as expected, various damages of bridge components depend on bridge configuration, seismic retrofit strategy, etc. Noting that in most previous LCC models, damage costs are formulated only as a function of system collapse which may produce unrealistic result, it may be reasonable to formulate LCC of a structure as a function of local damages of bridge components as well as system damage.
(b) And also realizing that previous studies on seismic reliability analysis, and LCC-effective optimum design, maintenance and seismic retrofit against seismic risk ignore the effects of environmental stressors and maintenance strategy, the effects of environmental stressors and maintenance strategy should be considered for realistic assessment of seismic reliability and LCC-effective seismic design, maintenance and retrofitting of a structure, etc.
(c) In conclusion, it may be positively stated that HYPER-DRAIN2DX-DS can be utilized as a useful numerical tool for LCC-effective optimum seismic design, maintenance, and retrofitting of bridges. And the basic concept for lifetime reliability analysis proposed in this study may be effectively utilized for other natural hazards such as typhoon, flood, etc.

REFERENCES

Enright, M.P. (1998), "Time Variant Reliability of Reinforced Concrete Bridge under Environmental attack," Ph.D. Dissertation, University of Colorado.
HAZUS (1999), "User's Manual," Federal Emergency Management Agency, Washington, D.C.
Hwang, H., Liu, J., and Chiu, Y. (2000), "Seismic Fragility Analysis of Highway Bridges," Center of Earthquake Research and Information, The University of Memphis, Memphis.
Korea Infrastructure Safety and Technology Corporation (2003), "Guideline for Inspection and Diagnosis of Bridge Structure," Final Report to Korea Road and Transportation Association.
Lee, K.M. (2006), "Life-Cycle Cost Effective Optimal Seismic Retrofit and Maintenance Strategy of Bridge Structures," Ph.D. Dissertation, Hanyang University.
Thoft-Christensen, P. (1997), "Life-cycle cost evaluation of concrete highway bridges," Structural Engineering World Wide 1998: Proceedings of Structural Engineers World Congress 1998, Elsevier Science, B.V., Amsterdam, The Netherlands, T132-6.
Yeh, C.-H. and Wen, Y.K. (1990), "Modeling of nonstationary ground motion and analysis of inelastic structural response," Structural Safety, 8, 281–298.

Implications of Life-Cycle Cost criteria for the design of structural components: The case of a simple steel-reinforced concrete beam

P.N. Christensen, M. Boulfiza, B. Sparling & G.A. Sparks
University of Saskatchewan, Saskatoon, Canada

ABSTRACT: Life-cycle cost (LCC) criteria hold important implications for the optimal design of structural components and systems. While a broad range of component designs, for instance, may satisfy local standards, the influence of environmental factors on structural deterioration – among other things – can cause wide variations in corresponding service life and related LCC forecasts. In order to select the 'best' design, therefore, engineers must: (a) select a satisfactory and defensible measure of deterioration, (b) link that measure to a LCC model, and (c) account explicitly for the uncertainty that naturally surrounds such complex problems resting on forecasts over often very long time periods. The authors of this paper elucidate a method of LCC analysis intended to satisfy these requirements and illustrate its implementation through a straight-forward case study. Subject to local standards, a range of steel-reinforced concrete beam designs are analyzed in the context of environmental exposure (including application of de-icing salts) typical of the Canadian prairies. Structural deterioration is measured in terms of bending moment capacity and related implications for safe use. Uncertainty is handled through an efficient, stepwise application of deterministic, sensitivity and risk analyses that leads naturally to stochastic characterization of demonstrably influential sources of uncertainty. This permits the derivation of time-varying structural reliability estimates that ultimately affect service life and related LCC forecasts. The results of the study demonstrate the sometimes surprising implications of LCC criteria for the practice of engineering design.

1 INTRODUCTION

The challenges surrounding the integration of design, performance, management and costs in the context of civil engineered artifacts is reflected in literature involving reliability and life cycle costing. As a logical means of merging inherent uncertainty and safety concerns, structural reliability estimates can serve as credible forecasts of the time-varying performance of infrastructure systems and components. Mated to relevant costs, therefore, structural reliability forecasts can reciprocally inform design and on-going management decisions facing engineers seeking to minimize life cycle costs (Stewart & Rosowsky 1998; Frangopol & Hendawi 1994; Tao, Corotis & Ellis 1994).

Although integral to the process, reliability assessment serves in this effort as a de facto check on a beam design generated through a novel method of life cycle engineering and costing (Christensen et al. 2004, 2005). Employing code-based design methods, chloride-induced deterioration modeling, and stepwise inclusion of stochastic data, the life cycle engineering process summarized herein seeks to select an optimal, steel-reinforced concrete beam design among a discrete number of possibilities. To validate the process from the standpoint of structural reliability, stochastic variation in material, load and deterioration parameters are employed, ex post, to estimate the probability of failure at the close of a forecasted service life. Although management decisions are limited to 'replace at end of service life', the exercise nonetheless illustrates the possible implications of life cycle cost criteria for the practice of civil engineering design.

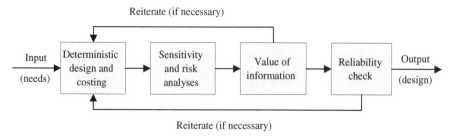

Reiterate (if necessary)

Reiterate (if necessary)

Figure 1. Iterative flow of beam design and costing process.

2 LIFE CYCLE ENGINEERING AND COSTING PROCESS

2.1 Method

The method of life cycle engineering and costing (LCE&C) employed herein reflects the iterative, comparative assessment process common to Decision Analysis (Clemen 1996; Matheson & Howard 1968). Figure 1 illustrates its basic steps in the context of the current design exercise. Canadian design standards (CSA 2004) are used to generate alternative, steel-reinforced concrete beam designs given a handful of initial parameter settings. In situ environmental factors and related models subsequently forecast the deterioration of the beams over time in terms of mid-span bending moment (Boulfiza et al. 2004; Thoft-Christensen 1998). In this initial, deterministic setting, end of service life occurs where time-varying bending resistance meets a pre-selected resistance threshold (based on factored loads). Combined with a range of initial cost parameters and a pre-selected discount rate, this furnishes sufficient information to estimate the annual worth of life cycle costs (AWLCC) attributable to all considered designs.

To reflect uncertainty across parameter estimates, one-way sensitivity analysis is then used to determine their relative impact on AWLCC across all beam design alternatives. The resulting, comparative ranking of parameters and associated variability then prioritizes the required assignment of corresponding probabilistic data. Sampling is then used to generate comparable risk profiles in terms of the cost-minimizing objective, AWLCC. Should one alternative demonstrate stochastic dominance, the costing process ends. Else, subsequent iteration – subject to value of information calculations (Raiffa 1968) – is needed.

To this point in the process, the presumed reliability of considered designs rests purely on partial safety factors. Yet this fails to account explicitly for the time-varying resistance of alternative designs crucial to the determination of an 'acceptably safe' service life and related life cycle cost estimates. A reliability assessment is therefore made at the forecasted end of service life of the chosen beam design. Should the check demonstrate adequate account of safe use, then further iteration is not required. Else, cyclical modification is required.

2.2 Models

2.2.1 General design and costing model

Let $\mathbf{x}$ and $\mathbf{c}$ stand, respectively, as matching vectors of beam design parameters (i.e., material specifications, dimensions, etc.) and costs. Then $C(\mathbf{x}, \mathbf{c})$ represents the total initial cost of any considered steel-reinforced concrete beam. Let T represent beam service life and i stand as a pre-selected discount rate common to all considered designs. Let A and P represent the annualized and present worth equivalents of a given monetary value. The annualized cost of any beam may then be expressed as $C(\mathbf{x}, \mathbf{c})(A|P,T,i)$. Let s_t and s_{thresh} be the bending resistance at any time t and a pre-selected resistance threshold based on factored loads. Finally, let λ be an arbitrary constant and $\tilde{\mathbf{x}}$ stand as a pre-selected subset of design parameters (reflecting functional requirements and client demands). Employing this notation, the guiding deterministic LCE&C model of the beam design exercise may be expressed as:

$$\text{Minimize}_{\mathbf{x}/\tilde{\mathbf{x}}} \quad C(\mathbf{x},\mathbf{c})(A \,|\, P,T,i) \tag{1}$$

$$\text{Subject to} \quad s_t \geq \lambda s_{thresh}, \; t = 0,1,2,...,T \; \text{where} \; \lambda \geq 1.0 \tag{2}$$

Given a vector of costs, a subset of design parameters and an arbitrary constant, the civil engineer's challenge is to select a range of design parameters, $\mathbf{x}/\tilde{\mathbf{x}}$, that minimizes the annualized cost of a beam over some service life, T (1). The service life, in turn, is determined by the initial design parameters, $\mathbf{x}$, time-varying resistance, $s_0, s_1, s_2, \ldots$, a resistance threshold, s_{thresh}, and an arbitrary constant, λ, of value 1.0 or greater (2).

2.2.2 Beam design and deterioration process

The annualized cost of a given beam rests on design decisions, temporal deterioration, and related requirements for safe use. Herein, design decisions comply with Canadian standards for a 50-year design (CSA 2004). It is assumed, furthermore, that the selected arrangement of reinforcing bars mirrors common practice where minimum initial cost often dictates the choice of bar diameter, d_b, and number, n_{bars}, needed to meet or exceed steel area requirements, A_s. Below, $A_s(0)$ represents the actual steel area at time 0 based on bar selection.

A stepwise series of equations describes the deterioration of a steel reinforced beam exposed to de-icing salts. Let $C(x, t)$ stand as chloride concentration at concrete depth x at time t, C_0 represent chloride concentration at the surface, and D represent a diffusion coefficient. According to Fick's second law, the process of chloride diffusion in concrete may then be expressed as:

$$C(x,t) = C_0 \left[1 - erf\left(\frac{x}{2\sqrt{Dt}} \right) \right], \text{ where } \quad erf(s) = \frac{2}{\sqrt{s}} \int_0^x e^{-\beta^2} d\beta \tag{3}$$

Corrosion of reinforcement, Cr, begins where chloride concentration reaches a predetermined threshold at depth of clear cover and (assuming constant temperature conditions) is described by:

$$Cr = k \, (\alpha_0 + a \, [Cl^-]) \, \phi \, S \, [O_2] \tag{4}$$

In (5): k, α_0 and a are empirical coefficients; ϕ and S stand as porosity and degree of saturation; and $[Cl^-]$ and $[O_2]$ are chloride ion and oxygen concentration in the pore solution phase. Steel loss per unit area of reinforcement at any time t, $W_r(t)$, is computed by integrating the corresponding corrosion rate, $q(s)$, from time of corrosion initiation, t_{ic}, to any time, t $(t \geq t_{ic})$:

$$W_r(t) = \int_{t_{ic}}^{t} q(s) \, ds \tag{5}$$

Let ρ_s and $\Delta m(t)$ stand, respectively, as the density of steel and temporal change in steel mass. The change in reinforcement area at any time t, $\Delta A_s(t)$, can then be derived as follows:

$$\Delta A_s(t) = \frac{\Delta m(t)}{\rho_s}, \text{ where } \quad \Delta m(t) = W_r(t) A_s(0) \tag{6}$$

The change in radius of sound steel at any time t is subsequently determined by:

$$\Delta R(t) = R(0) - R(t), \text{ where } \quad R(t) = \sqrt{\frac{A_s(0) - \Delta A_s(t)}{\pi}} \tag{7}$$

143

Imposed displacement at the steel-concrete interface caused by rusting, $ImpDisp(t)$, is then calculated as a function of rust diameter, $DrRust(t)$, a related expansion coefficient, η, and the change in radius of sound steel (7):

$$ImpDisp(t) = DrRust(t) - \Delta R(t), \text{ where } DrRust(t) = \eta\Delta R(t) \tag{8}$$

At the point-in-time, t_{spall}, where imposed displacement exceeds a critical threshold, IDc, the concrete spalls and, it is assumed, chloride concentration at the rebar, $C(x, t_{spall})$, immediately reaches surface chloride concentration, C_0. This affects the rate of corrosion (3, 4) which, in turn, affects temporal calculations involving the mass and area of sound steel remaining in the beam (5, 6).

The resulting temporal forecasts of remaining sound steel, $A_s(t)$, can then be used to determine the nominal bending moment resistance at any time t, s_t:

$$s_t = A_s(t) f_y \left[d - \frac{A_s(t) f_y}{2(\alpha_1 f_c b)} \right] \tag{9}$$

In (9): f_y and f_c stand, respectively for the yield strength of steel and nominal concrete strength; b and d represent the base width of the beam and effective depth of rebar; and α_1 is a stress block parameter specified in CSA 2004. Compared against extreme load-induced bending, the deterioration of nominal bending resistance ultimately derived through (9) provides the information required to specify a corresponding service life and annualized cost estimate (1, 2). Note that the absence of cracking models above reflects performance controls embedded in Canadian standards to avoid crack widths in excess of 0.3 mm.

2.2.3 Estimating failure probabilities

Let N stand as the number of steps in a discrete approximation of a continuous distribution and p_i represent the corresponding probability at step i. Assuming independence, the probability of a combined event is then $p_i p_j p_k \ldots$ For any combined event, let $z_{ijk}\ldots(T)$ be a binary variable equal to one if the corresponding factored resistance threshold, λs_{thresh}, exceeds moment resistance at time T, s_T (2). The failure probability for any considered beam design at end of service life, $p_f(T)$, may then be estimated as:

$$p_f(T) = \sum_{i=1}^{N}\sum_{j=1}^{N}\sum_{k=1}^{N} \cdots p_i p_j p_k \cdots z_{ijk\ldots}(T) \tag{10}$$

3 CASE STUDY AND RESULTS

3.1 A simple beam

A client demands a simple, steel-reinforced concrete beam of an engineer as a pedestrian bench. Pre-selected design parameters include length (L, 6.0 m), width (b, 250 mm), concrete quality (normal, $f_c \approx 35$ mPa) and concentrated design load (P_L, 25 kN). Typical of structures in the Canadian prairies, the beam will suffer periodic exposure to de-icing salts. It is assumed the beam is reinforced for bending only (i.e., stirrups are not included in the design) and the engineer is left to choose among only two design parameters: beam height (h) and clear cover (cc).

In this exercise, beam height selection is limited to 300 or 450 mm, while clear cover selection is limited to 40 or 65 mm. Hence, four mutually exclusive beam design alternatives are available to the engineer and subject to life cycle analysis. Once constructed, the beam is left to deteriorate until such time that the factored resistance threshold is met (marking end of service life). At that point, the beam is removed and replaced with a beam of identical specifications.

Table 1. Nominal data.

Var	Data Units	Var	Data Units	Var	Data Units	Var	Data Units
f_c	35 MPA	D	2.0E-12 m^2/s	S	0.90	i	5.0%
f_y	400 MPA	k	3.48	$[Cl^-]$*	0.90 kg/m^3	p_c**	350 \$/m^3
γ_c	2,400 kg/m^3	α_0	0.15 cm/day	ρ_s	7.85 g/cm^3	p_r	2.2–5.4 \$/m
g	9,807 m/s^2	a	169	η	3.5	p_f	450 \$
P_L	25 kN	$[O_2]$	9.50E-6 g/cm^3	IDc	0.0032	p_d	200 \$
C_0	3.5 kg/m^3	ϕ	11.0%				

*Presumed chloride corrosion threshold (otherwise, $[Cl^-]$ derived through equation 3).
**All dollar figures in \$CDN.

Table 2. Results of deterministic analysis.

Beam	h mm	cc mm	d_b mm	n_{bars}	IC* \$CDN	T years	AWLCC \$CDN	T** years	AWLCC** \$CDN
1	450	40	20	2	685	42	40.79	29	48.45
2	450	65	20	2	685	61	36.62	37	42.95
3	300	40	20	3	605	39	37.34	22	51.20
4	300	65	15	6	605	66	31.94	38	37.74

* Excludes disposal ahd hauling costs at end-of-service life.
**Results with $\lambda = 1.20$.

3.2 Data and results

3.2.1 Deterministic analysis
Nominal design, deterioration and costing data are listed in Table 1 (in addition to dimensions given previously). As can be seen, most parameters correspond directly to equations written above. Others are common material parameters familiar to engineers. Costing data includes a discount rate, i, and prices for concrete, steel reinforcement, formwork and disposal and hauling at end of service life (p_c, p_r, p_f and p_d, respectively). Note that the price of reinforcement varies by bar diameter, d_b. For the purpose of this study, three bar options are explored for each considered design alternative: 15, 20 and 25 mm (priced at approximately 2.2, 3.4 and 5.4 \$CDN/m, respectively) – where, subject to design standards, the least costly combination is chosen. Partial safety factors of resistance, ϕ_c and ϕ_s, are 0.65 and 0.85, respectively. Partial dead and live load factors, α_D, α_L, are 1.25 and 1.5. Two λ values are tested: 1.0 and 1.2. Note that the expected cost of failure is, in this case, assumed negligible.

Given beam height and clear cover selections, remaining dimension and unit cost parameters permit straight-forward derivation of initial cost estimates (see Table 2). Corresponding service life estimates, T, are found by comparing the pre-selected resistance threshold against time-varying moment resistance. The resulting AWLCC estimates are derived by combining initial cost, disposal and hauling costs, and service life predictions with a pre-selected discount rate, i (1). For sake of this exercise, the resistance threshold, s_{thresh}, is set equal to extreme load-induced bending based on the application of partial safety factors to dead and live loads.

Table 2 summarizes key results (for λ values 1.0 and 1.2). Recall that bar selection is based on standards and minimum bar cost. Hence, the diameter, d_b, and number of bars, n_{bars}, suggested for each unique combination of beam height and clear cover is not uniform (varying from 2, 20 mm bars to 6, 15 mm bars). While initial cost, IC, of options sharing identical height (beams 1 and 2, beams 3 and 4) are equal, predicted service lives, T, vary substantively due to clear cover and its implications for chloride diffusion as well as the selected λ value. Regardless of λ value, however, beam 4 posts the longest service life and, therefore, the lowest life cycle cost (AWLCC of \$31.94CDN and \$37.74CDN for λ values 1.0 and 1.2, respectively).

Table 3. Data and results of one-way sensitivity analysis.

		f_c MPA	f_y MPA	C_0 kg/m³	D^* m²/s	$[O_2]^{**}$ g/cm³	ϕ %	S	$[CL^-]^{***}$ kg/m³	ρ_s g/cm³	η
Low	Value	25	400	1.4	3.5	10.0	0.8	0.6	7.80	2.0	
	AWLCC	32.43	31.94	30.57	30.66	30.57	31.63	31.63	32.43	31.94	31.86
Nom	Value	35	400	3.5	2.0	9.5	11.0	0.9	0.9	7.85	3.5
	AWLCC	31.94	31.94	31.94	31.94	31.94	31.94	31.94	31.94	31.94	31.94
High	Value	47	470	8.8	3.9	12.0	13.0	1.0	1.2	7.90	8.0
	AWLCC	31.78	31.28	36.56	34.11	32.66	32.54	32.32	31.38	31.94	33.07

* Values to the power of 1.0E-12.
** Values to the power of 1.0E-6.
*** Threshold values.

In Table 2, note the relative difference in service life and life cycle cost results across initial design decisions, h and cc. Although height influences initial cost and, therefore, AWLCC, clear cover is the most crucial design decision given its concomitant influence on forecasted deterioration. As can be seen, an increase in clear cover from 40 to 65 mm reduces AWLCC by approximately 10 to 25 percent while initial costs remain the same. Note also the substantive impact of λ on service life, T, and AWLCC results. Across considered beam designs, an increase from 1.0 to 1.2 decreases service life by about 30 to 45 percent and concomitantly increases AWLCC by 15 to almost 40 percent. It is instructive to note that while beams 1 and 3 post the higher life cycle costs, they are – from a structural standpoint – the most efficient designs. Hence, in the context of deterioration, structural efficiency and economic efficiency need not lead to the same results.

3.2.2 *Sensitivity and risk analyses*
Uncertainty in parameter estimates is treated through systematic integration of sensitivity and risk analyses. The focus of sensitivity and risk analyses herein – reflecting the principal focus of this study – lies with design and deterioration parameters. Potential variation in costing parameters is therefore ignored.

Table 3 presents both the data and results of a one-way sensitivity analysis pertinent to the least life cycle cost alternative, beam 4 (where the nominal value of λ is 1.0 and all low-high ranges match, roughly, a 95 percent confidence interval). It is interesting to note that – despite wide variation in certain parameters (e.g., f_c, γ_c and $[Cl^-]$) – the influence over the objective, AWLCC, is generally slight (in the neighborhood of 2 to 3 percent). Since the key determinant of AWLCC is the corresponding service life estimate, T, the results suggest the minimal influence of the uncertainty associated with many parameters on subsequent deterioration forecasts. Clearly, uncertainty surrounding the diffusion coefficient, D, and chloride surface concentration, C_0, are exceptions to this general observation. For instance, at relatively high concentration ($C_0 = 8.8$ kg/m³) the AWLCC estimate jumps from a nominal value of \$31.94CDN to a value of \$36.56CDN – an increase of approximately 15 percent.

Of the parameters listed in Table 3, the five whose variation proved most influential on life cycle cost results were included in a risk analysis (D, C_0, $[O_2]$, $[Cl^-]$ and η) to permit stochastic comparison of all beam alternatives. The statistics assigned D, C_0 and $[Cl^-]$ are listed in Table 4 below. Both $[O_2]$ and η were assigned triangular distributions based on low, nominal and high estimates listed in Table 3 (statistics equally applied to reliability calculations below).

Figure 2 illustrates comparative risk profiles over AWLCC for all beam design alternatives based on assigned statistics. In all cases, note the degree of variance possible under each alternative. Depending on in situ material properties and environmental conditions, life cycle costs – reflecting a wide range of service life forecasts – may reach anywhere from approximately \$30CDN per year to almost \$130CDN per year. Nonetheless, the beam design favored under the deterministic analysis (beam 4) demonstrates stochastic dominance here – a fact that remains true under a λ value

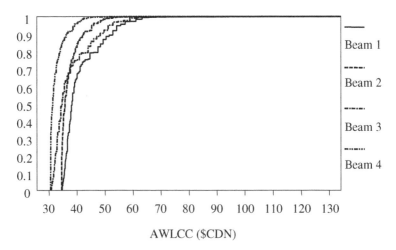

Figure 2. Comparative risk profiles for beam designs ($\lambda = 1.0$).

Table 4. Statistics for reliability assessment.

Var	Distribution	Mean	CoV	Source
f_c	Normal	$1.03f_c$	0.18	Stewart & Rosowsky 1998
f_y	Normal	$1.09f_y$	0.045	CSA 2000
P_L	Normal	$0.967P_L$	0.0323	CSA 2000
P_D	Normal	$1.05P_D$	0.10	CSA 2000
C_0	Lognormal	3.5	0.5	Stewart & Rosowsky 1998
D	Lognormal	2.0	0.75	Stewart & Rosowsky 1998
$[Cl^-]$	Uniform**	(0.6–1.2)		Stewart & Rosowsky 1998

*Presumed chloride corrosion threshold (otherwise, $[Cl^-]$ derived through equation 7).
** Defined in terms of low and high statistical parameters.

of 1.2. Hence, in terms of the design decisions at-hand, the expected value of additional information is zero (i.e., additional iteration of the LCC process appears unnecessary). Note that the annualized costs of beam 4 posts relatively less variability than beams with thinner covers. For instance, the likelihood of beam 4 ($cc = 65$ mm) posting an AWLCC value of \$35CDN or more is only about 10 percent, while for beam 3 ($cc = 40$ mm) the comparative probability is almost 40 percent.

The shape of the risk profiles in Figure 2 is reflected in expected value rankings. Beam 1 posts an expected AWLCC of \$41.13CDN (\$49.24CDN where $\lambda = 1.2$). For beam 2, the figure is \$37.33CDN (\$43.77CDN). For beams 3 and 4, the expected AWLCC is \$37.57CDN (\$51.08CDN) and \$32.58CDN (\$38.20CDN), respectively. Hence, regardless of λ, the expected life cycle costs of beam alternative 4 lie approximately 15 percent below its nearest competitor (beam 2). As with the deterministic results, clear cover is – in terms of life cycle costs – relatively more influential than beam height.

3.2.3 Implications for reliability
Based on (10) and the statistics of Table 4 (where P_D stands for dead load), failure probability estimates were calculated for beam design 4 at λ values 1.0 ($T = 66$ years) and 1.2 ($T = 38$ years). Note that live load statistics, P_L, are based on annual estimates for bridge loadings in Saskatchewan (admittedly less-than-ideal, yet useful for comparative purposes).

At a λ value of 1.0, the estimated failure probability, p_f, is 0.1703. Although expected failure costs, in this case, are assumed negligible, such high probability of failure clearly contravenes accepted safety standards. Where λ is set to 1.2, however, p_f falls to approximately 0.0030.

If the distribution of failure probabilities emulates, closely, a normal distribution, this translates to a reliability index, β, of almost 2.75. Although below an arguably desired target (e.g., $\beta = 3.50$), this is nonetheless a significant increase in reliability. Moreover, although the incremental decrease in service life over λ values is substantive (28 years), the incremental increase in life cycle cost for this 50-fold decrease in failure probability is only about six dollars per year, or 20 percent. Hence, where design decisions explicitly consider relevant deterioration processes, stringent safety standards may be met with little incremental life cycle cost.

4 CONCLUDING REMARKS

As demonstrated through the case of a simple beam, LCC criteria can markedly alter engineering design decisions. Indeed, in the context of likely environmental exposure and related deterioration, a structurally efficient design may prove relatively poor from the standpoint of economic efficiency. For this reason, engineers have little choice but to explicitly consider the life cycle performance of structural components and systems should they adopt LCC as guiding design criteria.

Some practical implications of this study include: (i) effective service life is highly contingent on the severity of the in situ environment and related deterioration mechanisms, (ii) at least in corrosive environments, substantive variability in service life, LCC and reliability forecasts involving steel-reinforced components are driven principally by statistics governing deterioration rather than materials or loads, and (iii) the latter observation suggests the inextricable nature of design, inspection and monitoring, and maintenance in the context of LCE&C. Since deterioration forecasts are highly uncertain, structures must be designed, observed and managed over time to minimize life cycle costs in practice.

REFERENCES

Boulfiza, M., Banthia, N. & Kumar A. 2004. Service performance of an RC beam in a chloride laden environment. *Canadian Society for Civil Engineering; Proc. 5th structural specialty conf., Saskatoon, 2–5 June*: 10 pp.
CSA 2000. *CSA Standard CAN/CSA-S6-00 Canadian Highway Bridge Design Code*. Mississauga, ON: Canadian Standards Association.
CSA 2004. *CSA Standard A23.3-04 Design of Concrete Structures*. Mississauga, ON: Canadian Standards Association.
Christensen, P.N., Sparks, G.A. & Kostuk, K.J. 2004. A proposed method for life cycle engineering and costing. *Advanced Composite Materials in Bridges and Structures; Proc. 4th intern. conf., Calgary, 20-23 July*: 8 pp.
Christensen, P.N., Sparks, G.A. & Kostuk, K.J. 2005. A method-based survey of life cycle costing literature pertinent to infrastructure design and renewal. *Canadian Journal of Civil Engineering* 32: 250–259.
Clemen, R.T. 1996. *Making hard decisions: an introduction to decision analysis*. Pacific Grove: Duxbury Press.
Frangopol, D.M. & Hendawi, S. 1994. Incorporation of corrosion effects in reliability-based optimization of composite hybrid plate girders. *Structural Safety* 16: 145–169.
Matheson, J.E. & Howard, R.A. 1968. An introduction to decision analysis. Reprinted 1989. In R.A. Howard & J.E. Matheson (eds.), *The Principles and Applications of Decision Analysis*: 17–55. Palo Alto: Strategic Decisions Group.
Raiffa, H. 1968. *Decision analysis: introductory lectures on choices under uncertainty*. Reading, Mass.: Addison-Wesley.
Stewart, M.G. & Rosowsky, D.V. 1998. Time-dependent reliability of deteriorating reinforced concrete bridge decks. *Structural Safety* 20: 91–109.
Tao, Z., Corotis, R.B. & Ellis, J.H. 1994. Reliability-based design and life cycle management with Markov decision processes. *Structural Safety* 16: 111–132.
Thoft-Christensen, P. 1998. Assessment of the reliability profiles for concrete bridges. *Engineering Structures* 20(11): 1004–1009.

Life-Cycle Cost and Performance of Civil Infrastructure Systems – Cho, Frangopol & Ang (eds)
© *2007 Taylor & Francis Group, London, ISBN 978-0-415-41356-5*

Bridge management system based on Life-Cycle Cost minimization

H. Furuta & T. Kameda

Department of Informatics, Kansai University, Japan

ABSTRACT: Recently, many bridge management systems have been developed in Japan to establish a rational maintenance program for existing bridges. Most of those bridge management systems use the concept of Life-Cycle Cost (LCC), because it is a useful concept in reducing the overall cost and achieving an appropriate allocation of resources. Using a bridge management system, it becomes possible to examine and compare many alternatives of maintenance plans. In this paper, an attempt is made to improve the bridge management system that was developed so far by introducing a new multi-objective genetic algorithm so as to discuss the superiority of several maintenance plans in a more detail manner. By introducing the new multi-objective genetic algorithm, it is possible to obtain several available solutions that have different structural life spans, safety levels and LCC values. Several numerical examples are presented to demonstrate the applicability of the bridge management system proposed here. Especially, the comparison of maintenance plans for the extension of service life and maintenance plans for renewal or rebuilding is made by changing the bridge span.

1 INTRODUCTION

Recently, maintenance work is becoming more and more important, because the number of structures requiring repair or replacement increases in the coming ten years, in Japan. In order to establish a rational and economical maintenance program, the concept of Life-Cycle Cost (LCC) has gained great attention, which minimizes the total cost of whole lives of structures (Liu and Frangopol, 2004a and 2004b).

So far, the authors have developed LCC based bridge maintenance systems for existing concrete bridge structures (Furuta et al., 2003 and 2004). The concrete bridges are deteriorating due to the corrosion of reinforcing bars and neutralization of concrete (Itoh et al., 2002). Then, it is necessary to achieve an optimal maintenance plan that can provide appropriate methods and times of repairing or replacement. However, the optimal maintenance problem is very difficult to solve, because it is one of combinatorial problems with discrete design variables and discontinuous objective functions. Furthermore, the problem may become tougher, when it becomes larger and more complex.

Although low-cost maintenance plans are desirable for bridge owner, it is necessary to consider various constraints when choosing an appropriate actual maintenance program. For example, the minimization of maintenance cost needs to prescribe the target safety level and the expected service life time. The predetermination of requirements may loose the variety of possible maintenance plans.

In this paper, it is intended to discover many alternative maintenance plans with different characteristics by introducing the concept of multi-objective optimization. When selecting a practical maintenance plan, it is desirable to compare feasible solutions obtained under the various conditions. This process is inevitable and effective for the accountability by the disclosure of information. Furthermore, another attempt is made to develop a new multi-objective genetic algorithm for the bridge management problems that have a lot of constraints. Several numerical examples are presented to demonstrate the applicability and efficiency of the proposed method.

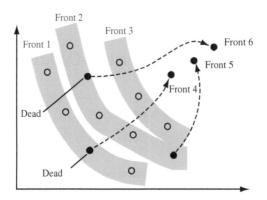

Figure 1. New sorting rules.

2 NEW MULTI-OBJECTIVE GENETIC ALGORITHM

Bridge maintenance planning has several constraints. In usual, it is not easy to solve multi-objective optimization problems with constraints by applying Multi-Objective Genetic Algorithm (MOGA).

In this study, a new MOGA is developed by introducing the sorting technique into the selection process. The selection is performed using so-called sorting rules which arrange the list of individuals in the order of higher evaluation values. Then, the fitness values are assigned to them by using the linear normalization technique. In usual, if the fitness values are calculated directly according to the evaluation values, the differences among every individuals decrease so that the effective selection can not be done. On the other hand, the linear normalization technique enables to keep the selection pressure constant so that it may continue the selection well. In this study, the selection procedure is constructed coupling the linear normalization technique and the sorting technique. Using the evaluation values, the individuals are reordered and given the new fitness values. Figure 1 presents the process of the selection proposed here. The individuals of satisfying the constraints are arranged first according to the evaluation values and further the individuals of non-satisfying the constraints are arranged according to the degree of violating the constraints. Accordingly, all the individuals are given the fitness values using the linear normalization technique.

In order to apply the sorting rules to the multi-objective optimization problems, the non-dominated sorting method is used (Kitano, 1995). In the non-dominated sorting method, the Pareto solutions are defined as *Front1*. Then, *Front2* is determined by eliminating the *Front1* from the set of solution candidates. Repeating the process, the new *Front* is pursued until the solution candidates diminish. Further, the *Fronts* are stored in the pool of the next generation. If the pool is full, the individuals in the *Front* are divided into the solutions to survive or die based on the degree of congestion.

In this study, the individuals are divided into the group of satisfying the constraints and the group without satisfying the constraints. The former is called as "alive individual", and the latter "dead individual". While the alive individuals are given the fitness values according to the evaluation values after the non-dominated sorting, the dead individuals are given the same fitness value. When implementing the non-dominated sorting, the *Pareto Front* may not exist at the initial generation, because there remain a lot of dead individuals after the non-dominated sorting. Then, the dead individuals are arranged in the order of degree of violating the constraints and some of them are selected for the next generation. Thus, the multi-objective optimization problems with constraints are transformed into the minimization problem of violation of constraints. The elite preserve strategy is employed for the selection of survival individuals (Kitano, 1995).

When the generation is progressed, alive individuals appear and then both the alive individuals forming the Pareto front and the dead individuals arranged in the order of violation degree exist together. In this case, appropriate numbers of alive and dead individuals are selected for the next generation.

150

3 APPLICATION OF NEW MOGA TO MAINTENANCE PLANNING

It is desirable to determine an appropriate life-cycle maintenance plan by comparing several solutions for various conditions. A new decision support system is developed here from the viewpoint of multi-objective optimization, in order to provide various solutions needed for the decision-making.

In this study, LCC, safety level and service life are used as objective functions. LCC is minimized, safety level is maximized, and service life is maximized (Furuta et al., 2004). There are trade-off relations among the three objective functions. For example, LCC increases when service life is extended, and safety level and service life decrease due to the reduction of LCC. Then, multi-objective optimization can provide a set of Pareto solutions that can not improve an objective function without making other objective functions worse.

Then, objective functions are defined as follows:

$$\text{Objective function 1} \quad : \quad C_{total} = \Sigma LCC_i \rightarrow min \tag{1}$$

where $LCC_i = $ LCC for bridge i

$$\text{Objective function 3} \quad : P_{total} = \Sigma P_i \rightarrow max$$

$$\text{Constraints} \quad : P_i > P_{target} \tag{2}$$

where $P_{target} = $ Target safety level

4 NUMERICAL EXAMPLES

A group of ten concrete highway bridges are considered in this study. Maintenance management planning for ten consecutive piers and floor slabs (composite structure of steel girders and reinforced concrete (RC) slabs) is considered here (Furuta et al., 2006).

In this study, environmental corrosion due to neutralization of concrete, chloride attack, frost damage, chemical corrosion, or alkali-aggregate reaction are considered as major deteriorations. Deterioration of a bridge due to corrosion depends on the concrete cover of its components and environmental conditions, among other factors. For each component, the major degradation mechanism and its rate of deterioration are assumed corresponding to associated environmental conditions. In this study, four environmental conditions are considered; neutralization, mild, middle and severe environmental conditions.

In the implementation of the proposed new MOGA, the GA parameters considered are as follows: number of individuals = 1000, crossover rate = 0.60, mutation rate = 0.05 and number of generations = 3000.

Figure 2 presents the results obtained by the new MOGA. This figure shows the evolution of the results from the 1st generation (iteration number) to the 100th generation. In Figure 2, the sum of penalty started to decrease from the 1st generation to around the 30th generation. The 20th generation was starting to increase the amount of alive solutions.

In Figure 3, the solutions at the 100th generation were not optimized. This means that the initial solutions can be generated uniformly. After the 120th generation, the solutions tend to converge to a surface, which finally forms the Pareto set as the envelope of all solutions. The number of solutions at the 3000th generation is larger than that at the 100th generation. This indicates that the proposed new MOGA could obtain various optimal solutions with different LCC values and safety levels. From Figure 4, it is seen that the new MOGA can find out good solutions, all of which evolve for all the objective functions, and the final solutions are sparse and have discontinuity. In other words, the surfaces associated with the trade-off relations are not smooth. This implies that an appropriate long term maintenance plan cannot be created by the repetition of the short term plans.

Figures 5 through 7 present the results for the mild environment. Pareto solutions converged after 30th generations, whereas the alive solutions became stable after 80th generations. Comparing with

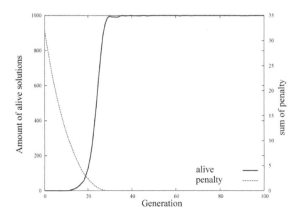

Figure 2.   Evolution of new MOGA under neutralization environment (alive solutions and penalty).

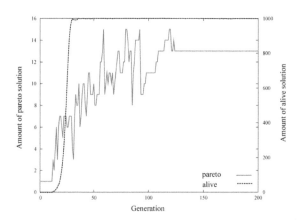

Figure 3.   Evolution of new MOGA under neutralization environment (Pareto and alive solutions).

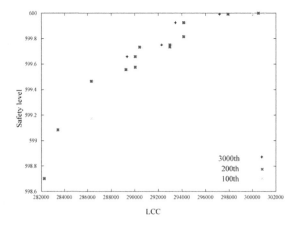

Figure 4.   Evolution of new MOGA under neutralization environment (safety level and LCC).

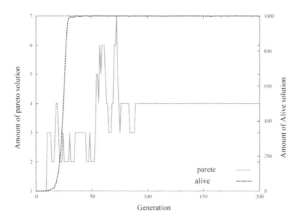

Figure 5.   Evolution of new MOGA under mild environment (Pareto and alive solutions).

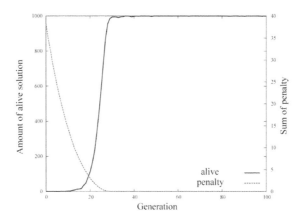

Figure 6.   Evolution of new MOGA under mild environment (alive solutions and penalty).

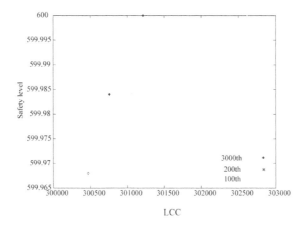

Figure 7.   Evolution of new MOGA under mild environment (safety level and LCC).

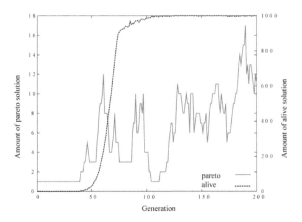

Figure 8.　Evolution of new MOGA under severe environment (Pareto and alive solutions).

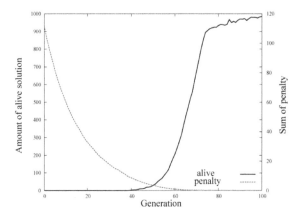

Figure 9.　Evolution of new MOGA under severe environment (alive solutions and penalty).

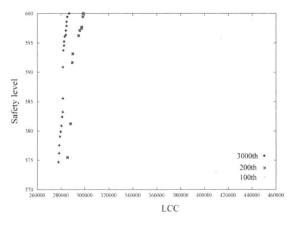

Figure 10.　Evolution of new GA under severe environment (safety level and LCC).

the case of neutralization environment, the Pareto solutions are sparser. Figures 8 through 10 show the evolution results for the case of severe environment. The evolution process required more generations to converge. This tendency can be understood by the fact that the number of alive solutions reduced gradually as shown in Figure 9. However, the Pareto solutions obtained showed

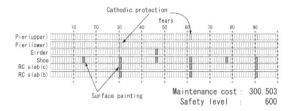

Figure 11.　Maintenance scenarios for mild environment.

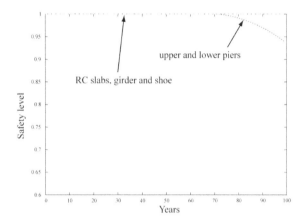

Figure 12.　Degradation curve under slight environment.

a good convergence to a smooth envelope curve of Pareto solutions. This means that it is easier to make a maintenance planning for the case of severe environment.

Figure 11 presents an optimal maintenance plan for the case of mild environment. It is noted that the simultaneous works are considered to be preferable to reduce the LCC value. Figure 12 shows the degradation curves of RC slabs, girders, shoes, and upper and lower piers used in this calculation.

Consequently, it is confirmeed that the proposed method can provide many useful solutions with different characteristics for determining an appropriate maintenance plan available for practical use. It is clear that LCC can be reduced by adopting simultaneous repair works. Finally, it is confirmed that the proposed method using linear normalization technique and sorting technique can provide many near-optimal maintenance plans with various reasonable LCC values, safety levels. Note that it is quite difficult to obtain such near-optimal solutions by the current MOGA.

5　CONCLUSIONS

In this paper, an attempt was made to formulate the optimal maintenance planning as a multi-objective optimization. By considering LCC and safety level as objective functions, it is possible to obtain the relationships among these two performance indicators and provide bridge maintenance management engineers with various maintenance plans with appropriate allocations of resources. Based on the results presented in this paper, the following conclusions may be drawn:

1. Since the optimal maintenance problem is a very complex combinatorial problem, it is difficult to obtain reasonable solutions by the current optimization techniques.
2. Although Genetic Algorithm (GA) is applicable to solve multi-objective problems, it is difficult to apply it to large and very complex bridge network maintenance problems. By introducing the

proposed new Multi-Objective Genetic Algorithm, it is possible to obtain efficient near-optimal solutions for the maintenance planning of a group of bridge structures.

3. The Pareto solutions obtained by the proposed method show discontinuity.
4. In the examples presented, the relation between safety level and LCC is non-linear. The increase of LCC hardly contributes to the improvement of safety level.
5. LCC can be reduced by adopting simultaneous repair works. The proposed method using linear normalization technique and sorting technique can provide many near-optimal maintenance plans with various reasonable LCC values and safety levels.

REFERENCES

Furuta, H., Kameda, T., Nakahara, K., & Takahashi, Y. (2003), "Genetic Algorithm for Optimal Maintenance Planning of Bridge Structures", *Proc. of GECCO, Chicago, US.*

Furuta, H., Kameda, T., & Frangopol, D.M. (2004). "Balance of Structural Performance Measures", *Proc. of Structures Congress, Nashville, Tennessee, ASCE, May,* CD-ROM.

Furuta, H., Kameda, T., Nakahara, K., Takahashi, Y., & Frangopol, D.M. (2006). Optimal Bridge Maintenance Planning Using Improved Multi-Objective Genetic Algorithm, *International Journal of Structure and Infrastructure, Vol.2, No.1.*

Ito, H., Takahashi, Y., Furuta, H., & Kameda, T. (2002). "An Optimal Maintenance Planning for Many Concrete Bridges Based on Life-Cycle Costs. *Proc. of IABMAS*, Barcelona, Spain, CD-ROM.

Kitano, H. (eds.), (1995). "Genetic Algorithm 3", *Tokyo, Sangyo-tosho (in Japanese).*

Liu, M., & Frangopol, D.M. (2004a). "Probabilistic Maintenance Prioritization for Deteriorating Bridges Using a Multiobjective Genetic Algorithm", *Proceedings of the Ninth ASCE Joint Specialty Conference on Probabilistic Mechanics and Structural Reliability, Hosted by Sandia National Laboratories, Omnipress, Albuquerque, New Mexico, July 26–28, 6 pages on CD-ROM.*

Liu, M., & Frangopol, D.M. (2004b). "Optimal Bridge Maintenance Planning Based on Probabilistic Performance Prediction", *Engineering Structures, Elsevier, Vol. 26, No. 7, pp. 991–1002.*

Optimal design and cost-effectiveness evaluation of MR damper system for cable-stayed bridges based on LCC concept

D. Hahm, H.-M. Koh & W. Park
Seoul National University, Seoul, Korea

S.-Y. Ok
University of Illinois at Urbana-Champaign, Illinois, USA

K.-S. Park
Dongguk University, Seoul, Korea

ABSTRACT: A method is presented for evaluating the economic efficiency of a semi-active magneto-rheological (MR) damper system for cable-stayed bridges under earthquake loadings. An optimal MR damper capacity maximizing the cost-effectiveness is estimated for various seismic characteristics of ground motion. The economic efficiency of MR damper system is addressed by introducing the life-cycle cost concept. To evaluate the expected damage cost, the probability of failure is estimated. The cost-effectiveness index is defined as the ratio of the sums of the expected damage costs and each device cost between a bridge structure with the MR damper system and a bridge structure with elastic bearings. In the evaluation of cost-effectiveness, the scale of damage cost is adopted as parametric variables. The results of the evaluation show that the MR damper system is highly cost-effective especially for the ground motion of moderate seismicity.

1 INTRODUCTION

Over the last few decades, studies of efficient vibration control systems that are applicable to cable-stayed bridges has been a major focus in research on structural control systems. One of promising control strategies, the semi-active magneto-rheological (MR) damper system has been extensively studied and applied to dynamic problems associated with cable-stayed bridges. However, most of these related studies focused on the control performance of the semi-active control system itself, while the economic efficiency of such a strategy has not yet been addressed. For the engineering community to fully embrace this technology, a semi-active control system should also establish the economic feasibility of its use in the cable-stayed bridge.

In previous studies (Wen & Shinozuka 1998, Koh 2002, Park et al. 2004a, b), the economic feasibility or cost-effectiveness of the control systems was investigated with respect to civil structures under the seismic events and a life-cycle cost (LCC) concept was introduced by summing up the initial construction cost and the expected damage cost during the lifespan of the structure. The assessment of the expected damage cost in the life-cycle cost concept can account for the probabilistic characteristics of the seismic event that can occur during the lifetime of the structural system. Therefore, an LCC-based cost-effectiveness evaluation of an MR damper system was performed for an earthquake-excited cable-stayed bridge. Furthermore, an optimal design method of MR damper is developed based on the cost-effectiveness of the seismic response control system.

Since economic damage varies very sensitively according to various kinds of surrounding conditions such as the regional economic status, the scale of damage to the bridge is adopted as a

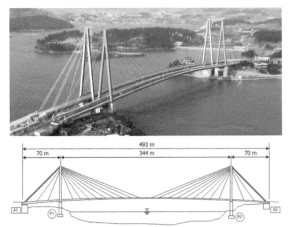

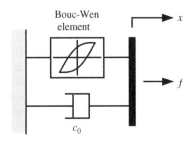

Figure 1.   2nd Jindo Bridge.

Figure 2.   Mechanical model of a parallel-plate MR damper.

parametric variable in the evaluation of cost-effectiveness of an MR damper system for a seismically excited cable-stayed bridge. To demonstrate the proposed approach, a cable-stayed bridge
is used as an example and the evaluation results of the cost-effectiveness of MR damper-based
semi-active control system compared to the previous elastic bearing system are presented.

2   SEMI-ACTIVE CONTROL OF CABLE-STAYED BRIDGE

2.1   *Modeling of Cable-Stayed bridge*

The cable-stayed bridge used in this study is the 2nd Jindo Bridge, which spans the Southern Sea
of Korea between Haenam and Jin-Island (Fig. 1). The cable-stayed bridge is modeled as a three-
dimensional finite element model with 560 DOFs. The original bridge employs elastic bearings in
the connections between the deck and the tower. MR damper systems are adopted in this study for
controlling the seismic response of the bridge. A number of MR dampers is varied from 4 to 12,
each of 2 to 6 between the deck and pier 1, the deck and pier 2.

2.2   *Dynamic model of MR damper*

A simple mechanical model consisting of a Bouc–Wen element in parallel with a viscous damper,
as shown in Figure 2, has been shown to accurately predict the behavior of a full-scale MR damper
(Yoshida & Dyke 2004).

Table 1. Parameters of the MR damper model.

Parameter	Value	Parameter	Value
α_a	1.087×10^5 N/cm	A_m	1.2
α_b	4.962×10^5 N/(cm·V)	n	1
c_{0a}	4.40 N·s/cm	β	$3\,\mathrm{cm}^{-1}$
c_{0b}	4.40×10^1 N·s/(cm·V)	γ	$3\,\mathrm{cm}^{-1}$
		η	$50\,\mathrm{s}^{-1}$

$$f = c_0 \dot{x} + \alpha z, \quad \dot{z} = -\gamma |\dot{x}| z |\dot{z}|^{n-1} - \beta \dot{x} |z|^n + A_m \dot{x} \tag{1}$$

where x = displacement of the device, z = evolutionary variable that accounts for the history dependence of the response. The parameters, γ, β, n, and A_m are adjusted to control the linearity in the
unloading and the smoothness of the transition from the preyield to the postyield region.

Device parameters α and c_0 are determined by the dependency on the control voltage u, as
follows.

$$\alpha = \alpha(u) = \alpha_a + \alpha_b u \quad \& \quad c_0 = c_0(u) = c_{0a} + c_{0b} u \tag{2}$$

To account for a time-lag in the response of the device to changes in the command input, moreover,
the resistance and inductance present in the driver circuit of MR damper introduce first-order filter
dynamics into the system. This dynamics is described by

$$\dot{u} = -\eta(u - v) \tag{3}$$

where v = command voltage applied to the control circuit and $1/\eta$ = the time constant of this first-
order filter. The parameters of the MR damper were selected so that the device has a capacity of
1000 kN, as shown in Table 1. These parameters were obtained via the identification process of a
shear-mode prototype MR damper tested at Washington University (Yi et al. 2001) and scaled up
so as to have a maximum capacity of 1,000 kN with maximum command voltage $V_{max} = 10$ V.

2.3 Design of MR damper system

As a strategy for controlling a seismically-excited cable-stayed bridge, we adopted the modified
clipped optimal control algorithm (Yoshida and Dyke 2004) which is able to continuously vary
the input voltage in the range of $[0 \sim V_{max}]$. a force feedback loop is appended to induce each MR
damper so as to approximately generate a desired optimal control force f_{ci}, which is determined
based on H_2/linear quadratic Gaussian (LQG) strategies in this study. Because the force generated
in the MR damper is dependent on the local responses of the structural system, only the control
voltage can be directly controlled to increase or decrease the damper force. The secondary controller
for continuously varying the command voltage is given by

$$v_i = V_{ci} H(\{f_{ci} - f_i\} f_i), \quad V_{ci} = \begin{cases} \mu_i f_{ci}, & \text{for } f_{ci} \le f_{max} \\ V_{max}, & \text{for } f_{ci} > f_{max} \end{cases} \tag{4}$$

where v_i = command voltage of i-th MR damper, V_{max} = maximum voltage to the current
driver (=10 V), $H(\cdot)$ = Heaviside step function, f_{max} = maximum force produced by the damper
(1,000 kN), μ_i = coefficient relating the voltage to the force (=V_{max}/f_{max}).

Determination of the command signal for the MR damper is graphically represented in Figure 3.
When the magnitude of the actual force produced by i-th MR damper f_i is larger than the magnitude

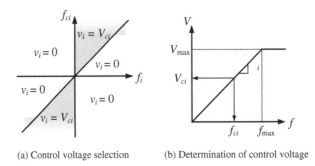

(a) Control voltage selection   (b) Determination of control voltage

Figure 3. Graphical representation of the modified clipped optimal control law.

of the desired optimal force required for i-th MR damper f_{ci}, or if these two forces have different signs, the control voltage applied to the current driver v_i is set to zero. If the magnitude of the force produced by the damper is smaller than the magnitude of the desired optimal force and the two forces have the same sign, the control voltage is determined to be V_{ci}, which is defined by a linear relationship between the applied voltage and the maximum force of MR damper. When the desired force is larger than the maximum force that the device can produce, the maximum voltage V_{max} is applied.

3 MODELING OF INPUT GROUND MOTION AND STRUCTURAL UNCERTAINTY

For the estimation of failure probability and expected damage cost, the uncertainty of input ground motion and structural system should be described properly. Since the seismic performance of an earthquake resistance system is highly dependent on the magnitude and frequency content of ground motion, input ground motion should be able to reflect the seismic characteristics. Frequency contents of ground motions are strongly related to the soil profile characteristics. In this study, we selected 20 earthquake ground motions from PEER Strong Motion Database. Each ground motion is assorted into the stiff-soil profile type and stiff-soil profile type according to its shear wave velocity. Ground motions corresponding to the soft-soil profile and stiff-soil profile are mainly composed by high frequency contents and low frequency contents, respectively. The peak ground accelerations of these earthquakes are scaled by 0.19 g representing the moderate seismicity, and 0.39 g for high seismicity.

Structural uncertainty often cause variations on stiffness of structures and affect to the structural responses significantly. In this study, we considered the structural uncertainty by constructing a global stiffness matrix from the element stiffness matrix which is considering the variation of stiffness.

$$
\left.\begin{aligned}
[K_1]_{12\times12} \times \alpha_1 &= [K_1]^*_{12\times12} \\
[K_2]_{12\times12} \times \alpha_2 &= [K_2]^*_{12\times12} \\
&\vdots \\
[K_n]_{12\times12} \times \alpha_n &= [K_n]^*_{12\times12}
\end{aligned}\right\} [K]^*_{m\times n}
\tag{5}
$$

where, α_i is a random variable corresponding to normal distribution of $N(1,\sigma)$, $[K_i]_{12\times12}$ is nominal stiffness matrix of 3-D beam element. $[K_i]^*_{12\times12}$ and $[K_i]^*_{m\times n}$ are element stiffness matrix and global stiffness matrix considering structural uncertainty, respectively.

160

Table 2. Mean and standard deviation of the maximum base-moment response of the tower.

# of dampers	4		8		12	
	μ_M	σ_M	μ_M	σ_M	μ_M	σ_M
0.19 g, Stiff-soil	11.4060	0.5018	11.3485	0.5113	11.3320	0.5008
0.19 g, Soft-soil	11.9316	0.3435	11.8774	0.2871	11.8482	0.2681
0.39 g, Stiff-soil	12.1192	0.5005	12.0478	0.5116	12.0189	0.5046
0.39 g, Soft-soil	12.6553	0.3480	12.5967	0.2977	12.5885	0.2994

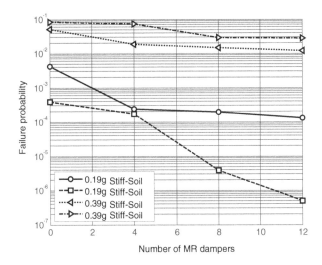

Figure 4. Failure probabilities of cable-stayed bridge with respect to the number of dampers.

4 FAILURE PROBABILITY ESTIMATION

4.1 Definition of limit state

To estimate the failure probability of a structure, the limit states of the structure should be predefined. Many structural components constituting the cable-stayed bridge may have their own limit states of failure. However, in this study, we assume that the base overturning moment is the most critical response to horizontal ground motion and consider only the failure caused by excessive base moment, because the base overturning moment is often the most important response of cable-stayed bridges that have a fixed pier-deck connection. In this study, the limit state of the base moment response is defined as 520,000 kN·m which induces the yield of longitudinal reinforcement bar.

4.2 Failure probabilities

The failure probability P_f of a cable-stayed bridge system is defined as the probability that maximum base-moment response of the tower (M_b) exceeds the limit state (M_{lim}). It is well-known that the probability distribution of the maximum structural response is generally correspondent to the lognormal distribution. In this study, we performed 100 time-history analyses to obtain the probabilistic parameters, i.e., mean (μ_M) and standard deviation (σ_M) of the maximum base-moment response of the tower. Each parameters corresponding to each seismic characteristics are listed in Table 2. Then, the failure probability is calculated from the cumulative probability density function of lognormal distribution. Figure 4 depicts the failure probabilities of cable-stayed bridge with respect to the number of dampers and seismic characteristics. For ground motion of moderate seismicity, it can be seen that the failure probabilities are significantly decreased as the MR dampers are installed and number of dampers are increased. On the other hand, for the ground motion of high seismicity, PGA of 0.39 g, we can found that the performance of MR damper on the viewpoint of a failure probability is not as good as the case of the 0.19 g.

5 COST-EFFECTIVENESS EVALUATION

In order to provide a quantitative basis for measuring an effectiveness of controls of MR damper systems, we proposed a cost-effectiveness index based on the expected life-cycle cost concept. Although the life-cycle cost function of a structural system under natural hazards is constructed by a sum of many cost items such as initial construction cost, expected damage cost, maintenance cost, retrofit cost, etc., for the evaluation of cost-effectiveness of semsic response control system, we do not have to estimate the exact values of all the cost items. The life-cycle cost function of bridge with MR damper (C_{MR}) or elastic bearing system (C_{EB}) can be formulated as following equations.

$$C_{EB} = C_I + C_M + C_{S.EB} + C_D \cdot P_{f.EB} \cdot \frac{v}{\lambda}(1 - e^{-\lambda t_{life}})$$ (6)

$$C_{MR} = C_I + C_M + C_{S.MR} + C_D \cdot P_{f.MR} \cdot \frac{v}{\lambda}(1 - e^{-\lambda t_{life}})$$ (7)

where, C_I is initial construction cost; C_M is discounted cost of maintenance; $C_{S.EB}$ and $C_{S.MR}$ are seismic retrofit/control cost related to the installation of elastic bearing and MR damper; $P_{f.EB}$ and $P_{f.MR}$ are failure probabilities of a cable-stayed bridge with elastic bearing and MR damper, respectively. C_D is total damage cost caused by failure of cable-stayed bridge. v is earthquake occurrence rate, λ is discount rate and t_{life} is lifetime of a structure. The term $v/\lambda(1 - e^{-\lambda t_{life}})$ is derived based on the assumption that earthquake occurrences can be modeled by a simple Poisson process with occurrence rate, v.

It is reasonable that the use of an MR damper system will increase the seismic retrofit/control cost compared with the case of the installation of an elastic bearing system. However, the expected damage cost due to a severe earthquake event through the lifespan of the structure will be reduced. The cost-effectiveness of the MR damper system can be explained by the balance between these two cost items, i.e., the additional seismic retrofit/control cost and the expected damage cost. Therefore, for the evaluation of cost-effectiveness of MR damper, we considered the summations of only these two cost items rather than evaluate all the cost terms in Eq. (6) and (7). A cost-effectiveness index J_{MR} is defined as follows:

$$\begin{aligned}
J_{MR} &= \frac{C_{S.EB} + C_D P_{f,EB} v/\lambda(1 - e^{-\lambda t_{life}})}{C_{S.MR} + C_D P_{f,MR} v/\lambda(1 - e^{-\lambda t_{life}})} \\
&= \frac{\alpha_{EB} + \beta \cdot P_{f,EB} v/\lambda(1 - e^{-\lambda t_{life}})}{\alpha_{MR} + \beta \cdot P_{f,MR} v/\lambda(1 - e^{-\lambda t_{life}})}
\end{aligned}$$ (8)

In this study, we normalized numerator/denominator of J_{MR} by the initial construction cost C_I. Hence, $\alpha_{EB} = C_{S.EB}/C_I$, $\alpha_{MR} = C_{S.MR}/C_I$ and $\beta = C_D/C_I$. Note that the MR damper is cost-effective when the value of J_{MR} is larger than 1.

Even though the cost-effectiveness index (J_{MR}) varies with respect to the scale of damage cost, β, a quantitative assessment of the subsequent effects of the failure of bridges due to the seismic events represents a delicate task, and may belong to the politic issues rather than the engineering issues. Since the damage scale defined as β cannot avoid some level of subjectivity and uncertainty, we performed a parametric study on the variation of this value and also examined the sensitivity of the results of a cost-effectiveness evaluation.

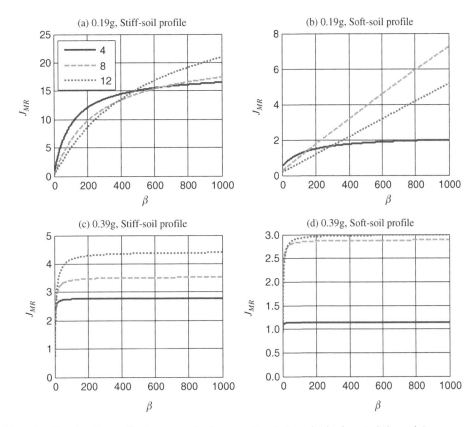

Figure 5. Results of cost-effectiveness evaluation according to the seismic characteristics and β.

If C_D or β is very large, i.e., a cable-stayed bridge performs a very important role on the traffic flow system and the cost induced by the failure of the bridge is nearly boundless, the variation of cost-effectiveness index will be governed by the ratio of failure probability between the case of installation of elastic bearing and MR damper. In that case, we can suppose that the employment of effective MR damper system will cost-effective if the failure probability will be reduced.

Figure 5 represent cost-effectiveness evaluation results of MR damper system for cable-stayed bridge subjected to the ground motions corresponding to the various seismic characteristics. The value of cost-effectivness index is larger than 1 in most cases and this results show that MR dampers are cost-effective compared to the elastic bearing systems. Figure 5(a) is a result for the ground motion of 0.19 g and stiff-soil profile type. We can found that the optimal damper capacity is varied according to the the damage scale, β. If β is smaller than 500, we can suppose that the smallest number of MR dampers is most effective. On the other hand, as the damage scale is increased, the optimal capacity of MR damper system is also increased. This result implies that we may increase the damper capacity if the cable-stayed bridge is very important and the failure of the bridge will cause a critical crisis on the regional economics. For the ground motion of moderate seismicity and low-frequency contents (Fig. 5(b)), we can found that the optimal number of MR damper is 8 if β larger than 100.

The results of the cost-effectivness evaluation for the ground motion of high seismicity, 0.39 g, are presented in Figure 5(c) and (d). It can be noted that the cost-effectiveness index is not varied according to the variation of β. And we can also found that the largest capacity of MR damper is most cost-effective. With this result, it can be inferred that the installation of large-capacity MR damper is profitable to resist the high magnitude of ground motion.

6 CONCLUSIONS

A method for evaluating the cost-effectiveness of an MR damper-based semi-active control system for cable-stayed bridge with respect to various seismic characteristics is proposed. The cost-effectiveness of the MR damper system is defined as the ratio of the life-cycle cost of a bridge structure with semi-active control devices to that of a bridge structure with elastic bearings. The proposed method was applied to the 2nd Jindo Bridge subjected to historical earthquake ground motions. For the ground motion of moderate seismicity, the evaluation results show that the cost-effectiveness is quite sensitive to the cost of damage induced by bridge failure. It is also found that the optimal capacity of MR damper system is varied according to the seismic characteristics of ground motion. We expect that the proposed method and the corresponding results on the cost-effectiveness of MR damper-based control system will serve as an aid to designers for reaching decisions on the use of such control systems.

ACKNOWLEDGEMENT

This work is a part of a research project supported by Korea Ministry of Construction & Transportation (MOCT) through Korea Bridge Design & Engineering Research Center at Seoul National University. The authors wish to express their gratitude for the financial support.

REFERENCES

Koh H.M. 2002. Cost-effectiveness analysis for seismic isolation of bridges. *In proceedings of the third world conference on structural control, vol. 1*. Como, Italy.
Park K.S., Koh H.M. & Song J. 2004a. Cost-effectiveness analysis of seismically isolated pool structures for the storage of nuclear spent-fuel assemblies. *Nuclear Engineering and Design* 231(3): 259–70.
Park K.S., Koh H.M. & Hahm D. 2004b. Integrated optimum design of viscoelastically damped structural systems. *Engineering Structures* 26(5): 581–91.
Shinozuka M. & Deodatis G. 1991. Simulation of stochastic processes by spectral representation. *Applied Mechanics Reviews* 44(4): 191–203.
Wen Y.K. & Shinozuka M. 1998. Cost-effectiveness in active structural control. *Engineering Structures* 20(3): 216–21.
Yi F., Dyke S.J., Caicedo J.M. & Carlson J.D. 2001. Experimental verification of multi-input seismic control strategies for smart dampers. *Journal of Engineering Mechanics* 127(11): 1152–64.
Yoshida O. & Dyke S.J. 2004. Seismic control of a nonlinear benchmark building using smart dampers. *Journal of Engineering Mechanics* 130(4): 386–92.

Cost evaluation in bridge inspection strategies by using real options method

T. Harada & K. Yokoyama
Department of Urban & Civil Engineering, Ibaraki University, Hitachi, Japan

ABSTRACT: To support expensive bridge inspection by the human being, bridge monitoring technique for damages such as crack, fatigue and corrosion has already been introduced. But, the cost evaluation of the bridge inspection strategies in consideration of the bridge monitoring has not discussed enough in current situation. In this study, the cost evaluation method for bridge inspection alternative by using Real Options method has been proposed. Then, by the proposed cost evaluation method considering the change of inspection cost in the future, the role of monitoring of bridge inspection has been discussed.

1 INTRODUCTION

Recently, due to the increasing inspection cost for bridge structures, there is an urgent need to find a more economical way to perform bridge inspection. In order to devise bridge maintenance strategies for the rest of the projected service life of existing bridge structures, we have to find a way to minimize the bridge inspection cost. On the other hand, to support expensive bridge inspection by the human being, bridge monitoring for damages such as crack, fatigue and corrosion has already been introduced (Sumitro et al. 2005, Wu 2003). But, the cost evaluation of the bridge inspection strategies in consideration of the bridge monitoring has not discussed enough in current situation.

The important point in planning inspection strategies is uncertainty in the future. The bridge inspection strategies might be modified in the future due to uncertainty like bridge degradation, budget, changing of social system, technology innovation and so forth. However, the rigid evaluation method, such as Discount Cash Flow (DCF) method, commonly used for alternative evaluation, does not consider uncertainty that may occur in future. On the other hand, Real Options (RO) method is a new technique that can take this into consideration providing the flexibility of alternatives. RO method applies financial options theory to real investments, such as manufacturing plants, product line extensions, and research and development. For example, a company that enters a new market may build a distribution center that it can expand easily if market demand materializes. This is the example of the expansion option. However, traditional valuation tools, including DCF method, cannot value the contingent nature of the exploitation decision. Therefore, these methods are called the rigid evaluation method (Amram et al. 1999, Harada et al. 2005, Kurino et al. 2000).

In this study, the cost evaluation method for bridge inspection alternative by using RO method has been proposed. The inspection cost increase caused by aging of the bridge, and the value of cost evaluation is quantitatively estimated based on some possibility that may cause change of inspection cost in the future. The bridge monitoring is introduced instead of some of the inspection. Therefore, instead of the bridge inspection cost being reduced, initial and running cost for bridge monitoring is occurred. By the proposed cost evaluation method considering the change of inspection cost in the future, the role of the monitoring in the bridge inspection can be discussed.

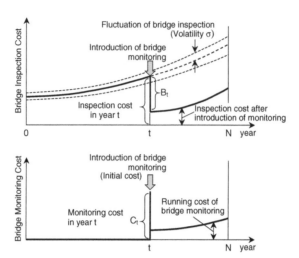

Figure 1. Modeling of bridge inspection and monitoring cost.

2 MODELING OF BRIDGE INSPECTION STRATEGIES IN CONSIDERATION OF BRIDGE MONITORING

Bridge inspection is the first work in order to plan repair and reinforcement of bridges. However, a bridge inspection cost increases due to decaying of the bridge, aggravation of the traffic environment, and so on. Then, to support expensive bridge inspection by the human being, it is necessary to introduce bridge monitoring. In this study, bridge inspection strategies in consideration of the bridge monitoring were modeled. Bridge inspection cost increases year by year in the future. Then, bridge inspection cost has volatility σ, and we supposed that they are defined by the geometry of Brownian motion.

This model can examine the reduction of the inspection cost by introducing bridge monitoring. When the bridge monitoring is introduced in the year t, inspection cost is reduced like the solid line in Figure 1. This is defined amount of inspection cost reduction B_t of the year t. If the bridge monitoring is not introduced, the amount of inspection cost reduction is zero ($B_t = 0$). On the other hand, when the bridge monitoring is introduced, we must estimate both the initial cost for setting sensor, hard and soft wear of monitoring system and its running cost. The cost C_t along with the introduction of these monitoring gives it minus value to the amount of bridge inspection cost reduction B_t.

3 COST ESTIMATION METHOD OF BRIDGE INSPECTION STRATEGIES IN CONSIDERATION OF BRIDGE MONITORING

3.1 Estimation of NPV by DCF method

In this study, Discount Cash Flow (DCF) method has been used as value estimate method of bridge inspection. In the DCF method, the fluctuation of the inspection cost in the future is not taken into consideration. Then, the net present value (NPV) in the DCF method can be expressed in the difference of the reduction of inspection cost and monitoring cost. The NPV is defined as the following equation.

$$NPV = \sum_{t=0}^{N} \frac{B_t}{(1+i)^t} - \sum_{t=0}^{N} \frac{C_t}{(1+i)^t} \tag{1}$$

166

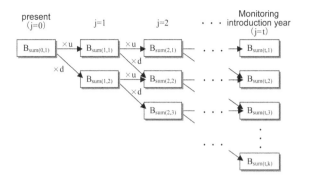

Figure 2. The event tree figure due to calculation of $B_{sum(j,k)}$.

where B_t is the inspection cost reduction (JPY) at year t, C_t is the monitoring cost (JPY) at year t, N is planning period (year), and i is discount rate (%).

3.2 *Estimation of ROV by Real Options method*

We used Real Options (RO) method as another value estimate method of the bridge inspection cost. In the RO method, the fluctuation of the inspection cost in the future can be taken into consideration. The calculated value by using real options method is called a real option value (ROV). The ROV estimation process is shown in the following steps:

Step 1: The bridge inspection cost reduction (B_{sum}) by introduction of bridge monitoring is calculated as the total amount from year t to year N. This value (B_{sum}) is shown in the first clause of the right sides of the equation (1) in the DCF method.

Step 2: The B_{sum} that it is calculated with step 1 is put as the $B_{sum(0,1)}$ which is value of the 0th year. $B_{sum(j,k)}$ of every year from the present ($j = 0$) until the monitoring introduction year ($j = t$) can be expressed with the equation (2) by giving the fluctuation in a year (increasing rate u or decreasing rate d) of the inspection cost. The increasing rate u is defined as the equation (3), and the decreasing rate d is defined as the equation (4). The flow of this estimation is shown as event tree figure in figure 2.

$$B_{sum(j,k)} = \begin{cases} B_{sum(j-1,k)} \times u \\ B_{sum(j-1,k-1)} \times d \end{cases} \quad (2)$$

$$u = e^{\sigma} \quad (3)$$

$$d = e^{-\sigma} \quad (4)$$

where j is the years until monitoring introduction, k is number of event at year j, and σ is volatility (%).

Step 3: $R_{sum(t,k)}$ shown in equation (5) can be estimated by reducing monitoring cost from $B_{sum(t,k)}$ at year t. Monitoring cost is the total amount of the initial cost and the running cost. When the value of $R_{sum(t,k)}$ becomes minus value, we can select the option that the monitoring has not been introduced, and $R_{sum(t,k)}$ is made zero.

$$R_{sum(j,k)} = B_{sum(j,k)} - \sum_{t=t}^{N} C_t \quad (5)$$

where, C_t is the monitoring cost (JPY) of the year t, and N is planning year.

Step 4: $R_{sum(j,k)}$ of every year shown in equation (6) can be estimated by using the $R_{sum(t,k)}$ that it is calculated with step 3, where i is discount rate (%). The increase probability p and the

167

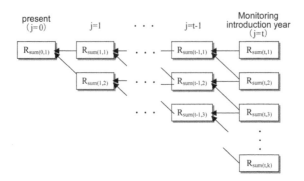

Figure 3.   The decision making tree figure due to calculation of $R_{sum(j,k)}$.

Table 1.   The basic information of the bridge made the target for verification.

Length of bridge	1,015 m
Effective width	11.5 m
Super structure	4 span continuous steel girder bridge + 3 span continuous steel girder bridge
Slab	Reinforced concrete slab

decrease probability q in equation (6) are shown respectively in equation (7) and (8). The flow of this estimation is shown as decision making tree figure in Figure 3. $R_{sum(0,1)}$ finally becomes ROV.

$$R_{sum(j-1,k)} = \frac{R_{sum(j,k-1)} \times p + R_{sum(j,k)} \times q}{\exp(i)} \qquad (6)$$

$$p = \frac{e^i - d}{u - d} \qquad (7)$$

$$q = 1 - p \qquad (8)$$

4   THE TARGET BRIDGE AND ITS INSPECTION AND MONITORING COST CALCULATION

4.1   *The basic information and inspection plan of the bridge made target for verification*

The basic information of real bridge, which is in Ibaraki Prefecture, Japan to examine the inspection plan, is shown in the Table 1. Then, the inspection alternative of the real bridge is shown in the Table 2. In the inspection alternative, we set up planning years with fifty years. The 1st bridge inspection is carried out at 10th year after its establishment, the bridge inspection of 2nd to 6th are carried out in every five years, and the other inspection are carried out in every two years.

4.2   *The settlement of the bridge inspection cost and the monitoring cost*

The bridge inspection cost is composed of the personnel costs and the scaffold cost. The personnel costs contain inspection cost, security guard cost, report writer cost, and so on. We defined that inspection period was thirty days, and the unit price of the scaffold cost was 10,000 JPY per square meter. From the above, the bridge inspection cost is calculated with 129.15 million JPY per one time.

On the other hand, to support expensive bridge inspection by the human being, in this study, the crack monitoring system of the reinforced concrete slab is introduced. The crack monitoring system of the reinforced concrete slab can measure the crack automatically by camera, and the measured data are collected in the computer. The monitoring system observes the occurrence and

Table 2. Inspection plan for the bridge made the target of the calculation.

Year	Inspection	Year	Inspection	Year	Inspection	Year	Inspection	Year	Inspection
1	No	11	No	21	No	31	No	41	inspection
2	No	12	No	22	No	32	No	42	No
3	No	13	No	23	No	33	No	43	inspection
4	No	14	No	24	No	34	No	44	No
5	No	15	inspection	25	inspection	35	inspection	45	inspection
6	No	16	No	26	No	36	No	46	No
7	No	17	No	27	No	37	inspection	47	inspection
8	No	18	No	28	No	38	No	48	No
9	No	19	No	29	No	39	inspection	49	inspection
10	inspection	20	inspection	30	inspection	40	No	50	No

growth of the crack by analyzing those data. We are considering that this monitoring system does not make bridge inspection unnecessary, but it makes up for bridge inspection. In other words, monitoring system of the reinforced concrete slab reduces the frequency of the bridge inspection. The cost of monitoring system is composed of initial costs and running costs. According to the catalog of the monitoring system of the reinforced concrete slab, initial cost is 252.53 million JPY, and running cost is 2.52 million JPY per year for the bridge made target of verification.

5 NUMERICAL EXAMPLES

5.1 The calculation condition of numerical examples

The usefulness of the proposed cost evaluation method and bridge monitoring in bridge inspection strategies were discussed through numerical examples. Through these examples, we can examine the appropriate introduction year of monitoring, the influence of the change in the inspection frequency by the monitoring introduction, and the influence of the fluctuations in the inspection cost in the future.

The bridge inspection alternative is evaluated in fifty years. A presumed inspection cost in the future is reduced by the introduction of the monitoring. The reduction of inspection cost is calculated due to the change in the inspection frequency as shown in Table 3. In other words, the introduction of the monitoring makes the change of inspection frequency and thus makes a reduction in inspection cost.

The future bridge inspection costs changes due to traffic environment. We defined the fluctuation of bridge inspection cost by geometry of Brownian motion. In the numerical examples, the volatility σ of the inspection cost changes as a parameter (3%, 5%, 7%, 10%). The discount rate i was set up 4%.

5.2 The result of numerical examples and verification

First, we estimated NPV and ROV supposing that monitoring has been introduced in the 0th-year (the present). Then, NPV and ROV in Figure 4 were estimated by introduction the monitoring change in every five years. In this numerical example, the volatility σ of the inspection cost is set up 5%. We can see a difference in the evaluation value on Case 1 between NPV by DCF method and ROV by using Real Options method in Figure 4(a).

By using DCF method, regardless of a year to introduce a monitoring, NPV is about fixed and minus. This shows that a monitoring cost is bigger than the reduction of the inspection cost by introducing monitoring. On the other hand, the ROV of about 300 million JPY occurs when a monitoring is introduced in the 30th year. We could confirm that the volatility of the inspection cost has the possibility to increase future inspection cost reductions greatly. When the bridge

Table 3. The change in the inspection frequency by the monitoring introduction.

Inspection cases	The change in the inspection frequency		
	From 0 year to 10 year	From 10 year to 35 year	From 35 year to 50 year
Case 1	1 time per 10 year	1 time per 6 year	1 time per 3 year
Case 2	1 time per 10 year	1 time per 7 year	1 time per 4 year
Case 3	1 time per 10 year	1 time per 8 year	1 time per 5 year
Case 4	1 time per 10 year	1 time per 9 year	1 time per 6 year
Inspection plan	1 time per 10 year	1 time per 5 year	1 time per 2 year

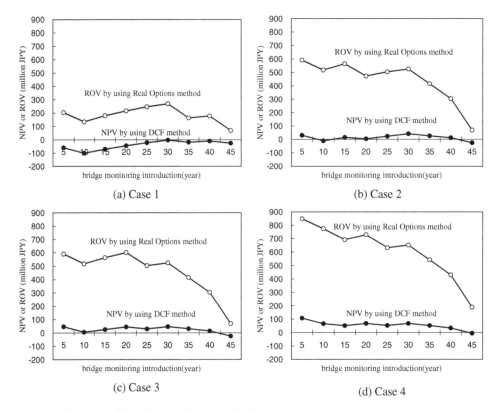

(a) Case 1   (b) Case 2

(c) Case 3   (d) Case 4

Figure 4. The result of the volatility 5% of numerical examples.

inspection strategies in consideration of the bridge monitoring are made, it is necessary to introduce a monitoring as early time as possible.

And, the influence of the change in the inspection frequency by the monitoring introduction can be examined from NPV and ROV that is calculated for all the cases from Case 1 to Case 4 shown in the Table 3. The calculation result of NPV and ROV are shown in the Figure 4(a)-(d). The DCF method could not estimate high NPV in every case. On the other hand, the ROV of the monitoring introduction grows bigger, when the frequency of the inspection decreases like Case 4. If the introduction of the monitoring can reduce the inspection frequency, the bridge inspection strategies in consideration of the bridge monitoring will get high evaluation.

Next, we estimated ROV of the different volatility (3%, 5%, 7% and 10%) to examine the influence of the fluctuations in the inspection cost in the future. Here, the condition of the change

Table 4. The ROV by the change of volatility in Case 2.

Introduction year of bridge monitoring	ROV by the change of volatility (million JPY)			
	3%	5%	7%	10%
5 year	590.69	590.69	592.11	593.36
10 year	517.17	517.17	519.17	527.72
15 year	563.20	563.20	564.98	573.12
20 year	472.10	472.21	473.84	487.59
25 year	503.52	503.60	505.83	514.97
30 year	510.88	513.88	524.06	529.50
35 year	400.06	405.15	415.27	421.74
40 year	293.77	295.41	303.66	310.20
45 year	66.20	69.04	71.00	78.07

in the inspection frequency by the monitoring introduction was made for Case 2. The ROV estimated by making volatility change is shown in the Table 4. As the result of estimating the ROV of the monitoring introduction of every five years by the volatility, we could confirm that the ROV increases due to the increase in the volatility. As the volatility increases, the influence of the fluctuations in the inspection cost in the future is estimated higher. However, the ROV is decided by the correlation of the inspection cost reduction and the monitoring cost.

6 CONCLUSIONS

The following conclusions may be made from this study:

(1) Generally, because the inspection cost in the future has the fluctuation, the appropriate introduction year of the monitoring can be found by using the real options cost evaluation method.
(2) To get the bridge inspection strategies in consideration of the bridge monitoring with high evaluation, it is necessary to reduce the number of bridge inspection by the introduction of the monitoring.
(3) Introduction of monitoring in the bridge inspection strategies has the possibility to reduce the bridge inspection cost.
(4) The proposed cost evaluation methods are available for decision making for effective bridge inspection strategies.

REFERENCES

Amram, M., and Kulatilaka, N. 1999. *Real Options: Managing Strategic Investment in an Uncertain World*, Harvard Business School Press.
Harada, T., and Yokoyama, K. 2005. Optimization of maintenance strategies for road pavement by considering flexibility of alternatives, *Proceedings of 4th International Workshop on Life-Cycle Cost Analysis and Design of Civil Infrastructure Systems*, pp. 307–315.
Kurino, M., Tamura, K., and Kobayashi, K. (2000). The optimal repairing rules for pavements under uncertainty, *Proceedings of Infrastructure Planning*, No.23(2), pp. 55–58. (in Japanese)
Sumitro, S., Nishido, T., Wu, Z.S., and Wang, M.L. 2005. Structural health assessment using actual-stress monitoring system, *Proceedings of International Symposium on Innovation & Sustainability of Structures in Civil Engineering - Including Seismic Engineering (ISISS'2005)*, pp. 1491–1502.
Wu, Z.S. 2003. Structural health monitoring and intelligent infrastructure, *1st International Conference on Structural Health Monitoring and Intelligent Infrastructure (SHMII-1'2003)* (Keynote paper), Vol. 1, pp. 153–167.

Life-Cycle Cost and Performance of Civil Infrastructure Systems – Cho, Frangopol & Ang (eds)
© 2007 Taylor & Francis Group, London, ISBN 978-0-415-41356-5

A case study: Evaluation for the replacement time of building equipment using building energy simulation and LCC analysis

Young Sun Jeong
Faculty of Architecture Urban Planning and Landscape Architecture, University of Seoul, Seoul, Korea
Building & Urban Research Department, Korea Institute of Construction Technology, Goyang-si, Korea

Seung Eon Lee
Building & Urban Research Department, Korea Institute of Construction Technology, Goyang-si, Korea

Jung Ho Huh
Faculty of Architecture Urban Planning and Landscape Architecture, University of Seoul, Seoul, Korea

ABSTRACT: This paper deal with case study on evaluating for the replacement time of building equipment during building operation and maintenance by using Life Cycle Cost. This research was to derive building energy performance analysis through building energy simulation with field measurement and documents (specification, drawing, et al.) about the case study building in Seoul Korea. From LCC analysis with making an investigation into initial cost, annual energy use cost and maintenance & repair cost, et al., the replacement time of the building equipment is calculated. The energy consumption unit of target building was 407.2 Mcal/m^2y in 2000 and 450.9 Mcal/m^2y in 2001. Energy consumption was on the increase gradually in comparison with average energy consumption of building in Korea. As the economic estimate result by LCC analysis, Given that discount rate is 8.0%, rate of price increase is 5.0% and the study period is 20 years, the economical replacement time of the target building's chiller is 7 years after completion of building construction work. With physical life and social life of building equipments, the calculation of economic service life will be applied to estimate reasonable replacement time for retrofit during building life-cycle.

1   INTRODUCTION

1.1   *Background and target*

Each element of buildings must properly have repairs and replacement according to its deterioration. Service life of building's equipment system is about 10~25years[1] and it become different according to the operating method and repairing maintenance. As the duration of use of building has been extending, the building systems can not avoid physical malfunction and buildings must have new function and the improvement of performance on demand of the rapid growth of society.

At present, the building operation & maintenance by repair and replacement of building elements in Korea has been using the limited method to evaluate energy efficiency of building, physical life span of building equipment and social life span of building without economic utility. In Korea, this condition can not been meeting the economic requirements of building owners and users.

Therefore, it is necessary to evaluate the economical time for replacement of building equipment by using life-cycle costing (LCC) during building operation and maintenance.

In this study, we would like to analyze energy performance of the case study building and to examine the method for evaluation of economic replacement time of building equipment.

1.2 *Scope and method of the research*

In this study, the computer simulation with a field audit and specification was performed to analysis of building energy performance on the target building. We investigated a initial cost, annual energy cost, the operation & maintenance (O&M) cost. The replacement time of building equipment was calculated by using LCC analysis.

Specially, in this study, we are here concerned with the replacement of chiller which is typical building equipment and has large energy consumption.

2 DETERIORATION AND ECONOMIC LIFE OF BUILDING EQUIPMENT

As the time is going from completion of a building construction, buildings are becoming super-annuated and performance of building is going down. The progress of deterioration is rapid from completion of building construction work to a certain extent elapsed time and very slow since the function of buildings maintains the least limitation. For the reasons mentioned above, the O&M cost goes on increasing continuously to maintain the function of buildings and is fixed since the function of buildings maintains the least limitation. This deterioration keep going continuously and finally when physical life span of building elements is over the improvement of building elements is carried out by the replacement.

The building and building equipment are used continuously until the employment of it become the impossible condition by physical deterioration that is undesirable from economic point of view because the function of building is lowering and O&M cost connected with it is increasing.

The economic life decided by annual equivalent worth of building is shown in Figure 1.[2] When the sum of annual equivalent worth of the initial investment cost for construction and annual equivalent worth of the O&M cost (energy cost + maintenance cost + replacement cost+repair cost + management expense+other service costs) become the least of year calculated by LCC analysis, this duration of use could be economic life.

In general, the initial plan and management plan is very important as regard of building life-cycle because the O&M cost is bigger than the initial cost. We have to consider that building equipment and interior is repaired periodically according to the use of building. For this reason, LCC analysis for time of replacement and repair in building is worth consideration.

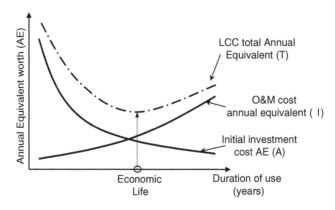

Figure 1.   Annual equivalent worth and economic life.

3 CASE STUDY OF EVALUATION FOR REPLACEMENT TIME OF BUILDING EQUIPEMENT

3.1 *Building summary and energy consumption*

The target building is office building located in the center of Seoul. This building was constructed at June 1992 and has eighteen floors above ground six underground. Table 1 shows building summary and general contents and Figure 2 shows the view of target building.

From field audit and specification, we became to know that this building was steel framed reinforced concrete structure and the envelope of building has curtain wall with closing large window. Also, we became to know that this building did not have a large-scale repair work and changed interior materials according to demand of building's users.

In order to look at the trends of energy consumption of building, we investigated the data of energy consumption between 1994 and 2001 with the exception of the early loss data managed carelessly. The data are collected from the specification and field audit. Figure 3 shows annual energy consumption of the target building.

Energy consumption of target building increased 2.3~12% until 1996. But because the condition of economy grew worse between 1998 and 1999 (The condition of Korea economy was on International Monetary Fund), the space of the target building was not used and target building spent few energy.

It has been reported that the energy consumption unit of office building in Korea is 405.7~412.5 Mcal/m^{2}yr[3][4]. The energy consumption unit of target building largely increased to

Figure 2.   View of target building.

Table 1.   Building summary.

Item		Description
Area	Site Area (Building Area)	3,351.83 m^2 (1,232.34 m^2)
	Gross Floor Area	37,708.38 m^2
Building	Bldg. Type	Office Building
	Exterior Materials	Granite, Double Clear Glass(6/12/6 mm)
	Floor Information	B2-6: plant, parking garage, engineering room
		B1: restaurant, office, F1: lobby, bank F2~F18: bank, tenant office

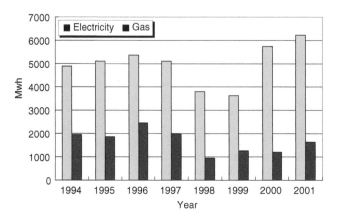

Figure 3. Yearly energy use profile of target building.

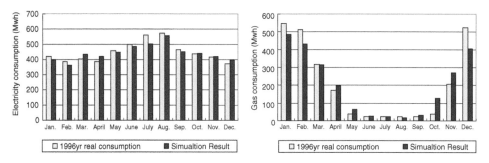

Figure 4. Comparison of monthly electricity consumption and gas consumption.

367.5 Mcal/m^{2}yr for 1994 from 407.2 Mcal/m^{2}yr for 2000 and 450.9 Mcal/m^{2}yr for 2001. Since the condition of Korea economy had greatly improved in 2000, the building expanded interior and the use of office appliances rapidly increased. So, as heating load decreased gas consumption did not increase. On the other hand, as cooling load increased electricity consumption rapidly increased. Because of it, energy consumption was increasing gradually.

3.2 *Analysis of annual energy consumption by simulation*

We estimated the future energy consumption and analyzed energy use of building components by using building energy simulation program. The building modeling for simulation was produced according to the building condition in 1996. In 1996 the target building was good management state and all room of building was used.

The simulation result of annual energy consumption with the weather data(1996 year) of Seoul, Korea showed 4.05% error as compared with the real energy consumption of the target building in 1996. As Figure 4 indicates, the properties of monthly energy consumption are showed similarly.

Figure 5 shows electricity consumption rate used by electrical use components and gas consumption rate used by gas use components of building. In case of electricity consumption, energy consumption by blowers and fans is 31% under the influence of air condition system. The cooling (18%) shows relatively much energy consumption. Most of gas consumption is used by heating, it is 96%.

3.3 *Life-cycle costing method and terms*

(1) Calculation method
Building LCC means all cost associated with the life cycle of a building, expressed in terms of equivalent dollars.

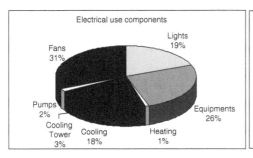
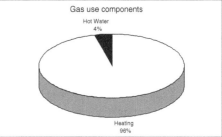

Figure 5.   Electrical use components and gas use components.

Table 2.   Cost data to perform LCC.

Cost item		Cost [×1000 Won]	Cost calculation
P	Initial investment	293,800	141,900×2(EA)
M	Energy annual cost	70,577	Total electrical use cost × 18% = 392,095 × 0.18
	M&R annual cost	4,407	Initial cost × 1.5%

Annual equivalent worth of the initial investment cost (P) by LCC analysis is that equal-payment should be paid equally at each term. Therefore We can represent annual worth (A_N) calculated by equal payment-series present-worth factor (UPV) in a numerical expression as follows[5][6]:

$$A_N = P \bullet \left(\frac{1}{UPV} \right) = P \bullet \left(\frac{i(1+i)^N}{(1+i)^N - 1} \right) \tag{1}$$

where i = discounting rate; e = price increase rate; and N = number of year.

Annual equivalent worth (I) of the O&M cost is future worth (I_N) of service life (year) of building. We can calculate this worth by using single-payment compound-amount factor (SCA) and annual O&M cost (M). Annual O&M cost (M) comes into existence every year.

$$I_N = M \bullet SCA = M \bullet (1+i)^N \tag{2}$$

We can represent LCC equal-payment (T_N) to calculate economic life of building as Figure 1 in a numerical expression as follows:

$$T_N = A + I \tag{3}$$

(2) Terms of Calculation
In order to do LCC analysis, we must set up cost and economic items that affect building LCC. Table 2 shows cost data to perform LCC for target building and next explanation is terms of calculation for LCC.

The chiller of the target building is a turbo refrigerating machine with 540 USRT. We produced the unit cost and the installation charge of the turbo refrigerating machine from Korea market price handbook and an estimate sheet of purchasing.

The target building do not have the data for repair and maintenance cost. Also the accurate data of it varies according to a scale, type, failure rate and operating process et al. For this reason, it is difficult to calculate the data for repair cost and maintenance cost. Therefore, in this study we calculated the repair and maintenance cost of target building supposing 1.5%[7] of the initial investment cost.

177

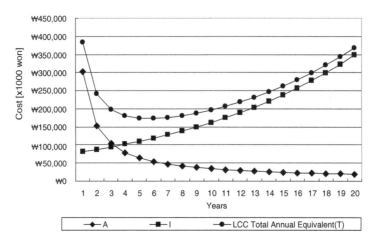

Figure 6.   LCC analysis for replacement time (i: 8.0%, e: 5.0%, N: 20).

Energy consumption cost used by two refrigerators of this building is calculated with electrical use rate (18%) for cooling according to Figure 5.

For the last 10 years, an interest loan rate of bank in Korea was 8.0~14.5% and rate of customer price increase was 4~9%.[8] The real social discount rate for public construction is 7.5% based on the report of Korea Ministry Construction & Transportation department.[9] In this study, discount rate is 8.0%, rate of price increase is 5.0%. The study period for an LCC is 20 years. The scrap value by replacing building equipments is ruled out because it can be offset by disposal cost for replacement. All costs were calculated at the each year's end time.

3.4   *Analysis of economical replacement time*

Figure 6 shows the results of life-cycle cost analysis in order to assess the replacement time of refrigerator as building equipment of target building. This result tells us that the economical replacement time of refrigerator is 6 years (2000) from completion of building construction work. At this time discount rate (i) was 8.0%, rate of price increase (e) was 5.0%.

Sensitivity analysis can help to assess the uncertainty of an LCC analysis. The critical inputs values to the LCC for assessing the replacement time of building equipment are rate of price increase, discount rate and energy consumption. We calculated annual equivalent worth by LCC when rate of price increase (e) were 0% and 9.0% and discount rate (i) was 12.0%. The results appear in Figure 7. The results show that annual equivalent worth is somewhat different but the replacement time is the same.

Assuming that real data is applied to electricity consumption cost in 1994~2001 and thereafter electricity consumption increases 5% every year as stated in paragraph 3.1, we can represent the result of the LCC analysis in Figure 8. In this case, it was found that the replacement time was 7 years (2001) from completion of building construction work.

There results lead us to the conclusion that the economical replacement time of building equipment could vary according to real energy consumption data with the properties of building operating and the simulation results for predicting the future event.

4   CONCLUSION

This research is the case study on evaluation for the replacement time with thought of the economic life of building equipments.

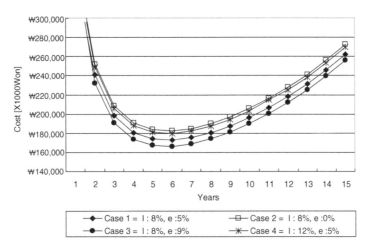

Figure 7. Annual Equivalent worth by LCC analysis for sensitivity analysis.

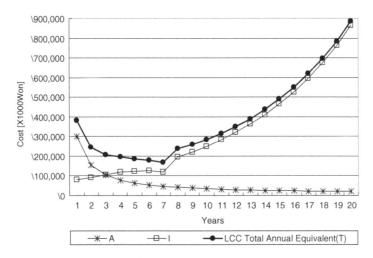

Figure 8. LCC analysis for replacement time.
(i: 8%, e: 5%, real electrical use cost, increase rate of annual electrical use: 5%)

The results are as follows:

(1) The energy consumption unit of target building was 407.2 Mcal/m²y in 2000 and 450.9 Mcal/m²y in 2001. Energy consumption was on the increase gradually in comparison with average energy consumption of building in Korea.
(2) The simulation results for target building with weather data (1996 year) of Seoul, Korea is Figure 4 and Figure 5. In comparison with annual real energy consumption, the error is 4.05%. According to these results, blowers and fans formed 31% of annual electricity consumption of target building because of FCU + VAV air condition system. The energy consumption by chiller for cooling is 18% of annual electricity consumption.
(3) Given that discount rate is 8.0%, rate of price increase is 5.0% and the study period is 20 years, the economical replacement time of the target building's chiller is 7 years after completion of building construction work.

179

(4) According to change of discount rate or rate of price increase, the replacement time do not change. But the more high discount rate is, the more high the LCC equal-payment is and the more high rate of price increase is, the lower the LCC equal-payment is.

(5) With physical life and social life of building equipments, the calculation of economic service life will be applied to estimate reasonable replacement time for retrofit during building life-cycle.

REFERENCES

ASTM, ASTM E917: Standard Practice for Measuring Life-Cycle Costs of Buildings and Buildings Systems, ASTM, US.[6]

Dell'Isola, A.J. & Kirk, S.J., 1995, *Life Cycle Costing for Design Professionals second edition*, New York: McGraw-Hill, Inc.[5]

Hwang, K. Y. and Koo, B. T. et al., 1989, A study on the economic replacement life of construction equipment, *Proceeding of Architectural Institute of Korea*, Vol. 9, No. 2, pp. 559–564.

Jeong, Y.S. and Huh, J.H. et al., 2001, Energy retrofit and estimate for small-to- medium office buildings, *Korean Journal of Air-Conditioning and Refrigeration Engineering*, Vol. 13, No. 4, pp. 279–287.

Jeong, Y.S. and Huh, J.H., 2001, Energy retrofit using economic analysis for a small-to- medium office building –with a focus of an existing office building, *Journal of The Architectural Institute of Korea*, Vol. 17, No. 10, pp. 209–216.

Kalpan, M.B. et al., 1990, "Reconcilation of a DOE2.1C Model with Monitored End-Use Data for a Small Office Building", *ASHRAE Transaction 1990*, pp. 981–992.

Korea Appraisal Board, 2000, *The table book of a new building unit cost*, The Research Institute of Korea Appraisal Board pp. 852–859.[1]

Korea Energy Economics Institute, Energy statistics database (http://www.keei.re.kr/).

Korea Institute of Energy Research, 1999, Final Report "A study on the typical energy consumption criteria in building": KIER-983522, KOREA Ministry Commerce, Industry & Energy.[3]

Korea Ministry Construction & Transportation, 1999, *The Study on Long-term Countermeasure of Energy Saving for Buildings in Korea*, Korea Institute of Construction Technology, pp. 53–58, pp. 77–106.[4]

Korea Ministry Construction & Transportation, 2002, *Manual for the Process and Method of LCC Analysis*, MOCT, Korea, pp. 34–36.[9]

Park, J.I., 1999, *Retrofit and Diagnosis of Building Equipments*, the publishing company Kun-ki-won, pp. 51–62.[2]

Park, S.H. and Choi, I. C. et al., 1999, A study on evaluation method for replacement time of heating equipment in apartment housing, *Journal of The Architectural Institute of Korea*, Vol. 15, No. 11, pp. 201–208.

Park, W. K., et al., 1993, Building equipment and L.C.C., *Monthly Journal of Equipment Technology*, No. 5, pp. 2–36.

SAREK, 2001, *SAREK Handbook Volume 1. Fundamentals*, The Society of Air-Conditioning and Refrigerating Engineers of Korea, Seoul Korea, pp. 2.6–11.[7]

The bank of Korea, Economic Statistics system database of the bank of Korea (http://ecos.bok.or.kr).[8]

Waltz, J.P., 1999, *Computerized Building Energy Simulation Handbook*, The Fairmont Press, Inc.

Life-Cycle Cost and Performance of Civil Infrastructure Systems – Cho, Frangopol & Ang (eds)
© 2007 Taylor & Francis Group, London, ISBN 978-0-415-41356-5

A probabilistic methodology for sustainable bridge management

N.H. Jin, M.K. Chryssanthopoulos & G.A.R. Parke
University of Surrey, Guildford, UK

ABSTRACT: A new methodology for sustainable bridge management is proposed which encompasses whole life costing and life cycle assessment objectives under conditions of uncertainty. Direct impacts from maintenance works themselves, as well as indirect impacts from traffic disruption and delay, can be taken into account in calculating whole life costs and environmental scores of alternative maintenance plans. Moreover, the multi-criteria decision analysis technique is used in order to calculate the most preferable maintenance plan in terms of a linear combination of whole life cost and environmental score based on each decision-maker's preference. The proposed methodology is demonstrated through a simple example of a deteriorating reinforced concrete bridge element.

1 INTRODUCTION

For some years, it has been advocated that bridge management decisions should be based on minimum whole-life cost subject to safety or reliability constraints, i.e. minimise the total cost over the remaining service life provided that performance, in terms of a load capacity index or reliability indicator, remains above an acceptable/tolerable level. However, from the 1990s onwards, another strategy for decision making, termed sustainable development, is gaining ground among public policy makers. At the heart of sustainable development is the simple idea of ensuring a better quality of life for everyone, now and for generations to come. This implies achieving social, economic and environmental objectives at the same time. (Figure 1).

There has been some effort towards the definition and elaboration of methodologies and practical measures to underpin sustainable construction principles in recent years but only few research studies have attempted to treat sustainability issues related to bridge structures (Steele 2000). Therefore, this study attempts to develop a theoretical framework and a decision-making support tool for bridge management based on sustainability concepts.

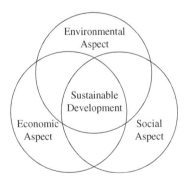

Figure 1. Three dimensions of sustainable development.

Table 1. The decision making process in this study.

Step	General decision making process	Its application to this study
1	Identifying objectives	Finding sustainable bridge maintenance plan
2	Identifying options for achieving the objectives	Developing feasible bridge maintenance plans
3	Identifying the criteria to be used to compare the options	Using whole life cost and environmental score as economic and environmental performance indicators, respectively
4	Analysis of the options	Calculating whole life costs and environmental scores of maintenance plans using WLC and LCA tools
5	Making choices	Applying MCDA techniques

Among the three aspects of sustainable development, economic and environmental performance can be represented by costs incurred and environmental parameters, such as quantities of used energy and/or emission, etc., respectively. On this basis, it is relatively straightforward to quantify and calculate economic and environmental performance. It is much more difficult to quantify social factors and criteria influenced by bridge management activities. In this context social criteria may include different indicators of noise, nuisance, inconvenience and job opportunity, etc. associated with different maintenance activities. There is generally a lack of objective tools to measure such indicators and a case-by-case examination is often merited. Therefore, social impacts arising from bridge maintenance activities are not considered in this study, and only economic and environmental factors are calculated and integrated.

A bridge is a structure with a long lifespan, and it is generally accepted that life cycle thinking is suitable for underpinning the sustainability of construction and life cycle management. Analytical tools for economic and environmental performance based on life cycle thinking include whole life costing (WLC, or life cycle costing) and life cycle assessment (LCA). Therefore, these two tools, namely WLC and LCA, are adopted in order to evaluate the economic and environmental performance indicators of bridge maintenance options. Additionally, multi-criteria decision analysis techniques are used for integrating economic and environmental performance indicators and making choices. Table 1 compares a general decision making process advocated for multi-criteria analysis. (Office of the Deputy Prime Minister 2001) and its application to this study. The methodological framework developed, together with relevant background theories, is described below in more detail.

2 DEVELOPMENT OF METHODOLOGY

2.1 General framework

In order to embody the concepts of sustainability in bridge management under uncertain conditions, a computer based tool was developed. Monte Carlo simulation was used for the generation of random samples, in terms of the uncertain variables which include effective life of maintenance actions as well as deterioration rate and cost parameters. Figure 2 shows the main steps in the methodology in terms of a flowchart. The program carries out deterministic and/or probabilistic analysis according to available input data and the selected option for analysis type. In probabilistic analysis, the concept of a global plan number is used, so that if the same maintenance plan is generated in different iterations representing different samples, it can be uniquely identified. In deterministic analysis, the procedure is basically the same, except that the number of iterations is equal to one.

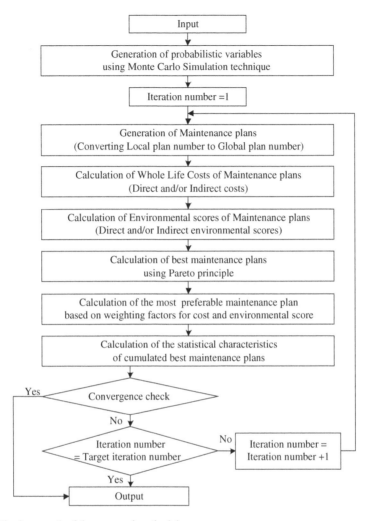

Figure 2. The framework of the proposed methodology.

2.2 *Generation of bridge maintenance plans*

Bridge maintenance plans can be developed by considering different maintenance options and their combination. Three approaches may be used to determine the time of application of maintenance options. They are:

(a) time-based approach: applicable primarily to preventative maintenance options;
(b) performance-based approach: applicable primarily to essential maintenance options;
(c) time- and performance-based approach: applicable to both types of maintenance.

The time-based approach uses two variables: time of first application and time of subsequent applications, independently of predicted or measured profiles of any performance indicators. In the case of a performance-based approach, maintenance actions are applied when a performance threshold is violated, which implies that some prediction/estimation of performance is needed. Time- and performance-based approach is a mix of these two approaches. Typically, the timing of preventative maintenance actions may be determined by a time-based approach but, in addition, essential maintenance is applied if/when some performance threshold is violated. The computer program developed in this study can generate maintenance plans based on either time-based approach, i.e.

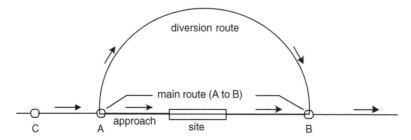

Figure 3. Basic elements of the network system used in QUADRO.

based on effective life concept (Vassie 1997, Rubakantha 2001), or time- and performance-based approach, i.e. based on reliability theory (Frangopol 2001), according to the availability of input data and analysis type.

2.3 *Whole life costing (WLC)*

For assessing public sector investments, future costs are converted into their present value (PV) at a given base year using the expression:

$$PV = \frac{C}{(1+r)^t} \tag{1}$$

where, C is cost at current price levels; r is the test discount rate (TDR) and t is the time period in years (Tilly 1997). In this study, the total costs for maintenance plans are calculated using the above formulae and by summing up the present values of all the component maintenance options. The total costs consist of direct costs, which the bridge owner should pay for, and indirect or notional costs, which bridge users may incur due to traffic disruption/delay during the maintenance activities. The components of direct and indirect costs are as follows.

(a) Direct costs: maintenance costs such as repair and upgrading, traffic management;
(b) Indirect costs: user delay cost, vehicle operating cost and accident cost.

Generally, it is difficult to quantify the maintenance cost data because there are huge variations in their values depending on timing, location and other factors. In the UK case, direct costs are determined based on data collected from bridge maintaining agents, whereas indirect costs are assessed using the DETR computer program QUADRO (Queues And Delays at ROadworks) (The Highways Agency 2002).

The QUADRO program uses a simplified network system which has a single main route and a single diversion route (Figure 3). The QUADRO program requires a number of input data to specify the types of main and diversion route, traffic volume and its mix, work type (traffic management method), etc. Furthermore, the QUADRO program includes a wealth of statistical data for costs, accident rates, fuel consumption, vehicle proportions, vehicle occupancy, traffic growth, etc. which change with time. The QUADRO program calculates net traffic delay costs as a difference between two cases of without and with site works based on these input data and internal statistical data. This study adopted/modified the principles used in QUADRO program in order to quantify the indirect costs; further details may be found in Jin (2006).

2.4 *Life cycle assessment (LCA)*

Life Cycle Assessment (LCA), a scientific tool to quantify and evaluate the environmental impact of a product system, can be used to select environmentally friendly construction options on a

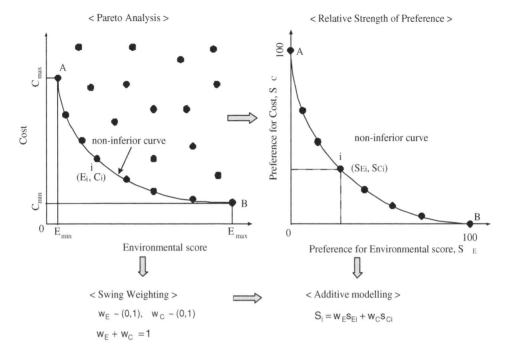

<Pareto Analysis>

<Relative Strength of Preference>

<Swing Weighting>

$$w_E \sim (0,1), \quad w_C \sim (0,1)$$

$$w_E + w_C = 1$$

<Additive modelling>

$$S_i = w_E s_{Ei} + w_C s_{Ci}$$

Figure 4.   Concepts in multi-criteria decision analysis.

quantitative basis. The phases of an LCA consist of goal and scope definition, inventory analysis, impact assessment and interpretation (ISO 1997). This study uses Eco-indicator 99 methodology (Goedkoop & Spriensma 2001) and evaluates single environmental scores which represent the total environmental impacts of bridge maintenance options. Similar to total costs, total environmental scores are composed of direct and indirect environmental scores, i.e.

(a) Direct environmental scores: representing environmental impacts from maintenance works themselves such as production of materials, delivery, machinery operation, electricity use and waste disposal;

(b) Indirect environmental scores: representing environmental impacts associated with additional fuel consumption of vehicles when traffic disruption happens.

When the environmental scores of maintenance options are calculated, this study depended on databases stored in the SimaPro program (Goedkoop). Furthermore, the principles in QUADRO and COBA. (The Highways Agency 2002) program were adopted in order to quantify the additional fuel consumption of vehicles due to traffic disruption; the amount of fuel consumption was then converted into an indirect environmental score.

2.5   Multi-criteria decision analysis (MCDA)

The results from separate WLC and LCA analyses have different characteristics since economic or environmental quantities are calculated independently. Multi-criteria decision analysis techniques may be employed to combine them. Pareto analysis complemented by the relative strength of preference and swing weighting concepts are chosen herein (Office of the Deputy Prime Minister 2001). The relationships between these techniques are schematically shown in Figure 4.

2.5.1   Pareto Analysis

If one represents the two values of whole-life cost and environmental score as a point in a two-dimensional graph for all bridge maintenance plans, one can draw a non-inferior curve; only points on this curve can represent the best maintenance plans. In other words, whole-life cost and environmental performance score of all other points are worse than those of one of the points on the non-inferior curve.

2.5.2 Relative strength of preference and swing weighting

The main idea of 'relative strength of preference and swing weighting' is to construct scales representing preferences for the consequences, to weigh the scales according to their relative importance, and then to calculate weighted averages across the preference scales. In this study, considering that the lowest cost and environmental score best meet the two decision criteria, the most preferred option is assigned a preference score of 0, and the least preferred a score of 100. Scores are assigned to the remaining options so that differences in the numbers represent differences in strength of preference.

If a decision-maker determines the weighting factors for environmental score and total cost, then the overall weighted score can be calculated by the expression below. Thus, a maintenance plan with the smallest value of S_i becomes the best maintenance plan.

$$S_i = w_E s_{Ei} + w_C s_{Ci} \qquad (2)$$

where, S_i: overall weighted score for ith maintenance plan; w_E, w_C : weighting factors for environmental score and whole-life cost, respectively; s_{Ei}, s_{Ci}: relative strength of preferences for environmental score and whole-life cost for ith maintenance plan, respectively.

2.6 Uncertainty in bridge management

Many factors involved in bridge management are subject to considerable uncertainty. Their modelling varies depending on the purpose of the analysis, the methodology used and the availability of probabilistic data. Monte Carlo simulation is a general method used for random uncertainty modelling, and is also used in the present study in order to quantify the impact of the randomness in the input data on the choice of the best maintenance plan.

3 EXAMPLE

In order to demonstrate the validity of the proposed methodology, both a deterministic and a probabilistic analysis for a simple example of reinforced concrete slab structure are executed. Concrete repair, waterproofing, cathodic protection and replacement of element are selected as four alternative maintenance options for a slab structure which might suffer from corrosion deterioration. For simplicity, the time-based approach is used for developing maintenance plans and only direct impacts are considered in this example. The data used for effective lives, unit costs and environmental scores for maintenance options are extracted from several recent studies (Maunsell Ltd. 1999, FaberMaunsell Ltd 2003).

3.1 Input data

The input data for deterministic analysis are shown in Tables 2 and 3. Furthermore, the basic assumption for probabilistic analysis is described below.

For probabilistic analysis, it is assumed that the variables associated with effective life, cost and environmental score in Table 3 have a triangular distribution shape and its minimum, mode and maximum values are 0.75, 1.0 and 1.25 times of the above values, respectively.

3.2 Deterministic analysis results

Figure 5 shows the distribution of deterministic analysis results. The total number of generated maintenance plans is 241. Three maintenance plans on the non-inferior curve are obtained by

Table 2. Basic assumptions adopted in the example.

Input data	Value
Functional unit	1m^2 of reinforced concrete bridge deck
Present year	2002
Reference year for discount	2002
Maintenance required year	60 years
Discount rate	3.5% until 30 years; 3.0% from 31 years onwards

Table 3. Characteristics of maintenance options.

Name (Abbreviation for Name)	Effective life (years)	Cost (GBP/m^2)	Environmental score (/m^2)	Repair Area
Concrete repair (CR)	10	429	10.2	10%
Waterproofing (WP)	12	27	1.84	100%
Cathodic protection (CP)	30	100	4.89	100%
Replacement of element (RE)	50	2106	61	100%

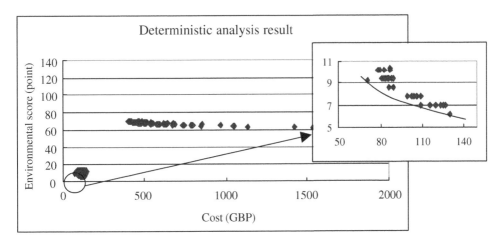

Figure 5. Deterministic analysis results.

Table 4. Details of optimal maintenance plans on the non-inferior curve.

No	Maintenance plan	Original value		Relative strength of preference		Range of WF$_{cost}$
		Cost	E-score	Cost	E-score	
115	WP-WP-WP-WP-WP	70.5	9.20	0.0	100.0	0.533 ~ 1.0
182	CR-CR-CR-CR-CR-CR	108.8	6.94	64.2	26.6	0.426 ~ 0.533
118	WP-CR-CR-CR-CR-CR	130.2	6.12	100.0	0.0	0.0 ~ 0.426

Pareto analysis, and their details are shown in Table 4. These three plans can be chosen as optimal according to decision maker's preference. The values in column 7 of Table 4 represent the range of weighting factor associated with cost which will make each of the maintenance plan shown become the most preferred. The corresponding weighting factor for the environmental score can be calculated by the relation of W.F$_{envi}$ = 1-W.F$_{cost}$.

Table 5. Results of probabilistic analysis.

Maintenance identification no. and plan		Frequencies for maintenance plans to become best (%)		
		Based on Whole life Cost	Based on Environmental score	Based on Both criteria
139	WP-WP-WP-CR-CR	–	–	12.7
143	WP-WP-WP-WP-CR	–	–	13.6
144	WP-WP-WP-WP-WP	52.1	–	2.0
194	WP-CR-CR-CR-CR-CR	–	6.5	7.8
210	WP-WP-CR-CR-CR-CR	–	–	29.9
214	CR-CR-CR-CR-CR-CR-CR	–	25.6	–
568	CR-CR-CR-CR-CR-CR	–	49.8	–
593	WP-WP-WP-CR-CR-CR	–	–	14.8
600	WP-WP-WP-WP-WP-WP	42.6	–	0.2

3.3 Probabilistic analysis results

Table 5 shows the frequencies of optimal maintenance plans based on cost, environmental score and their combination, respectively. The three optimal maintenance plans in deterministic analysis also appear with high frequencies in the probabilistic analysis. Furthermore, new maintenance plans such as WP-WP-CR-CR-CR-CR and CR-CR-CR-CR-CR-CR-CR also appear with relatively high frequency. Table 5 shows clearly that the optimal maintenance plan varies according to the preference for cost and environmental score.

4 CONCLUSIONS

The purpose of this paper is to highlight the framework and an associated decision-making support tool for bridge management based on sustainability concepts.

It is shown that the most preferable maintenance plan depends on the decision-maker's preference for economic and environmental impacts. Probabilistic analysis can help a decision-maker to improve the confidence in his/her choice by quantifying relative frequencies of the plans chosen and to provide a broader selection of alternatives.

Although the indirect impacts from traffic disruption are not treated in the example, unless the traffic volume is very small and works duration is short, it has been found that the indirect impacts are much larger than direct impacts (Jin 2006), and hence indirect impacts should be considered as a dominating factor in calculating total impacts.

Finally, in order that the methodology proposed in this study is satisfactorily applied to real decision making situations, continuous efforts should be made to collect the cost and environmental data associated with bridge maintenance options.

REFERENCES

FaberMaunsell Ltd (2003). DRAFT BD 36 Application of Whole Life Costs for Design and Maintenance of Highways Structures. Design Manual for Roads and Bridges. Birmingham.

Frangopol, D. M. and J. S. Kong, et al. (2001). Reliability-Based Life-Cycle Management of Highway Bridges. Journal of Computing in Civil Engineering 15(1): 27–34.

Goedkoop, M., SimaPro Database Manual – Methods library, PRé Consultants: Downloaded from http://www.pre.nl/download/manuals/DatabaseManualMethods.pdf

Goedkoop, M. and R. Spriensma (2001). The Eco-indicator 99: A damage oriented method for Life Cycle Impact Assessment – Methodology Report. Downloaded from http://www.pre.nl/eco-indicator99/ei99-reports.htm, PRé Consultants.

ISO (1997). Environmental management – Life cycle assessment – Principles and framework. International Standard ISO 14040. Geneva, International Organisation for Standardisation.

Jin, N.H. (2006). A risk-based decision making tool for sustainable bridge management, Department of Civil Engineering. PhD thesis. Guildford, University of Surrey.

Maunsell Ltd. (1999). Serviceable life of highway structures and their components, Final report. Birmingham, The Highways Agency.

Office of the Deputy Prime Minister (2001). DTLR Multi-criteria analysis manual. UK.

Rubakantha, S. (2001). Risk-Based Methods in Bridge Management. Department of Civil Engineering. PhD thesis. Guildford, University of Surrey.

Steele, K. N. P., G. Cole, et al. (2000). Bridge maintenance strategy and sustainability. Bridge Management 4. J. E. Harding. London, Thomas Telford: 361–369.

The Highways Agency (2002). The COBA manual (Design Manual for Roads and Bridges: Volume 13).

The Highways Agency (2002). The QUADRO manual (Design Manual for Roads and Bridges: Volume 14).

Tilly, G. P. (1997). Principles of whole life costing. Safety of Bridges. P. C. Das. London, Thomas Telford: 138–144.

Vassie, P. R. (1997). A whole life cost model for the economic evaluation of durability options for concrete bridges. Safety of Bridges. P. C. Das. London, Thomas Telford: 145–150.

Life-Cycle Cost and Performance of Civil Infrastructure Systems – Cho, Frangopol & Ang (eds)
© 2007 Taylor & Francis Group, London, ISBN 978-0-415-41356-5

Life-Cycle Cost analysis in bridge structures: Focused on superstructure

Jihoon Kang, Seunghoon Lee, Taehoon Hong, Kyo-jin Koo & Chang-taek Hyun
Department of Architectural Engineering, University of Seoul, Seoul, South Korea

Duheon Lee
Korea Institute of Construction Technology, Goyang, Gyeonggi-Do, South Korea

ABSTRACT: As social infrastructures including bridges require relatively high initial costs and long service lives, the systematic feasibility study is highly recommended at the early phase of projects. This study performs a life cycle cost analysis (LCCA) focused on the bridge superstructures. A cost breakdown structure was developed and also the data required for analysis such as analysis periods and unit costs, are collected. Recently, an on-going railroad bridge project was applied for LCCA as a case study.

1 INTRODUCTION

Increasing attention on operating and maintaining social infrastructure such as bridges and tunnels several studies have attempted to predict precise life cycle cost from the planning phase to the maintenance phase. New construction technologies have been introduced in some advanced countries. However, there are a couple of problems to be solved; (i) domestic construction technology of bridges is quite limited and (ii) the economic feasibility and durability of the bridges built by using the construction technologies have not been verified due to the lack of the preceding case studies. Maintenance costs including repair, rehabilitation, and replacement cost might be different depending on each bridge super-structural type as well as initial const. Therefore, it leads to difficulties in selecting an optimal bridge structure.

This study focuses on four different bridge superstructures; (i) Pre-stress Concrete (PSC) Beam (ii) PSC Box Girder (iii) Steel Box Girder and (iii) Re-prestressed Preflex (RPF) Girder, and also analyzes their life cycle costs systematically. It is expected to enhance burdensome decision-making process of selecting optimal bridge superstructures in terms of LCC.

2 LIFE CYCLE COST ANALYSIS FOR BRIDGES

This study itemized the costs required for LCCA of the four bridge types mentioned above and developed cost breakdown structure (CBS), and collected appropriate cost data related to the CBS. Then, LCC analysis was performed using Present-worth method. In the LCCA, there were two basic assumptions such as the discount rates and analysis periods. Figure 1 shows the research framework for this study.

The cost items considered for the LCCA in this research mainly consist of initial costs, maintenance costs, and waste deposit cost as shown in Table 1. As it is difficult to predict the precise traffic density, the user cost was not considered in the research (An & Cha 2001).

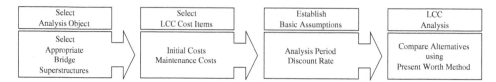

Figure 1.   Research framework.

Table 1.   LCC cost items.

Cost items	Sub-items	Note
Initial cost	Construction cost	C
	Design cost/Supervision cost	C
	Compensation cost for site and existing structure	N/C
Maintenance cost	Inspection/Diagnosis cost	C
	Repair/Rehabilitation/Replacement cost	C
	User cost	N/C
Waste deposit cost	Waste deposit cost	N/C

Note: C = Considered and N/C = Not Considered.

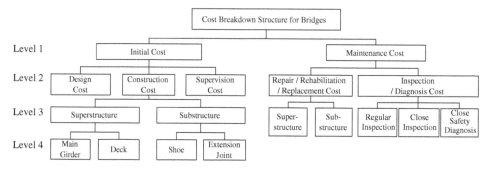

Figure 2.   Cost breakdown structure for bridges.

2.1   Cost breakdown structure

As shown in Figure 2, the CBS for this research was developed based on the fact that the elements of main bridge structures such as shoe and expansion joint are different depending on the types of bridge (i.e., normal bridges and railroad bridges).

2.2   Initial costs

The initial costs considered in this study are construction cost, design cost, and superintendent cost. The existing historical data of construction unit cost per meter or square meter were applied according to bridge superstructural types. The "criteria for engineering service fee" (Official Announcement from the Ministry of Science and Technology) and the "criteria for construction work supervision fee" (Korea Construction Consulting Engineers Association) were applied for design cost and supervision cost, respectively.

2.3   Maintenance costs

2.3.1   Inspection cost and diagnosis cost

Regular inspection, close inspection, and close safety diagnosis were estimated by the official announcement from the Ministry of Construction and Transportation No.2003-195.

Table 2. Estimated construction cost for superstructures.

	Drawing	Length (m)	Construction Cost (10,000 won / m)	Application
PSC BEAM		20.0~25.0	-Superstructure: 1,144 -Substructure: 1,036 -Total: 2,180	Standard Section
STEEL BOX GIRDER		40.0~50.0	-Superstructure: 3,172 -Substructure: 1,351 -Total: 4,523	Crosscut Section over a river and road
PSC BOX GIRDER		40.0~50.0	-Superstructure: 2,136 -Substructure: 1,351 -Total: 3,488	Crosscut Section over a river and road
RPF BEAM		30.0~35.0	-Superstructure: 2,640 -Substructure: 948 -Total: 3,588	Section required under-bridge height and double span
PPC GIRDER		20.0~35.0	-Superstructure: 1,405 -Substructure: 948 -Total: 2,353	Section required under-bridge height and double span
PF BEAM		30.0~35.0	-Superstructure: 2,725 -Substructure: 948 -Total: 3,673	Section required under-bridge height and double span

2.3.2 *Repair/Rehabilitation/Replacement cost*

This study used the repair cost, rehabilitation cost, and replacement cost data that Korea Highway Corporation (refer to Table 3) has developed for different superstructure types that are typically being used in 2003 (Korea Highway Corporation 2003). Roughly estimating the annual maintenance cost by using the specified percentage out of the construction cost is the most common method in calculating maintenance costs due to the lack of historical cost data. For better estimation, the maintenance data (i.e., the rate, the first application, the frequency thereafter and the unit cost) of repair, rehabilitation and replacement are highly requested.

As shown in Table 3, the maintenance cost for bridge superstructures were categorized into deck, main girder, cross beam, coating, shoe, and expansion joint (Ministry of Construction and Transportation 2002). For instance, PSC beam has a twenty one percentage of repair rate, and the first application of PSC beam repair is performed in eighteen years. After the eighteen years, PSC Beams is repaired every fifteen years and the unit costs of the repair cost 132, 900 won per square

Table 3. Maintenance data for superstructures.

Item description		Maintenance data	Rate (%)	First application (Year)	Frequency thereafter (Year)	Unit cost (1,000 won/m^2)
Deck	PSC Beam	Repair	21	18	15	132.9
		Rehabilitation	22.3	25	23	337.6
		Replacement	–	–	–	–
	Steel Box Girder	Repair	21	18	13	89.4
		Rehabilitation	22.3	22	21	337.6
		Replacement	–	–	–	–
	PSC Box Girder	Repair	21	18	13	89.4
		Rehabilitation	22.3	22	21	337.6
		Replacement	–	–	–	–
	RPF Beam	Repair	21	18	13	89.4
		Rehabilitation	22.3	22	21	337.6
		Replacement	–	–	–	–
Shoe	PSC Beam	Repair	18.4	10	8	132.5
		Rehabilitation	21.6	19	11	346
	Steel Box Girder	Repair	–	–	–	–
		Rehabilitation	19.8	20	12	321.5
	PSC Box Girder	Repair	–	–	–	–
		Rehabilitation	19.8	20	12	321.5
	RPF Beam	Repair	–	–	–	–
		Rehabilitation	19.8	20	12	321.5
Cross Beam	PSC Beam	Repair	18.4	10	8	132.5
		Rehabilitation	21.6	19	11	346
	Steel Box Girder	Repair	–	–	–	–
		Rehabilitation	19.8	20	12	321.5
	PSC Box Girder	Repair	–	–	–	–
		Rehabilitation	19.8	20	12	321.5
	RPF Beam	Repair	–	–	–	–
		Rehabilitation	19.8	20	12	321.5
Coating	PSC Beam	Repair	–	–	–	–
		Recoating	–	–	–	–
	Steel Box Girder	Repair	19.1	10	7	49.4
		Recoating	100	15	15	49
	PSC Box Girder	Repair	19.1	10	7	–
		Recoating	100	15	15	–
	RPF Beam	Repair	19.1	10	7	–
		Recoating	100	15	15	–

Note: N/A (Not Available).

meter (about $132.90/m^2). The maintenance cost data (refer to Table 3) of shoe and expansion joint were estimated using the arithmetic mean value for them because normal road bridges and railroad bridges use different type of shoes and expansion joints and all their database systems are also not systematically developed.

2.4 Basic assumptions for LCC analysis

2.4.1 Analysis period

After conducting surveys on the existing researches in order to identify the end of service life for railroad facilities, it was found that no standardized databases on the analysis period has been established yet in South Korea. Accordingly, this study used sixty years, the end of service life of

Table 4. Lifetime for railroad facilities.

Division	Facilities						
	Roadbed						
	Civil engineering	**Bridge**	Tunnel	Station	Rail	Building	Signal · Communication Track · Power · Machine
The end of service life (Year)	80	**60**	60	60	25	60	20

(Source : Korean Society of Transportation (2003).

bridges (refer to Table 4), as the analysis period which was the result of a research from Korean Society of Transportation on the life of railroad facilities in 2003.

2.4.2 *Real interest rate*

In order to analyze the life cycle cost of infrastructures, discount rates are used to convert present and future costs to a single point in time. The nominal discount rate which is required for figuring out the real interest rate might have the same meaning as market discount rate and it is the interest rate of long-term public bonds generally used as a nominal discount rate. However, it is tended to use the bank interest rate in domestic situations; the domestic market forces of long-term public bonds are not significant enough to fully affect discount rate (Korea Research Institute for Human Settlements & Korea Institute of Construction Safety Technology 2000). Calculating the real interest rate in this research, the discount rate for loans to corporations was used as the nominal discount rate because it was frequently used by most private construction companies. Consumer price index(CPI) was also used as a measure of inflation (Korea Research Institute For Human Settlements, Korea Institute of Construction Safety Technology. 2000).

Based on these assumption values, the real interest rate could be calculated as shown in equation (1).

$$i_r = \frac{(1+i_n)}{(1+f)} - 1 \tag{1}$$

where i_n = Real Interest Rate; i_r = Nominal Discount Rate; and f = Inflation.

As shown in Table 5, the nominal discount rate and inflation between 1998 and 1999 are much bigger than the mean value of the bottom of the Table 5. From 1998 to 1999 was a historically difficult duration of time going through the IMF (International Monetary Fund) economic crisis. For this reason, many existing studies have excluded the specific period of time when calculating the real interest rate. As discount rate is an uncertain value itself, it might be quite important to the feasibility analysis for private investment projects. Thus, the values in this period of time were not excluded calculating real interest rate in this research.

3 LIFE CYCLE COST ANALYSIS

In South Korea, with the government's strong intention to introduce the private fund to the public projects, including various types of Public-Private-Partnership(PPP) such as BTL(Build-Transfer-Lease, BTO(Build-Transfer-Operate) and BOT(Build-Operate-Transfer), many social infrastructures have been built by PPP. A certain section of a recently on-going railroad bridge project was used for the purpose of a case study. Originally, the 800 meter long bridge was supposed

Table 5. Real interest rate.

Year	Nominal discount rate (i_n)	Inflation (f)	Real interest rate (i_r)
1996	10.98	4.90	5.80
1997	11.75	4.40	7.04
1998	15.20	7.50	7.16
1999	8.91	0.80	8.05
2000	8.18	2.30	5.75
2001	7.49	4.10	3.26
2002	6.50	2.70	3.70
2003	6.17	3.60	2.48
2004	5.92	3.60	2.24
2005	5.65	2.70	2.87
Mean	8.68	3.66	4.83

(Source : Bank of Korea, http://ecos.bok.or.kr/).

Table 6. Superstructures applied for the original plan and the alternative.

Division	The original plan	The alternative
Superstructure Combination	PSC Beam + Steel Box Girder 20@25 + 4@50 + 110 = 800 m	PSC Beam + PSC Box Girder + RPF Beam 19@25 + 1@40 + 4@40 + 110 = 785 m

to consist of two different kinds of superstructure; PSC Beam and Steel Box Girder. In order for a group of specialists from many different field of bridge-related industry to seek better alternatives, they performed group brainstorming on each superstructure's characteristics, such as durability, feasibility, and maintainability. Going through this process, an alternative consisted of three different superstructures; PSC Beam, PSC Box Girder and RPF Beam, was developed. In order to check the economic efficiency in terms of life cycle cost, a comparison between the original plan and the alternative plan with the same length and width was made. The basic information about the original and the alternative for a case study is shown in table 6.

Even though the alternative was found to require more maintenance cost than the original plan, the original plan required much more initial cost as compared to the alternative(Refer to Table 7). Therefore, it may be said that the alternative is more advantageous than the original plan in terms of life cycle cost.

In order to predict the change of cumulative life cycle cost on the two plans for the next sixty years, based on the assumption of the analysis period, their cumulative graph for the results of their LCC analysis was developed as shown in Figure 3. It seems reasonable to interpret that no break-even point might be generated because the difference in their initial costs was incomparably bigger than those of their maintenance costs.

4 CONCLUSION

As social infrastructures including bridges are more likely to require relatively high initial costs and long service lives, the systematic feasibility study is highly recommended at the early phase of projects. In order to analyze life cycle cost for various superstructures, a cost breakdown structure was developed and the data required for analysis, such as periods and unit costs, were collected. As the result of a case study from recently an on-going railroad bridge project, the alternative (PSC Beam + PSC Box Girder + RPF Beam) needed more maintenance cost and the original plan (PSC Beam + Steel Box Girder) required much more initial cost as compared to the alternative.

Table 7. The Result of the LCC analysis.

Cost item			The original (A) (1,000 won)	The alternative (B) (1,000 won)	A-B (1,000 won)
Initial cost	Construction cost		56,340,000	54,140,000	
	Design / Supervision cost		2,721,222	2,614,962	
	Initial cost (Subtotal)		59,061,222	56,754,962	2,306,260
Maintenance cost	Deck	Repair	561,947	553,642	
		Rehabilitation	905,928	893,084	
		Replacement	–	–	
	Main Girder	Repair	426,985	405,636	
		Rehabilitation	1,056,607	1,072,274	
	Coating	Repair	140,548	–	
		Recoating	346,694	–	
	Shoe	Repair	50,824	50,824	
		Replacement	1,404,593	1,404,593	
	Expansion Joint	Joint	Repair	280,878	280,878	
		Replacement	1,991,756	1,991,756	
		Filler	Repair	280,878	280,878	
		Replacement	1,991,756	1,991,756	
	Substructure	Abutment	Repair	128,338	128,338	
		Rehabilitation	2,868	2,868	
		Pier	Repair	603,523	693,525	
		Rehabilitation	158,154	181,739	
	Safety Diagnosis Cost	Close safety diagnosis cost	1,709,101	2,076,062	
		Cost inspection cost	413,774	498,679	
		Regular inspection cost	231,778	272,582	
	Maintenance cost (Subtotal)		12,686,940	12,779,122	−92,181
Total life cycle cost			71,748,162	69,534,084	2,214,078

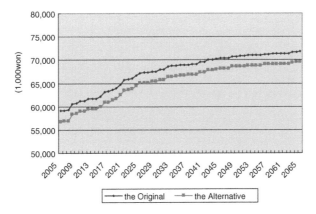

Figure 3. Cumulative graph for the result of LCC analysis.

REFERENCES

An, J. & Cha, K. (2001). "A Case Study on the Life Cycle Cost Analysis of Steel Box Girder and Prestressed Concrete Box Girder Bridge," Journal of Korea Institute of Construction Engineering and Management, KICE, 2(2), 59–66.
Korea Highway Corporation. (2003). "A Study on Life Cycle Cost Analysis for Highway Bridge Structures".

Korea Research Institute For Human Settlements, Korea Institute of Construction Safety Technology. (2000). "A Study on developing infrastructure safety system introducing Life Cycle Cost", A interim report of Korea Infrastructure Safety & Technology Corporation.

Korean Society of Transportation (2003). "A Study on Lifetime for Railroad Facilities".

Ministry of Construction and Transportation. (2002). "A Study on Life Cycle Cost Analysis Technique and System for Optimum Design and Economical Maintenance of Steel Bridges". Annual Report of Hydrologic Data 2002.

Life-Cycle Cost and Performance of Civil Infrastructure Systems – Cho, Frangopol & Ang (eds)
© 2007 Taylor & Francis Group, London, ISBN 978-0-415-41356-5

A new concept of performance-based maintenance for infrastructures

A. Kawamoto
Aratani Construct Consultant Inc, Tottori, Japan

W. Shiraki
Kagawa University, Kagawa, Japan

K. Yasuda
Newjec Inc, Tokyo, Japan

N. Ito
CAE Inc, Tottori, Japan

M. Dogaki
Kansai University, Osaka, Japan

ABSTRACT: A huge number of infrastructures have been constructed with economic growth after the 2nd World War in Japan. Hereafter, it is an important issue how to maintain such a huge stock of infrastructures effectively. A lot of studies on the life cycle cost management (LCCM) and asset management (AM) of infrastructures are urged onward. Most of management systems developed up to the present, however, is not able to deal with the deterioration of performance such load-carrying capacity of structures and the change of service conditions. In this study, a new concept based on the performance-based maintenance of infrastructures is developed according to the concept of live design introduced for a method of the soft disaster prevention after the terror attacks in New York on 11th Sept. 2001. A new management process for maintenance of structures is described for meeting multiple demands during the reference time.

1 INTRODUCTION

Up to the present, the maintenance of infrastructures has been performed by conventional specification-based design rule coded referring to the load history subjected in the past. However, the maintenance under the specification-based design rule may be exchanged to a performance-based design one near future. Therefore, not only prescribes definitely about ranges from serviceability limit state to ultimate limit state, the performance have to be prescribed for security of safety when the makeup fell into the states out of assumptions. We mean the critical state of the structures, namely the destruction state of the structures, with a state out of the assumption in this article.

The infrastructures are constructed so as to provide the function being demanded by many citizens. To minimize user's damage when infrastructures receive damage, not only reduce user's damage by improving seismic performance and the disaster prevention of facilities, but also idea whether to make it quickly take shelter becomes extremely important. The soft disaster prevention is one of ideas so as to reduce user's damage in unexpected circumstances.

The idea of soft disaster prevention attempts the reduction of damage by promptly understanding the damage situation, securing the escape route, and inducing the user appropriately when the catastrophe occurs. Such idea is an extremely important concept that starts shifting from the idea of the current hard disaster measures that give priority to the reinforcement of facilities to soft disaster prevention that values effective evacuation measures. The idea in this new disaster prevention measures is called "Live Design (lifesaving design)" for present idea "Dead Design" by G. Dasgupta (Columbia University, United States) advocated. This word "Dead Design" is the one compared with "Live Design", and our coinage.

The simulation using the soft-computing technology such as cellular automata (CA) etc. for the problems of people shelter action are studied by Chikada et al.[1], Matsuda et al.[2], Shiraki et al.[3][4], and Yasui et al.[5], in order to analyze how to take shelter when the disaster has happened. Moreover, Mitsutaka et al.[6] examined the optimum problem on the arrangement of shelter exit by using Genetic Algorithm (GA). The shelter simulation based on the soft idea of disaster prevention is new research field, that is started in these several years. The problem on the maintaining and managing of infrastructures during life cycle is also the same as soft disaster prevention. It is of problem considering how to control when the disaster happens. Nevertheless, it has been thought by having separated from the design of facilities and equipment up to the present. In Japan, it has misgivings about the occurrence of a large-scale earthquake such as Tokai, Nankai, and Tonankai earthquakes at a high probability of about 50% in these 30 years. In such a situation, in order to maintain and manage a large amount of retrofitting existing infrastructures over a long period in the future, the idea of soft disaster prevention to the maintenance and management measures should be introduced immediately.

In this paper, the design methodology named as Live Design for the damage reduction, and a new maintenance idea that puts in soft disaster prevention to Dead Design, i.e., present design methodology is proposed. This idea differs from present maintenance/management and Asset Management, and is not measures to give priority to a strength increase that has been done up to the present either. Moreover, it is not a concept in the narrow sense how it is necessary to take shelter in a time of disaster either. A live design in the maintenance and management of infrastructures proposed by this research consists of four core elements. That is, (1) collection and analysis of information, (2) transmission of information, (3)education for disaster prevention with learning function of participation type, and (4) definition of safety expressed by performance matrix.

2 WHAT IS A LIVE DESIGN?

The decrease in safety becomes a serious problem in the United States since the terrorism had occurred at World Trade Center in New York on 11th Sept. 2001. And, rescuing people using the facilities in the emergency and securing their life become important social issues. In such a situation, Dasgupta advocated Live Design as a design concept of achieving shelter that suited the situation in the emergency. The concept of Live Design[7] is also introduced in this paper.

The parameter of the structure was decided by provided strength and the economy in the design basis in Dead Design, and the examination concerning measures for safety after it had designed was not done at all. However, since September Eleven Terror Attacks, the manager in facilities came to examine measures for safety in the emergency by the shelter simulation that effectively uses the information technology is used, and develop the prototype of Live Design. Especially, at the building that a lot of citizen use, the safe escape routes are requested to be decided instantaneously by analyzing huge sensor information when the disaster of a fire etc. occurs. Live Design is a method of calculating huge sensor information at high speed by using the probability statistics such as Bayes' theorem processing technique, and executing measures for safety instantaneously in the emergency. Live Design consists of the following main three stages. And, it is considered that the best means of escape is induced while communicating directly with Live Design Database with the engine handling various kinds of information (it is named as IT engine) at each stage.

(1) Acquisition of initial data such as design method and legal restriction condition etc. and the real-time data brought from the sensors

(2) Setting of scenario (The threat pattern is assumed by a statistical combination of indices that compose the scenario of the threat)
(3) Detection of optimal solution

New information is instantaneously analyzed by always communicating with the data base, and the judgment is updated in requesting the optimal solution. Live Design presented by G. Dasgupta is one of ideas preventing the disaster that aims at the necessity of such real-time correspondence.

Now, the Live Design concept begins to attract notice also in Japan. Especially, the effectiveness of Live Design is expected in the disaster prevention field where the correspondence in the emergency is the most important[3,4]. However, the application to the maintenance and management field has not been done yet.

In this paper, the method so as to reduce the damage is defined as Live Design, to execute efficient maintenance and management under the performance-based design, the new framework of maintenance and management that puts in soft disaster prevention to Dead Design is proposed.

3 CHANGE IN DESIGN SYSTEM AND IDEAL MAINTENANCE/MANAGEMENT

It is necessary to think about the performance of the maintenance and management that should satisfy with the movement of the change in the design system in recent years from the present specification-based to performance-based regulations additionally. Maintenance and management in both the specification-based and performance-based regulations are arranged as follows:

3.1 *Specification-based maintenance and management*

The situation of maintenance and management in a present specification-based concept is as follows:

(1) Maintenance is not described clearly, and it has been executed by the administrator's original judgment.
(2) At time when finance was abundant, remedial correspondence is executed, and it has been managed by administrative each different standard.
(3) As the systematized practice activity to manage the infrastructure effectively, because fiscal resources were insufficient recently, Asset management (AM) is main currents.

3.2 *Performance-based regulations*

The content of maintenance and management demanded by the performance regulations is enumerated as follows.

In the performance-based design, it is necessary to keep the performance demanding the structure, to carry out the maintenance and management so as to satisfy the demand performance. Therefore, to set the demand performance to all of events that might happen on service of structures, it is necessary to refer the collapse as a limit state of the structure. That is,

(1) Setting the performance level in the collapse of the structure in the limit state, and considering the correspondence in that case become important.
(2) It is necessary to think about the action when the decrease is caused in the performance when designed due to deterioration etc.

However, the in existence a large amount of infrastructure is designed by the specification before regulations of the performance, and it is difficult to evaluate the performance quantitatively. Therefore, it is requested to maintain it to can endure using long though the damage of the in existence infrastructure stock is allowed. Therefore, it is requested to maintain it to be able to use long allowing the damage of existing infrastructures.

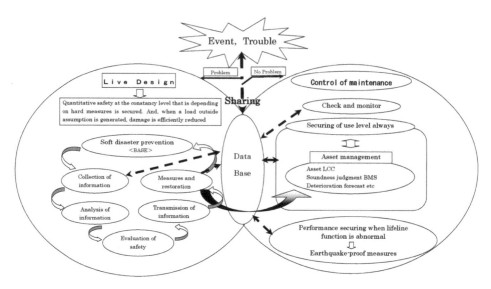

Figure 1.   The intelligent constructed system to have considered live design.

3.3   *Ideal maintenance and management of infrastructures in the future*

The maintenance and management concept is not clearly referred in the method based on the conventional specification-based design rule as shown in paragraph 3.1 and 3.2. However, in maintenance and management under the performance-based design, it is necessary to show the performance of infrastructures clearly at both of the serviceability and ultimate limit states and to refer even by the action when the state not assumed at the time of the design is happened. Therefore, it is indispensable to clearly definite the relation between Live Design and maintenance/management. Figure 1 shows the relation between Live Design and maintenance/management. In this figure, the earthquake that occurs happening by accident while using the structure as an event is assumed, and the state that became unavailability as trouble by the progress of deterioration and damage of the structure is assumed.

In the present maintenance and management concept, seismic measures that attempts the improvement of seismic performance of the existing bridge for prior of the event to correspond to the Asset management[8] (the bridge management[9] and the life cycle cost[10] are included) that executes maintenance and management strategic based on the management data that administrator has accumulated is thought by separating.

In a word, it is not unified original maintenance over wide range, and the management of the service level at usualness is executed chiefly. And, it thinks about the correspondence when the event is happened by separating.

On the other hand, a main purpose of Live Design is to achieve the damage reduction when a load outside assumption acts effectively by a software method in addition to securing quantitative safety by hard measures at a certain level. In Live Design, the education for disaster prevention that thinks about the action when the event happens is executed for that beforehand. When the event occurs, the reduction of damage and the activity of security are executed according to the following procedures.

(1) Information is collected
(2) Information is analyzed
(3) Safety level is evaluated
(4) Information is communicated
(5) Measures are executed

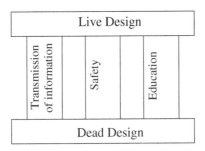

Figure 2.   Three pillars of Live Design.

Both of hardware and software measures are in implications, and thinking by separating in management is irrational. And, separately corresponding is irrational for the execution of efficient management.

Therefore, it is possible to correspond smoothly at the period of normal and abnormal circumstances by exchanging data mutually by Live Design Data Base (LDDB), i.e., a common data base to maintenance, management, and Live Design.

4   PROBLEM ON APPLICATION OF LIVE DESIGN

In general, Live Design is supported by main three pillars, i.e., (1) safety, (2) transmission of information, and (3) education as the basis of Dead Design, as shown in Fig. 2.

The present paper describes the point and problem on how to consider each of these three pillars, in order to unite Live Design with present maintenance/management technique, and to smoothly correspond to the management in normal and abnormal circumstances.

4.1   *Transmission of information (Collection and analysis of information)*

The act of corresponding to this item is the setting of the management scenario of existing structures at the normal circumstances and at the event. In many existing structures, there is neither architecture nor a strategic scenario for maintenance and the management made when they are designed.

Therefore firstly, it is necessary to arrange the design parameter, to analyze the current condition (design policy, present traffic, and damage conditions etc.), and to make the deterioration model according to the deterioration history up to the present. And, it is necessary to make the management scenario at the normal circumstances by using deterioration model.

When the deterioration of structures is permitted by the management scenario at the normal circumstances, there is a possibility that the performance expected when designing cannot be satisfied for the lower performance of structures. Then, at the same time, it is necessary to set the event curve that show state of the load of the structure at the event related to the management scenario of normal circumstances, and to make the management scenario at the event.

4.2   *Education for disaster prevention*

The act of corresponding to this item is the correction and the verification of the scenario. The state of the object structure after measures is regularly checked out, and it is verified whether to achieve the effect of assuming it when measures are done, whether there is an improvement. And, it proceeds to the next management step.

It is necessary to establish the consensus building between the administrator and citizen so that the maintenance scenario and escape action at the event, etc. may become more important. So, the education preventing the disaster[11] for all of the administrators, inspection engineers, and citizen become important. For instance, it is necessary to achieve the design that stands in user's aspect by the consensus building technique of the resident participation type about the shelter road, the place in the escape facility, and the use, etc., and to settle on the management scenario similarly at the normal circumstances and at the event.

4.3 *Setting of safety*

The act of corresponding to this item is a setting of the measures level. In Dead Design, the specification for keeping the same performance level for all of the structures at the use condition has been decided. However, each structure comes to have an individual characteristic by site condition, environment, and construction circumstances after construction. Therefore, maintaining all the structures to an initial setting (range of the detailed specification) will spend a large labor and cost.

Then, it is necessary to set an appropriate diagnosis and the measures level according to the service state of structures, and to satisfy the function as the road network. Then, it is necessary to set an appropriate diagnosis and the measures level according to the service state of structures, and to satisfy the function of the highway network. So, it is necessary to satisfy the performance of structures at service condition, to decide the performance level that should be maintained in addition at the event, and to set an appropriate performance matrix of each structure.

4.4 *Maintenance of data base*

It is necessary to unite the maintenance and management with Live Design, in order to establish maintenance and management in the performance-based regulations as to be described previously. In this study, Live Design Data Base (LDDB) is considered as a strategic location shown in Fig. 1, in order to achieve uniting of both.

The data base concerning maintenance and management is individually constructed by each administrator among LDDB. But, the technique to manage effectively and to use a large amount of accumulated data has not been mounted yet, and the effective usage of the data base for maintaining and managing[12] is studied. Moreover, the research aimed at the data base of maintenance, management, and the Live Design sharing does not exist.

In this study, the data items stored in LDDB is as follows:

(1) Data needed Asset management:
   (a) bridge ledger, (b) data at construction (c) inspection data,
   (d) traffic investigation data, (e) accident investigation data, (f) cost data etc.
(2) Road disaster prevention check (earthquake) data
(3) Ground information data
(4) Management scenario data at normal circumstances and at event etc.
(5) Data relating to seismic measures[13]
   (a) Factor data of characteristic of main shock, aftershock, tsunami, and meteorological conditions, etc. (b) Geographical features and land use data concerning characteristic of area of distress (c) Factor data concerning possibility of secondary disaster generation (d) Geography data concerning relation to facilities outside jurisdiction (e) Urgent investigation and monitor data of infrastructure (road, cutting, slope, banking, bridge, tunnel, other road structures, and excluding the road structure).

These data items have been individually constructed and managed up to the present. Though there is a lot of overlapped content of data, an efficient communication of data becomes possible because these are managed in frame work of LDDB, and the execution of maintenance and management due to Live Design is facilitated. Therefore, the technique for effectively managing and operating these data becomes more important.

5 EFFECT OF MAINTENANCE DUE TO LIVE DESIGN TECHNOLOGY

In the performance-based design, the performance is provided for corresponding to various limit states of structures. Therefore, various correspondences from the slight maintenance and management that repair or reinforce to secure the provided performance to the urgent safety management for lifesaving are needed.

In this chapter, enumerates the matter that should be prepared beforehand to correspond to various demands quickly, and shows the outline.

5.1 *Making of administrative map*

In the maintenance and management based on Live Design, the infrastructure in the targeted wide region should be able to be used without the inconvenience in normal circumstances. Moreover, it is necessary to be secured the function as the main route of resident's shelter road and disaster recovery at the event, and to guarantee securing resident's safety. The administrative map that consolidates the object region information is made beforehand, and the scenario that executes strategic maintenance and management based on this map data is made. If it is prepared like this, even when not only abnormal circumstances assumed when designing but also unexpected emergencies, the execution of effective maintenance and management measure becomes possible.

5.2 *Making of maintenance and management network in infrastructures*

Because the object region of maintenance and management is usually wide, it is necessary to arrange the position of the infrastructure and the dwelling area in the management region beforehand. Then, the infrastructure (airport, harbor equipment, dam, and bridge, etc.) scattered in the management region and the inhabitable areas are modeled on the network as the nodes and the infrastructure such as the highway, rail way etc. are modeled as the edges. The disaster prevention trunk road that functions as a trunk line for disaster prevention is selected based on the network and, the branch line that divides from the trunk route are treated as the resident's life route and escape route. Strategic maintenance and the management of the social infrastructure facilities become executable because it sets weight to the route like that.

5.3 *Arrangement of information on infrastructures on highway network*

Administrative information as area can be arranged by the administrative map and maintenance and management network previously shown. However, the highway composes of various structures such as the bridge, crossing structures, and tunnels, and can carry out a function as a network only after all those facilities function effectively. Therefore, if peculiar information to the object region is consolidated, arranged, and the management scenario of each region is set based on that information, effective maintenance and management measures can be executed.

5.4 *Making of management scenario in region*

In this paper, the idea of maintenance and management concept in the performance-based design is proposed. It aims at strategic maintenance and the management in normal circumstances, and the achievement of resident's security and an immediate restoration activity when an event such as earthquakes happens by accident. For that, it is necessary to assume various events that might occur in normal circumstances, and to plan the action when there happens. The base of these plans making is the management scenario. Therefore, it is necessary to make the management scenario to suit a regional characteristic, and to show normal circumstances and the event uniformly.

The forecasts and analyzes by the simulation of situation in which infrastructure fall at event is one method to make the management scenario.

5.5 *Visualization of struck situation at event*

Management strategy from 5.1 to 5.4 is for the manager. However, if the event actually occurs, nobody can control following the administrator's guideline. It is important to inform the management scenario at the event planned by the administrators to the public. It is necessary for the administrators to prevent and reduce the disaster in union cooperating with citizen.

The struck situation at the event may be predictably simulated by current soft-computing technique above-mentioned by 5.4. What the administrator and the resident examine together based on the disaster situation that is forecasts by such a technology, and is visualized is effective though executes maintenance and management based on common consideration of the administrator and the resident.

5.6 *Update of management scenario in region*

In general, the performance of infrastructures varies day by day. Therefore, it is impossible to correspond to initial management scenario at current state for a long term, even if it is carefully planned at the construction. Then, it is indispensable to update management scenario. It is necessary to properly reconstruct the management strategy described by 5.5 from 5.1 so as to correspond to the current maintenance/management data obtained by monitoring and inspection. It is appropriate to change in various situations such as the reform of service condition in the region and the construction of a new route, etc. It is also important to evolve the strategic management scenario so as to effectively maintain and manage valuable infrastructures in the region. It is dominant to secure the serviceable performance at normal circumstances, and to reduce the damage at the event.

6 CONCLUSIONS

In this study, as an idea of maintenance and management under the performance-based design, maintenance and management at each performance level for both of the serviceability limit state and ultimate limit state, and management under the situation in which the performance that had been expected when designing was not able to be secured is described. Live Design concept, that is paid attention in the field of disaster prevention, is introduced into the conventional maintenance and management technology, the management both at normal circumstances and at the event was unified. The idea is proposed for the maintenance and management to secure the serviceable performance at normal circumstances and the safety at the event.

Firstly, the concept of Live Design proposed by Dasgupta, G., Connor, J., and Sutner, K. is introduced. Next, the ideal maintenance/management method according to the change in the design system from the specification-based design to the performance-based design is described. And, the Live Design concept necessary for maintenance and management in the performance-based design system is clarified. In addition, the problem in the Live Design application in maintenance and management is extracted, and shown to be able to unify the management scenario that ware treated separately up to the present both at normal circumstances and at the event by uniting present maintenance and management with Live Design. Finally, by uniting of both, it is shown to be able to minimize human damage by not only improvement of the seismic performance and the disaster prevention of facilities by the hardware means, but also prompt shelter inducement when facilities with performance at a certain level where receive damage, and arranged a necessary management strategy for that.

This research is just started, and it is scheduled that present maintenance and management data base are extended to Live Design Data Base (LDDB), and the study on the three pillars sustaining Live Design, i.e., the transmission of information (collection and analysis of information), disaster prevention education, and setting of safety is advanced in the future.

Firstly, construction of the Live Design data base is carried out. Next, the prototype model is set. Thirdly, the guarantee performance, measures, and the limit performance at the event are set based on them. Finally, a unified management scenario both at normal circumstances and at the event is proposed.

In addition, it reviews the affectivity and the problem of the new intelligent constructed system which has the present maintenance and live design. And it thinks of the harmony of the maintenance at the time with doing the usual to have thought much of the efficiency and the cost performance, and the maintenance at the time of the event to have thought much of the safety and functionality. Then, it plans to proceed with the research to build a new intelligent constructed system about the performance-based design system.

REFERENCES

Dasgupta G., Connor, J., and Sutner, K.: Information-based security for civil infrastructure : Deep Domain Ba,yesian Models-case studies of office buildings, NSF Proposal No.0331054, 2002.3.

Guidance of road bridge management, Japan Bridge Engineering Center, 2004.8(Japanese).

Imasa, H., and Furuta, H.: Shelter action simulation of underground shopping center by artificial life, The 58th Annual Meeting of Japan Society of Civil Engineers, Part IV, pp.737–738, 2003(Japanese).

Matsuda, Y., Otsuka, H., Otiki, T., and Oono, M.: Research on shelter action simulation of the crowd who considered man's individual difference and interaction in god underground shopping center, Symposium on Underground Space, Japan Society of Civil Engineers, Vol.8, pp.19–28, 2003(Japanese).

Mitsutaka, M., Furuta, H., Hirokane, M., and Gotou, H.: The best in shelter entrance arrangement of underground shopping center by genetic algorithm and simulation, The 55th Annual Meeting of of Japan Society of Civil Engineers, pp.220–221, 2000(Japanese).

Mizuno, Y., Abe, M., Fujino, Y., Meret Sandy, and Abe, M.: Proposal of maintenance management data management technique concerning social infrastructure structure by data base technology, Expressway and car The third volume 49 issue, 2006.3(Japanese).

Nishikawa, K.: Longevity and control of maintenance of road bridge, Civil Engineering association thesis collection, No.501/I-29, pp.1–10, 1994.10(Japanese).

Road earthquake measures handbook (chapter of earthquake restoration), Japan Road Association, 2002.4(Japanese).

Shiraki, W., Inomo, H., Ishikawa, H., Yasuda, K., and Aritomo, H.: Simulation of pedestrian dynamics at occurrence of disaster using CA-model, Proceedings of the 2004 International Conference on Intelligent Mechatronics and Automation (ICIMA), IEEE, pp.191–195, 2004.8.

Shraki, W., Inomo, H., Ishikawa, H., and Aritomo, H.: Simulation of pedestrian dynamics in emergency for live design of buildings, Proceedings of ICOSSAR2005, Milltress, CD-ROM, pp.823–829, 2005.6.

The Ministry of Land, Infrastructure and Transport, Chapter of advisory committee concerning ideal ways of management and renewal, etc. of road structure in the future: Proposals of how for being of management and renewal, etc. of road structure in the future, 2003.4(Japanese).

Tikada, Y., Asaji, K., and Shiroto, T.: Consideration concerning shelter simulation by CA, Proc. of Structural engineering, Japan Society of Civil Engineers, Vol.49A, pp.217–224, 2003.3(Japanese).

Yamori, K., Atsumi, K., Suwa, S., Yoshikawa, C., Koshimura, S., Kouda, K., Imamura, F., Gotou, R., and Miura, F.: Feature article Frontier of disaster prevention education, Natural disaster science, J.JSNDS 24-4, pp.343–386, 2006(Japanese).

Life-Cycle Cost and Performance of Civil Infrastructure Systems – Cho, Frangopol & Ang (eds)
© 2007 Taylor & Francis Group, London, ISBN 978-0-415-41356-5

Optimum design of seismically isolated bridges in a region of moderate seismicity based on the minimization of LCC

D.S. Kim, S.J. Kim, D. Hahm & H.-M. Koh
Seoul National University, Seoul, Korea

W. Park
Korea Bridge Design & Engineering Research Center, Seoul, Korea

ABSTRACT: In this paper, a nonlinear approach for the life-cycle cost (LCC) analysis is proposed. Computational procedure for the analysis of LCC for seismically isolated bridges is established. To compute the failure probability of critical structural components, most probable failure modes for the structure-isolator system are defined as unseating failure of a superstructure, local shear failure of an isolator, and damage of a pier. Multi-level damage state for a pier is introduced according to the level and type of visual damages and the level of corresponding damage result. A relationship between the damage index and the damage state of a pier structure is established by performing quasi-static cyclic loading test. To calculate the failure or damage probability, instead of a large number of nonlinear dynamic analyses this study uses method that generates a sufficient number of response sets which preserve the correlation between the responses of interest by using the result of small number of nonlinear analyses. The proposed procedure is applied to the optimal design of seismically isolated bridge based on the minimum LCC. The cost-effectiveness of seismically isolated bridge designed optimally is investigated.

1 INTRODUCTION

Seismic isolation is often used for bridges in a region of low to moderate seismicity in order to reduce construction cost usually caused by seismic performance requirements. However the available design specifications are generally following the design concept of the regions of high seismicity. This design concept may be conservative in a region of low to moderate seismicity. Therefore a new design approach is required, which fits a balance between the seismic risk and corresponding cost in a lifetime perspective. For this purpose, several researches for the life-cycle cost (LCC) analysis of seismically isolated bridges in a region of low to moderate seismicity have been performed (Koh 2000, Koh 2002, Hahm & Koh 2004). As the extension of previous researches, in this study a nonlinear approach for the life-cycle cost (LCC) analysis is proposed. Computational procedure for the analysis of LCC consists of the following steps: modeling of input ground motion; structural modeling; computing failure probability of critical structural components; defining and evaluating LCC function. To compute the failure probability of critical structural components, the most probable failure modes for the structure-isolator system are defined as unseating failure of a superstructure, local shear failure of an isolator, and damage of a pier. Since the level and type of damage of a pier strongly affects the expected damage cost, multi-level damage state for a pier is introduced according to the level and type of visual damages. The relationship between the damage states and the damage index of a pier structure was established by performing quasi-static cyclic loading test for identifying "visual damage" and numerical simulation of the same test for computing "damage index" (Hahm & Koh 2004).

The failure or damage probability exceeding the limit state can be calculated by Monte Carlo simulation method in nonlinear dynamic analysis. However Monte Carlo simulation method requires a large number of time history analyses in order to obtain a reliable result. This process is more time-consuming as the structure model is more complex. To reduce the calculation time various methods can be used. Among the methods this study uses method that generates a sufficient number of response sets which preserve the correlation between the responses of interest by assuming the distribution of response from the results of nonlinear dynamic analyses for some tens of earthquake records (Moehle 2005).

The proposed method will be applied to numerical models of seismically isolated bridges to derive optimal design variables for a region of moderate seismicity. Additionally, the cost-effectiveness of seismic isolation system designed optimally will be investigated.

2 MODELING OF INPUT GROUND MOTIONS

In general, seismic performance of the structural system highly depends on the magnitudes and the frequency characteristics of ground motion. In the assessment of seismic reliability, therefore, using an appropriate excitation model, capable of reflecting the specific characteristics of the construction site, is necessary for performing a satisfactory estimation. In this study, a collection of real earthquake records that obtained from PEER strong motion database are used. Since seismic isolation has been known as being more effectively in stiff soil condition than soft soil condition, USGS A type soil classification was only considered. A collection of real earthquake records were selected in the range of magnitude 5.0 to 7.0, distance 10 to 30 km respectively. To represent seismic intensity of low and moderated seismicity the peak acceleration of selected earthquake records was scaled to 0.154 g that is corresponding to Korean seismic design code.

3 MODELING OF ISOLATED BRIDGE

Ductility of pier structures constitutes an essential seismic capacity in bridges, particularly in non-isolated bridges. Therefore, ductility of pier must be considered to calculate the failure probability and evaluate the cost-effectiveness since piers of isolated and non-isolated bridges may show a relatively different nonlinear behavior. Thus it is desirable that nonlinear model considering the ductility of pier is used for pier structure. Also it is desirable that isolator is modeled as nonlinear model such as bilinear model.

In the present analysis, seismically isolated bridge is modeled as multi-DOFs and the pier structure is modeled by using fiber element that can consider material nonlinear properties of steel and concrete. Girder and coping are modeled by using beam element and isolator is modeled as bilinear model which is characterized by the pre-yielding stiffness and the post-yielding stiffness. In our numerical simulation, we took the pre-yielding stiffness as one tenth of pier's stiffness from results of previous researches and post-yielding stiffness was defined as one tenth of pre-yielding stiffness.

4 FAILURE/DAMAGE PROBABILITY ESTIMATION

4.1 *Definition of limit states*

We defined three different failure modes most likely to occur for the structure-isolator system, namely unseating failure of a superstructure, local shear failure of an isolator, and multi-level damage state of a pier. As a simple failure-safety model for the unseating failure and the isolator's shear failure, we defined the limit states of the responses of superstructure and isolator, respectively, in terms of displacement at relevant DOFs. However, since the expected damage cost of a seismically isolated bridge strongly depends on the level and type of damage of a pier, it is desirable that the

Table 1. Relationship between damage state and damage index.

Damage state	Damage index (DI)	Visual damage	Damage result
None	0.00~0.10	None or small number of light cracks, either flexural (90 deg) or shear (45 deg)	No loss of utility or need not structural repair
Minor	0.10~0.20	Widespread light cracking; or a few cracks >1 mm wide; or light shear cracks tending to flatten toward 30 deg	Minimum loss of utility, need of a little repair for recovery of design strength
Moderate	0.20~0.50	Significant cracking, e.g. 90 deg cracks >2 mm; 45 deg	No use for main repair in a term
Severe	0.50~1.00	Very large flexural or shear cracks, usually accompanied by limited spalling of cover concrete	Irreparable damage state, dismantlement
Collapse	>1.00	Very severe cracking and spalling of concrete; buckling, kinking or fracture of rebar	Complete or partial collapse of structure

limit state of pier structure is defined according to the level and type of damage of a pier structure. Therefore, in this study the damage state of a pier structure was classified into 5 categories, according to the level and type of damages such as the size and distribution of cracks, spalling of concrete, and failure of rebar, and the level of corresponding damage result or repairability. A certain type of damage index, such as the Park and Ang's damage index (Park 1985), was then used to correlate the computed response of the pier to the damage state. A relationship between the damage index and the damage state of a pier structure was established by performing quasi-static cyclic loading test (Hahm 2004). In this study such a relationship is used for the multi-level damage states. The result of the correlation for the pier structure is presented in Table 1.

4.2 Calculation of failure/damage probability

The failure or damage probability exceeding the limit state can be calculated by Monte Carlo simulation method in nonlinear dynamic analysis. However Monte Carlo simulation method requires a large number of time history analyses in order to obtain a reliable result. This process is more time-consuming as the structure model is more complex. To reduce the calculation time various methods can be used. Among the methods, in this study we will use the method that generates a sufficient number of response sets which preserve the correlation between the responses of interest by assuming the distribution of response from the results of nonlinear dynamic analyses for some tens of earthquake records (Moehle 2005). This section shall briefly describe the procedure of generating an additional correlated response vector.

With selected ground motion, a series of nonlinear dynamic analyses are used to determine response matrix (X). In this study, Open System for Earthquake Engineering Simulation (OpenSees) software is used for the analyses. From the results of nonlinear dynamic analyses, peak structural responses are identified and summarized in a matrix. Because each row of the response matrix is calculated from a single ground motion, the correlation among responses is preserved. The response matrix can be extended by considering any number of responses in interest and any number of ground motions. Instead of running additional nonlinear dynamic analyses to obtain more response realizations, a joint lognormal distribution is fitted to the response matrix. Additional correlated response vectors are generated using the correlation matrix (R) and artificially generated standard normal random variables (u). Figure 1 shows the procedure of obtaining additional correlated response vectors.

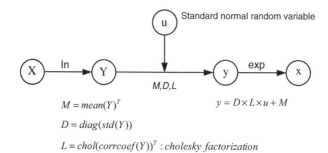

$$M = mean(Y)^T$$

$$y = D \times L \times u + M$$

$$D = diag(std(Y))$$

$$L = chol(corrcoef(Y))^T : cholesky\ factorization$$

Figure 1.   Procedure of generating additional correlated response vector.

5   EXPECTED TOTAL COST FUNCTION AND COST-EFFECTIVENESS INDEX

Generally, life-cycle cost includes all cost such as initial construction cost, damage cost, maintenance cost and repair cost that can be occurred for the life time of the structure. However in this
study initial construction cost and damage cost which is expected when earthquake with seismic
intensity of design level occurs are only considered as total life-cycle cost. The expected value of
total life-cycle cost of isolated bridge structure can be evaluated as a sum of the initial construction
cost and expected damage cost throughout the life-time of the structure, and expressed as

$$E\big[C_{Ti}(k_{pi})\big] = C_{Ii}(k_{pi}) + E\big[C_{Di}(k_{pi})\big] \qquad (1)$$

where k_{pi} is the stiffness of pier of i-th design level; $E[C_{Ti}]$ and C_{Ii} are the expected life-cycle cost
and the initial construction cost for the i-th design level of pier. Note that both the initial construction
cost and the expected damage cost are expressed in terms of the same design variables, which in
this case is the stiffness of a pier.

Initial construction cost was estimated by the sum of various cost items such as material cost,
labor cost and general cost induced by transportation, insurance, etc. In this study, the proportional
ratios of each cost items were evaluated by investigating previous construction costs. Then, total
initial construction cost was formulated as a function of direct material cost which could be modeled
by using design variables such as stiffness of pier.

Expected damage cost function can be represented as a sum of the expected cost due to the
failure of superstructure/isolator and the expected cost due to the damage of a pier as follows:

$$E\big[C_{Di}(k_{pi})\big] = \left[\sum_{u=1}^{2} DS_u P_{ui}(k_{pi}) + \sum_{k=1}^{4} DS_k P_{ki}(k_{pi}) + C_h + C_R + C_{IR} \right]$$
$$\cdot \frac{v}{\lambda}(1 - \exp(-\lambda t_{life})) \qquad (2)$$

where $E[C_{Di}]$ is the expected damage cost of i-th design level of pier; DS_u and P_{ui} are respectively
the damage cost induced by failure of superstructure/isolator and the failure probability of super-
structure/isolator; DS_k and P_{ki} are the pier's damage cost of k-th damage state and the probability
of k-th damage state; C_h, C_R and C_{IR} are the costs due to human loss, traffic congestion delays and
indirect local economic loss, respectively; v is the occurrence rate of earthquake, λ is the discount
rate and t_{life} is the life-cycle of the bridge. In this study, the costs arising from human loss and
traffic congestion delays were formulated as a function of damage index.

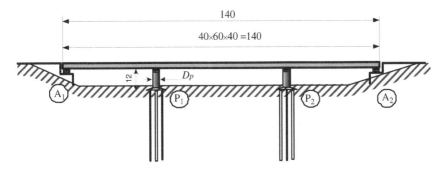

Figure 2.   Example of seismically isolated bridge.

Table 2.   Properties of the structure.

Span length	40-60-40 m	Height of pier	12 m
Superstructure weight	1.60×10^5 N/m	Reinforcement ratio of pier	2.0%
Damping ratio of pier	5%	Diameter of pier	variable

To evaluate the relative cost-effectiveness of a seismically isolated bridge against a non-isolated bridge, the cost-effectiveness index can be defined as follows.

$$E_{iso/non} = \frac{E[C_{iso}]_{min}}{E[C_{non}]_{min}} \qquad (3)$$

where $E[C_{iso}]_{min}$ and $E[C_{non}]_{min}$ are the minimum LCCs optimized for an isolated bridge and a non-isolated bridge, respectively. According to the definition of the cost-effectiveness index, the smaller the index is the higher cost-effectiveness of seismically isolated bridges. We can also conclude that a seismically isolated bridge is more cost-effective than a non-isolated system if the cost-effectiveness index is less than 1.

6   NUMERICAL EXAMPLES

The proposed procedure was applied to the following example (Figure 2) to verify the procedure and investigate the cost-effectiveness of isolated bridge in a region of low to moderate seismicity. The properties of the structure are shown in Table 2. To calculate social damage cost such as human loss, traffic congestion delays and indirect local economic loss this bridge is assumed to be located in a region near to the city such as Seoul.

For the example bridge, life-cycle cost is investigated according to stiffness of pier ranging from 3.186e6 to 5.098e7 N/m which is corresponding to diameter of pier ranging from 1.5 to 3.0 m. Figure 3 shows the life-cycle cost of isolated bridge with respect to diameter of pier.

Isolated bridge and non-isolated bridge have minimum LCC when diameter of pier is 2.0 m and 2.5 m respectively. Such a result shows that initial construction cost can be saved by using seismic isolation system. Since cost-effectiveness index ($E_{isol/non}$) is very small as 0.16, a seismically isolated bridge is greatly more cost-effective than a non-isolated bridge. Such a high cost-effectiveness is because social damage cost such as human loss, traffic congestion delays and indirect local economic loss as well as structural damage cost are considered. Therefore it must be noted that optimal design values and cost-effectiveness can be changed according to definition of damage cost function. But it must also be remembered that seismic isolation system is still more cost-effective than non-isolation system in a region of stiff soil condition and low to moderate seismicity.

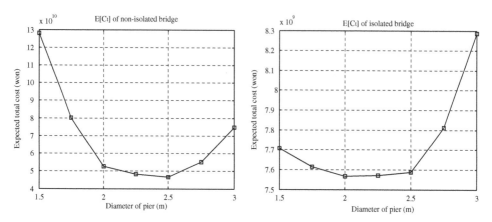

Figure 3.   LCC of isolated bridge and non-isolated bridge with respect to diameter of pier.

7   CONCLUSIONS

In this paper we presented a procedure for the cost-effectiveness analysis and optimal design for the seismic isolation of bridges based on the Life-Cycle Cost concept. The procedure uses a nonlinear time history analyses, including modeling of input ground motion; nonlinear structural modeling for pier and isolator; computing failure probability of critical structural components; and defining and minimizing LCC function. Design and analysis examples showed that the present method is capable of being effectively used not only to provide rational basis for the cost-effectiveness investigation on the use of isolation but also to optimize the design variables of the isolated structural system. From the results of numerical example, it was reconfirmed that the seismic isolation system with more flexible isolator than in a region of high seismicity is cost-effective in a region of low to moderate seismicity and a pier in such a region may be designed to behave elastically or to have small cross section.

ACKNOWLEDGEMENT

This work was supported by the Ministry of Construction and Transportation (MOCT) through the Korea Bridge Design and Engineering Research Center at Seoul National University.

REFERENCES

Hahm, D., Koh, H.M., Shin, J.H. & Park, Y.S. 2004. Life-cycle cost analysis for the seismic isolation of bridges in a region of low to moderate seismicity. *Proc. of the 13th World Conference on Earthquake Engineering.* Vancouver, Canada.

Koh, H.M. 2002. Cost-Effectiveness Analysis for Seismic Isolation of Bridges. *Proc. of the Third World Conference on Structural Control.* Como, Italy.

Koh, H.M., Song, J. & Ha, D.H. 2000. Cost Effectiveness of Base Isolation for Bridges in Low and Moderate Seismic Region. *Proc. of the 12th World Conference on Earthquake Engineering.* Auckland, New Zealand.

Moehle, J., Stojadinovic, B., Der Kiureghian, A. & Yang, T.Y. 2005. An application of PEER performance-based earthquake engineering methodology. *PEER research digest 2005-1.*

Open System for Earthquake Engineering Simulation. Pacific Earthquake Engineering Research Center. University of California, Berkeley.

Park, Y.J. & Ang, A. H.-S. 1985. Mechanistic Seismic Damage Model for Reinforced concrete. *Journal of Structural Engineering.* Vol. 111, No. 4, April.

Life-Cycle Cost and Performance of Civil Infrastructure Systems – Cho, Frangopol & Ang (eds)
© 2007 Taylor & Francis Group, London, ISBN 978-0-415-41356-5

LCC analysis of structures on a network level in The Netherlands

H.E. Klatter
Ministry of Transport, Public Works and Water Management, Civil Engineering Division, Utrecht, The Netherlands

J.M. van Noortwijk
HKV Consultants, Lelystad, The Netherlands
Delft University of Technology, Delft, The Netherlands

ABSTRACT: The paper describes the methodology for a probabilistic life-cycle cost approach to management of structures applied to the national road network in the Netherlands. A large fraction of these structures, mainly concrete highway bridges, were built in the seventies of the previous century. The annual maintenance costs of these bridges form a substantial part of the total budget. A long-term prognosis of maintenance costs is made. Using renewal theory, the expected cost of maintenance is computed as a function of time by modeling repairs, replacements, inspections and routine maintenance of the bridge elements as "renewals". Uncertainty in maintenance intervals is taken into account. Two different scenarios are used, with and without backlog of maintenance.

1 INTRODUCTION

The Dutch Directorate General for Public Works and Water Management is responsible for the management of the national road infrastructure in the Netherlands. Maintenance is one of the core tasks of this directorate. Structures such as bridges and tunnels are important objects in the road network. They largely determine the functionality of the road network as well as the necessary maintenance budgets.

Structures such as bridges are characterized by large investments and a long service lifetime of 50 to 100 years. Although the annual maintenance cost is relatively small compared to the investment cost (less than 1 percent), the sum of the maintenance cost over the service lifetime is of the same order of magnitude as the investment costs. A reliable prognosis for long-term development of maintenance costs is needed. The development of future budget need is influenced not only by the expected deterioration of the bridges, but also by maintenance carried out in the past. Questions as, for example, how backlog of maintenance affects future costs arise. A fundamental solution to these problems is a life-cycle cost (LCC) approach, while taking into account the uncertainties involved.

2 THE NATIONAL ROAD NETWORK LEVEL IN THE NETHERLANDS

The Dutch national main road network consists of 3200 km of road, including 2200 km of motorway. It serves mainly one function, mobility, while traffic safety and environmental aspects are also taken into account. The network divides assets into four object categories: pavements, structures, traffic facilities and environmental assets. The total number of structures in the network is 3236. The structures are categorized into generic types, each having their own maintenance characteristics. An overview of the types, number of structures and deck area is given in Table 1.

Table 1. Structures in the main road network.

Object	Number [−]	Deck area [m^2]
Concrete bridge	3,112	3,436,754
Steel bridge (fixed)	38	248,291
Movable bridge	60	351,780
Tunnel	18	475,558
Aqueduct	8	114,654
Total	3,236	4,627,037

3 LIFE-CYCLE MAINTENANCE COSTS

Effects of deterioration and maintenance strategies influence the maintenance costs of the bridge stock. The age distribution of the bridge stock plays a role. A prognosis of the life-cycle costs should take into account all these effects. Uncertainties in such a prognosis are large and have to be dealt with. The next sections describe the methodology to determine the expected maintenance cost of the bridge stock and to assess the effect of uncertainties. The methodology consists of assessing the maintenance costs of a single bridge, which is extended to bridge types. Next, the effect of the age distribution of the bridges and the current backlog of maintenance are discussed.

3.1 Maintenance costs of a single bridge

The first step is to determine the expected maintenance measures over the lifetime of a single bridge. They are based on all planned maintenance measures for the bridge elements during the lifetime of the structure. Maintenance strategies were drawn up for frequently used elements, e.g. concrete elements, preserved steel, extension joints, and bearings. Such a strategy requires a description of the minimally acceptable quality or condition, or a description of acceptable defects. In addition, standardized maintenance measures were formulated with an estimate of the maintenance intervals and the cost of standardized measures. This results in a so-called preservation plan for a structure. For operational management the preservation plans are updated with results of inspections. For a specific type of bridge a reference preservation plan can be drawn up. Such a reference preservation plan represents the expected life-cycle costs of an average bridge. A typical outcome of a maintenance plan for a concrete bridge is given in Figure 1.

3.2 Maintenance cost of bridge types

From the previous step the annual average maintenance costs over the lifetime can be determined. The actual costs in a specific year will differ from this average due to the effect of the number of structures that actually needs maintenance. The effects of uncertainties on the costs have to be dealt with also. Sources of uncertainties are the characteristics and extent of actual maintenance measures, variation in costs and variability of maintenance intervals. Although the uncertainty in the cost of individual measures can be large, it is small in comparison with the uncertainty in the maintenance interval. Therefore, we merely focus on the uncertainty in maintenance intervals in this research. The uncertainty in the cost is treated as a separate item in further assessment of the cost of the maintenance programs.

Possible maintenance actions are inspections, routine maintenance, repairs and replacements of the bridge elements. To include the cost of maintenance of the bridge stock, three types of concrete viaducts and bridges have been identified: (i) concrete viaducts in the highway and small concrete bridges shorter than 200 m, (ii) concrete viaducts over the highway, and (iii) large concrete bridges longer than 200 m. For the three types of concrete viaducts and bridges, the respective stock is 2116, 951, and 45 (summing up to a total of 3112). For 117 viaducts and bridges, the year of construction

216

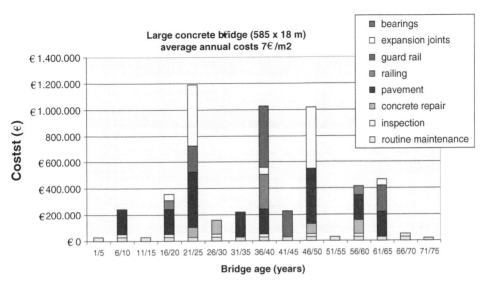

Figure 1. Maintenance costs of a large concrete bridge (Summarized over units of time of five years).

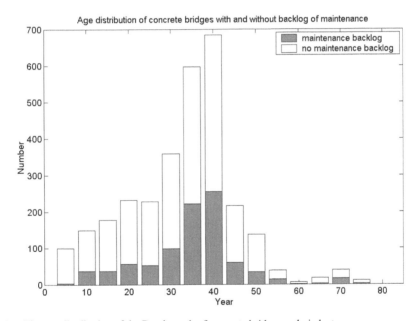

Figure 2. The age distribution of the Dutch stock of concrete bridges and viaducts.

was unknown (i.e., was not stored in the database). The age distribution of the remaining stock of concrete bridges is shown in Figure 2.

To include maintenance measures into the life-cycle cost analysis, the cost of repair and replacement have been quantified for the following bridge elements: kerbs; bridge deck and pavement; piers, abutments and main carrying element; railing and guard rail; asphalt top layer (i.e., pavement maintenance); joints; and bearings. Large concrete bridges are inspected every ten years, smaller concrete viaducts and bridges every six years. Routine maintenance is performed every year. To compute the expected cost of maintenance as a function of time, renewal theory is used for the ten

217

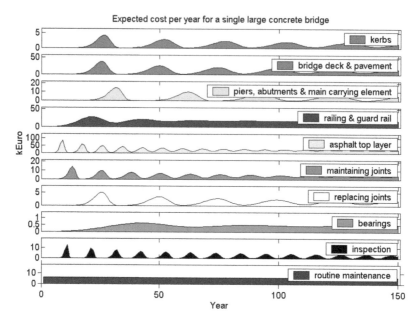

Figure 3.   Expected cost per year for a single average large bridge.

maintenance actions listed in Figure 3 for all the three types of concrete viaducts and bridges (for the mathematics, see van Noortwijk and Klatter, 2004).

3.3   Costs of maintenance of the bridge stock

For computing the expected cost of maintenance, the following four assumptions are made. First, maintenance actions that are combined in operational practice are assumed to be completely dependent; that is, the times of maintenance are judged to be completely dependent for the bridge deck and pavement, for the piers, abutments and main carrying element, and for the railing and guard rail. Second, for those maintenance actions that could not meaningfully be combined beforehand, the times of maintenance are judged to be independent. Third, the times between two subsequent maintenance actions of the same type have a Weibull distribution. Because no observations were available, the Weibull parameters have been assessed using (informal) expert judgment. Fourth, the age of a bridge element is the age of the bridge minus the time at which the last maintenance action was scheduled. Because the maintenance history is not available in the database, the time of the last maintenance action is taken from the maintenance schedule in Figure 1. For example, the age of a bearing of a fifty-year old bridge is assumed to be ten years.

For the purpose of illustration, the annual expected costs of maintaining the bridge elements of a new large concrete bridge are shown in Figure 3. For this type of bridge, the predicted evolution of the maintenance costs over time is displayed. It can be seen that the more uncertain the times between identical maintenance actions, the faster the corresponding cost spreads out. For example, compare the expected cost of maintaining the bearings to the expected cost of inspection. When the first time of maintenance is quite uncertain, the second time of maintenance is even more uncertain, et cetera. In the long run, the expected cost per year approaches the quotient of the expected cycle cost (cost of a single maintenance action) and the expected cycle length (length of the interval between two identical maintenance actions).

The long-term expected annual cost of maintenance of the bridge stock without taking account of bridge replacements is 38 million Euro. To avoid double counting of maintenance actions in the event of bridge replacement, the long-term expected cost of maintenance per year should be adjusted to about 32 million Euro. This adjustment is based on subtracting the sum of all maintenance activities

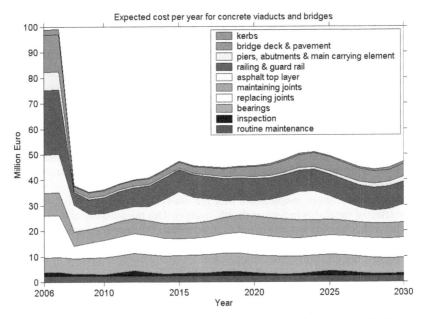

Figure 4.   Expected costs of maintenance per year for the bridge stock with maintenance backlog.

from the long-term expected average maintenance cost per year due to building a new bridge (see van Noortwijk and Klatter, 2004).

Due to the averaging over all bridges in the bridge stock as well as over all its bridge elements, the expected cost of bridge maintenance nearly finds itself in an equilibrium value. Other reasons are that the maintenance intervals are relatively short and the uncertainties relatively large. The annual cost average can be used for reserving future maintenance budgets.

4   BACKLOG OF MAINTENANCE

In the Netherlands, the past decade a large backlog in maintenance of structures was built up. The budget need for large investments in infrastructure made budget for maintenance scarce. The effects of postponing maintenance were underestimated. The past five years these effects became apparent in a rise in incidents that caused many traffic delays. As a result of the public discussion, budget became available to solve the backlog in maintenance. In 2006 and 2007, approximately 200 million Euro will be spent on maintenance of structures in the road network in addition to the regular budget.

In mathematical terms, solving the backlog of maintenance is formulated as follows. The structures are subdivided into concrete bridges with and without a backlog in maintenance (935 and 2177 bridges, respectively). The average age shows that the backlog does not depend much on the age of the bridge: the average age is 31 year for all bridges, and 33 and 30 year for bridges with and without a maintenance backlog, respectively. This can also be seen in Figure 2.

The expected maintenance cost of bridges without a backlog in maintenance is computed as described in Section 3.3. On the other hand, the expected maintenance cost of bridges with a backlog in maintenance is computed by forcing the maintenance action to be performed immediately (in the period 2006–2007) after which the renewal processes of the individual bridge elements proceed. For this analysis it is assumed that there is no backlog of maintenance for the bearings, because backlog of bearing replacements is apparent in only a relatively small part of the bridge stock. The expected cost per year of solving the backlog of maintenance as well as performing planned maintenance for the stock of concrete bridges is shown in Figure 4.

It can be seen that solving the backlog of maintenance temporarily decreases the necessary expected average maintenance cost with about 10 million Euro per year. Due to averaging the cost over the age and stock of the individual bridges, a second maintenance peak after about ten years is therefore not expected. The uncertainty in the individual maintenance actions is simply too large.

5 CONCLUSIONS

A life-cycle cost analysis of the maintenance of the stock of concrete bridges has been set up. The input for this model was standardized maintenance measures and maintenance intervals for bridge elements. These were specified for the different types of concrete bridges. For the cost calculations three types were used.

The variability of costs of maintenance over the life-cycle of structures was studied in more detail. The effect of variability in time on the life-cycle cost for maintenance of the bridge elements was modeled with renewal theory. By introducing the effect of the age distribution of the structures, the maintenance costs for the bridge stock turned out to be fairly constant.

The effect of a discontinuity in operational maintenance, causing a backlog of maintenance of almost one third of the bridges, was studied. This backlog is currently solved in two years. The effect on the total needed budget of the bridge stock was fairly moderate. The first years after clearing the backlog a substantial lower budget is needed. Because all backlog bridge maintenance actions are synchronized, a second maintenance peak after ten to twenty years does appear, but is nearly negligible on the scale of the total budget need. This is caused by the variability in maintenance intervals and the bridge stock.

REFERENCE

van Noortwijk, J.M. and Klatter, H.E. The use of lifetime distributions in bridge maintenance and replacement modeling, *Computers & Structures*, 82(13–14):1091–1099, 2004.

Life-Cycle Cost and Performance of Civil Infrastructure Systems – Cho, Frangopol & Ang (eds)
© 2007 Taylor & Francis Group, London, ISBN 978-0-415-41356-5

Performance regression models for the optimal maintenance evaluation of steel box bridges

Jung S. Kong
Department of Civil and Environmental Engineering, Korea University, Seoul, Korea

Kyung-Hoon Park
Korea Institute of Construction Technology, Goyang-Si, Gyeonggi-Do, Korea

Jong-Kwon Lim
Infra Asset Management Co., Seoul, Korea

Hyo-Nam Cho
Department of Civil and Environmental Engineering, Hanyang University, Ansan-Si, Korea

ABSTRACT: Most of current structural management systems make decisions based on previous maintenance records. In these systems Markov evaluation with transition probabilities between different condition states are used to predict future performance of deteriorating systems. However, the values of transition probabilities have to be modified according to material and structural differences. As a result, the same deterioration transition rule cannot be used for different types of structures. Besides the transition probabilities are very subjective when it is combined with condition states based on unreliable assessment methods such as visual inspection. Recognizing this problem, a new method has been developed to construct more objective performance models for steel box bridges. The new method also gives enough flexibility and extendibility for the next generation of BMS. Based on this new performance regression model, a bridge management system has been developed also.

1 INTRODUCTION

Modern structure management systems (SMSs) such as a bridge management system (BMS) need to evolve according not only to their functions but also to the environment in which the systems are developed. Fundamental function of SMS would be the assessment and prediction of structure condition, the evaluation of cost flow, and finding optimal management solution for a structure or an inventory of structures. From the structure management policy of structure owners to the degree of quantification of structure condition profile models, the environment for SMSs is comprehensive, moreover, the environment may change over time and so does SMS.

However, condition/performance profile models have not changed much even though the analysis methods and inspection techniques have been evolved significantly. One of the reasons would be that the early condition profile model is easy to construct and use. Besides, many applications and decision making processes already have been developed based on the early model.

The condition or the performance profiles are the most important perspectives among many factors and parameters consisting SMSs. Reasonable results and the confidence of the results depends on the quality of the condition/performance profiles. As a matter of fact, the decision making process and solutions provided by the modern management systems such as the optimal maintenance scenario are meaningless if the quality of condition or performance profiles as fundamental data cannot be assured to some extent. Therefore, providing higher quality condition/performance profile models are very important.

It's been commonly accepted that condition/performance profiles can be constructed based on existing inspection data. Based on this, significant investments have been made for last a couple of decades to establish databases storing inspection results. However, the result is not so satisfactory. In Korea, only 20–25% of inspection data for highway bridges have been collected for 7 years since the Korea Expressway Corporation has started to collect inspection data, about 25% data is not continuously gathered and 50% of data are not even available. Besides, experts' opinions on condition/performance profiles of bridges are so subjective that it is doubt to be used with some quantitative accuracy. To overcome this inefficient data problem, we may use some additional information such as numerical or experimental analysis results.

In many cases, the decision making process of a SMS need to predict the future condition of target structures, however, it is inevitable that the future condition of civil infrastructures includes significant uncertainties. To estimate the effect of uncertainties, probability-based methods have been developed in structure management areas. However, the conventional discrete condition/performance profile models with 4 or 5 rates are not good enough to be collaborated with advanced structural management techniques. This urges developing more refined classification for the condition/performance assessment.

Modern SMSs require to provide decision supporting solutions such as optimal lifetime scenarios for target structures. Complicated civil infrastructures such as bridges are associated with various inspections and maintenance activities, therefore, finding an optimum solution of this kind of structures requires a very complicated process and serious computation time.

As we consider these aspects involved with a SMS, it is evident that a condition/performance profile model of SMS should be not only elaborate to insure the accuracy including uncertainties but also simple enough to guarantee the efficiency in computation. In this study, a methodology of constructing quantitative condition/performance profile models based on the Response Surface Method (RSM) was developed. The most important process of the suggested method was to identify essential design variables to reduce the order of response surface. It was found that the number of essential design variables to produce a good performance profile is relatively small for steel box bridges because design variables are highly correlated to each other and its design process is well established without excess.

Once a set of design variables have been selected, a lot of computations were taken place for different sample points to create the response surface. Based on this surface a regression function was constructed. This process has to be repeated for different group of members and bridges subject to different deterioration process and maintenance interventions.

The suggested method was applied to a group of steel box bridges subject to Corrosion and fatigue. It was found that the suggested method is very effective to produce quantitative condition/performance profiles. Especially, quantitative performance profiles was accurate and reliable than the condition profile based on inspection data, if the quality of analysis can be guaranteed. Updating performance profiles including the effect of a maintenance intervention was also available. The variation of performance as a function of properties of a maintenance intervention and environment effects such as time of application were computed and corresponding response surfaces have been constructed.

2 QUANTIFICATION OF DETERIORATION

Quantification of the effect of deterioration of members is a difficult and tedious process because of the diversities in properties of individual bridges depending on designer's selections and loading history. Therefore, it is very important to develop a simple but reliable enough method to compute the lifetime performance of the target structure.

The best and exact solution would be using imbedded analysis module to compute the variation of bridge performance overtime based on given environmental properties and expected loading condition. Advantage of the direct analysis module is to get a relatively accurate result reflecting the properties of each individual bridge. But it takes long computational time even with high performance computers to include various possible scenarios that the bridge will experience during its lifetime. As a result real time decision supporting is not possible until we have a computer with very high capability.

Another solution may lie on a regression model. This regression model can also be established based on numerical and/or experimental results associated with all the expected scenarios of loading conditions and environmental conditions. Therefore, the model also requires enormous time to be established. But the difference to the imbedded module is that the computation and analysis can be done in advance to create a regression model, so decision supporting process will not take long time. In this study, the Response Surface Method (RSM) was used successfully to construct a performance profile of deteriorating bridge members including the effect of maintenance intervention.

Eq. (1) shows a sample of response surface equations, where α_i are unknown coefficients to be determined, X_i is design variable i, k is the number of design variables, $\hat{g}(\mathbf{X})$ is an approximate representation of $g(\mathbf{X})$, and $H.O.T.$ is the higher order terms (Haldar & Mahadevan 2000).

$$\hat{g}(\mathbf{X}) = \alpha_o + \sum_{i=1}^{k} \alpha_i X_i + \sum_{i=1}^{k} \alpha_{ii} X_i^2 + \sum_{i=1}^{k-1} \sum_{j>i}^{k} \alpha_{ii} X_i X_j + H.O.T \tag{1}$$

Among 1,954 steel box bridges in Korea, 633 steel box bridges maintained by the Ministry of Construction and Transportation have been analyzed to decide the design variables. Bridge types have been classified depending on length of bridges, number of spans, span length, width of bridge, different combinations of different span lengths, number of lanes, number of steel boxed, height of steel boxes, variation in thickness of steel plate, support conditions and so on.

Different response surface models were computed in advance according to the types of members, deterioration processes, and maintenance interventions. For instance, among possible deterioration models for girders, deterioration of painting, corrosion of steel surface, fatigue accumulation by overly weighted vehicles, and associated crack propagation have been selected as major deterioration factors.

To reduce the design variables, dependency between design variables was investigated. For instance, it is well known that the height of steel box is dependent on the length of bridge. Moreover, bridge owners suggest standard dimensions and properties for certain standard bridges. For instance, most of steel box bridges in Korea can be divided into two groups in terms of the width of bridges. One group has the width of 10 to 13 m and another group about 19.5 m. Therefore, two different response surface models were constructed for these two groups. Table 1 shows grouping results of steel box bridges and important design variables considered for grouping.

Table 2 shows the variation of upper and lower flange thickness for bridges with width of 19.5 m. Mean values were used for the thickness since the effect of variation on the performance of bridges is not significant. The number of ribs of the steel box and their size were also investigated and it was found that the effect of variation is relatively small. So it was not used as design variable for the response surface model.

Similarly, many dependent variables could be eliminated and important design variables were identified. For simply supported one-span steel box girder bridges, it was found that only five design variables such as the length of bridge, representative web thickness at center of bridge, thickness of concrete slab, the width of girders, and age are of significant importance to decide the performance profile of bridges. (See Table 3) For two-span bridges, two additional design variables such as the length of the first span and the web thickness at the middle of span were added for the response surface models. Time dependent performance of three-span bridges was analyzed by using similar variables as two-span bridges.

Effect of corrosion and fatigue were considered. It was assumed that web surface and bottom flange were corroded according to Komp' corrosion rate. Ang and Mouse's fatigue model was adapted to simulate the performance reduction by fatigue.

Table 1. Grouping results of steel box bridges and design variables considered for grouping.

Bridge groups	Simple span		Two-span		Three-span
Width	10.5 m	19.5 m	10.5 m	19.5 m	19.5 m
Typical bridge length	40 m, 45 m, 50 m		80 m, 90 m, 100 m		100 m, 140 m, 180 m
Span length ratio	–	–	1.0		0.4
L_1/L_2			1.2		0.7
			1.4		1.4
Main web thickness	0.0212 m		0.024 m		0.0212 m
	0.0224 m		0.027 m		0.023 m
	0.0235 m		0.030 m		0.024 m
Concrete slab thickness	0.215 m		0.20 m		0.215 m
	0.265 m		0.25 m		0.265 m
	0.315 m		0.30 m		0.315 m
Width of steel box	1.839 m	1.968 m	1.839 m		2.115 m
	2.046 m	2.349 m	2.046 m		2.210 m
	2.253 m	2.731 m	2.253 m		2.305 m
Point in time	0, 55, 70 years				

Table 2. Variation of upper and lower flange thickness (mm) for bridges with width of about 19.5 m.

Range	14%	22%	7.5%	13%	7.5%	22%	14%
Variation of upper flange thickness	10~14	12~22	18~20	24~30	18~20	12~22	10~14
Variation of lower flange thickness	10~14	14~22	16~18	24~30	16~18	14~22	10~14

Table 3. Sample point values of design variable for the response surface of simply supported steel box bridge (Width of 19.5 m, units: m, years).

Design sets	Bridge length (x_1)	Web thickness At center (x_2)	Concrete slab thickness (x_3)	Width of steel box (x_4)	Time (x_5)
1	45	0.0256	0.265	2.046	55
2	50	0.0256	0.265	2.046	55
3	40	0.0256	0.265	2.046	55
4	45	0.0215	0.265	2.046	55
5	45	0.0315	0.265	2.046	55
6	45	0.0256	0.215	2.046	55
7	45	0.0256	0.315	2.046	55
8	45	0.0256	0.265	1.839	55
9	45	0.0256	0.265	2.253	55
10	45	0.0256	0.265	2.046	0
11	45	0.0256	0.265	2.046	70

Eq. (2) shows the response surface function of the reliability index (β) for the group of simply supported steel box bridges with average width of 19.5 m, located at urban area. Tables 4, 5, and 6 show the values of coefficients associated with design variable of the response surface functions for the reliability index of steel box bridges located at urban area. They are

Table 4. Coefficients of response surface function associated with the reliability index of simply supported steel box bridges subject to corrosion (with Average Width of 19.5 m, Located at Urban Area).

Coefficient	Value	Coefficient	Value
α_0	-9.8883	α_{31}	-1.8320
α_{11}	0.6584	α_{32}	-1.6000
α_{12}	-8.48×10^{-3}	α_{41}	0.9391
α_{21}	111.75	α_{42}	-0.1050
α_{22}	-1171.3	α_{51}	-0.0018
		α_{52}	1.1601×10^{-6}

Table 5. Coefficients of response surface function associated with the reliability index of simply supported steel box bridges subject to fatigue (with Average Width of 19.5 m, Located at Urban Area).

Coefficient	Value	Coefficient	Value
α_0	7.2292	α_{41}	0.1671
α_{11}	-0.1416	α_{42}	-0.0710
α_{12}	0.0011	α_{51}	-4.37×10^{-4}
α_{21}	120.09	α_{52}	1.68×10^{-8}
α_{22}	-1780.1	α_{61}	-0.0863
α_{31}	0.4674	α_{62}	6.44×10^{-4}
α_{32}	-0.1207		

Table 6. Coefficients of response surface function associated with the reliability index of simply supported steel box bridges subject to both corrosion and fatigue (with Average Width of 19.5 m, Located at Urban Area).

Coefficient	Value	Coefficient	Value
α_0	6.9316	α_{41}	0.6275
α_{11}	-0.1547	α_{42}	-0.07225
α_{12}	1.30×10^{-8}	α_{51}	-4.41×10^{-4}
α_{21}	164.12	α_{52}	1.70×10^{-8}
α_{22}	-2633.8	α_{61}	-0.0878
α_{31}	0.4557	α_{62}	6.5×10^{-4}
α_{32}	-9.82×10^{-2}		

corresponding to deterioration models due to corrosion, fatigue, and both corrosion and fatigue, respectively.

$$\beta = \alpha_0 + \alpha_{11}x_1 + \alpha_{12}x_1^2 + \alpha_{21}x_2 + \alpha_{22}x_2^2 + \alpha_{31}x_3 + \alpha_{32}x_3^2 + \alpha_{41}x_4 + \alpha_{42}x_4^2 + \alpha_{51}x_5 + \alpha_{52}x_5^2$$
$$= \alpha_0 + \sum_{i=1}^{5} \alpha_{i1}x_i + \alpha_{i2}x_i^2 \tag{2}$$

3 QUANTIFICATION OF MAINTENANCE EFFECT

Quantification of maintenance effect is basically the same as the quantification of deterioration. Maintenance interventions causing variation in the structure performance such as essential maintenance interventions were analyzed numerically and their effects on structure performance were quantified by a response surface model. The number of design variables to construct the response

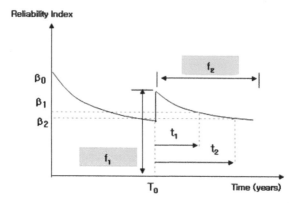

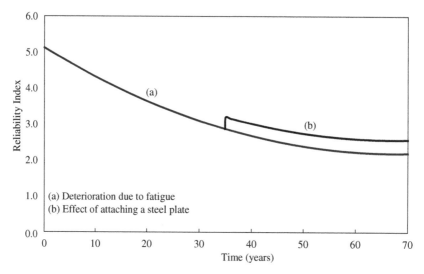

Figure 1.   Quantification model of maintenance effect.

Figure 2.   Reliability index profile of simply supported steel box bridges due to fatigue (a), and Effect of steel plate attaching (b).

surface model changes depending on the characteristics of maintenance interventions. For instance, two more variables, that is, time of application and thickness of steel plate were added to the initial design variables to compute the response surface equation for the effect of steel plate attaching method. Fig. 1 shows the quantification model of a maintenance effect. Two functions, f_1 and f_2 were used to quantify the effect of maintenance intervention on the reliability index for a bridge. This model is applied to steel box bridges for maintenance interventions associated with corrosion scenarios. Fig. 2 shows the reliability index profile of a simply supported steel box bridge corroded due to fatigue and also the improvement of reliability by a steel plate attaching at 35 years. The thickness of steel plate was 1.2 cm.

4   CONCLUSIONS

This paper introduced the method of constructing performance profiles based on the response surface models for steel box bridges. This could be possible by reducing the number of design variables significantly based on the correlations between design variables. It was proven that the response surface method is very effective to construct a regression function for bridge performance profile. This model was fast and accurate enough to be used for advanced structure management systems. In the future, performance profile models of various bridges.

ACKNOWLEDGMENTS

This study was partially sponsored by Smart Infra-Structure Technology Center of Korea Science and Engineering Foundation and by the Korea Research Foundation Grant funded by the Korean Government (MOEHRD) (KRF-2006-331-D00564). The supports received are gratefully acknowledged.

REFERENCES

Cho, H-N., Kim, J-H., Choi, Y-M. and Lee, K-M., (2003), "Practical Application of Life-Cycle Cost Effective Design and Rehabilitation of Civil Infrastructures", Proc. of JSSC03

Furuta, H. and Hattori, H. (2005). "Health monitoring system using AdaBoost technique", Proceedings of the 2nd International Conference on Structural Health Monitoring of Intelligent Infrastructure, Nov. 16-18, Shenzhen, China

Haldar, A. and Mahadevan, S., (2000), Reliability Assessment Using Stochastic Finite Element Analysis, John Wiley & Sons, Inc.

Kong, J.S. and Frangopol, D.M. (2003). "Life-cycle reliability-based maintenance cost optimi-zation of deteriorating structures with emphasis on bridges." J. Struct. Eng., ASCE, 129(6), 818–828.

Kong, J.S. and Frangopol, D.M. (2004). "Prediction of Reliability and Cost Profiles of Deterio-rating Bridges under Time- and Performance-Controlled Maintenance," J. Struct. Eng., ASCE, 130(12), 1865–1874.

Ministry of Construction and Transportation, MOCT (2004) Interim Standard Guidance Manual of Life Cycle Cost Analysis Technique and System for Cost-Effective Optimum Design and Economical Maintenance of Steel Bridges, Interim Research Project Report, MOCT 2003

Ministry of Construction and Transportation (2005) White Paper of the States of Bridges in Korea 2005

Life-Cycle Cost and Performance of Civil Infrastructure Systems – Cho, Frangopol & Ang (eds)
© 2007 Taylor & Francis Group, London, ISBN 978-0-415-41356-5

Strategies based on Life-Cycle Cost to reduce fatalities and economic losses from seismic and wind hazards in Mexico

David De Leon
Universidad Autónoma del Estado de México, Facultad de Ingeniería, Cerro de Coatepec, Toluca, Estado de México, Mexico

Alfredo H-S. Ang
University of California, Irvine, California, USA

ABSTRACT: Optimal safety levels are obtained for two structures located on seismic and eolic zones in Mexico. The formulation is based on the estimation of expected life-cycle costs for the individual risks of the structures and the uncertainties on the seismic and wind loadings are taken into account according to the probabilistic characterization of the corresponding hazards. The criterion is illustrated for two typical structures: a reinforced concrete building located on the soft soil of Mexico City and a bridge built on Tampico, a city on the coast of the Gulf of Mexico.

1 INTRODUCTION

The decision regarding the appropriate safety level for a given facility is largely an economic decision. Usually, it is assumed that a higher safety will require a higher initial cost. However, in the long-term, the life-cycle cost (consisting of the initial cost plus the life-time maintenance cost) is most relevant and must be considered in order to achieve an optimal balance in the whole life cost of the facility. On the other hand, given the uncertain nature of some design parameters, such as the seismic and wind loading, the safety level of a structure cannot be estimated with certainty. Moreover, because of imperfect models in the estimation of the load effects as well as in assessing the structural capacity to a given loading, there is further uncertainty (of the epistemic type) in the estimated safety measure (safety index or failure probability). Information that is specific for a given site can serve to reduce the epistemic uncertainty in the loading from a particular hazard, and should be used to calculate the risk for a given facility at a specific site. From these standpoints, new strategies for reducing the risks of fatalities and economic losses against strong seismic and hurricane events may be formulated.

The principal objective of the proposed study is to apply the tools of risk and reliability analyses to obtain optimal safety levels for facilities located in seismic and eolic zones. Two specific examples, namely a reinforced concrete building on soft soil in Mexico City and a bridge in Tampico on the Atlantic coast will be examined in detail.

The assessment of the risks of fatalities and economic losses from future earthquakes and hurricane winds in urban areas serves to estimate the reduction in the respective risks as functions of incremental investments in strengthening specific infrastructures.

The development of appropriate cost-benefit functions and relationships leads towards the determination of policies to improve the allocation of funds in the planning stage of infrastructure to reduce long-term costs within acceptable risks.

Cost benefit ratios are calculated for the building and the bridge and some strategies are recommended to obtain the optimal safety level that reduces the risk of fatalities and economic losses, in the long term.

2 GENERAL FORMULATION

The expected life-cycle cost $E[C_L]$ is composed by the initial cost C_i and the expected damage costs $E[C_D]$ and it depends on the structure failure probability for the specific hazard governing the facility performance

$$E[C_L] = C_i + E[C_D] \tag{1}$$

The expected damage costs include the components of damage cost: expected repair or reposition $E[C_r]$, expected loss due to business interruption $E[C_{bi}]$, expected cost of injuries $E[C_{inj}]$ and expected fatality costs $E[C_{fat}]$ where each one depends on the particular use of the structure and failure consequences.

The reposition cost, for example, is defined:

$$E[C_r] = C_r(PVF)P_f \tag{2}$$

where:
C_r = reposition cost, which includes the contents loss,
PVF = present value function (Ang and De León, 2005).

$$PVF = \sum_{n=1}^{\infty} [\sum_{k=1}^{n} \Gamma(k,qL)/\Gamma(k,\upsilon L)(\upsilon/q)^k](\upsilon L)^n / n! \exp(-\upsilon L) \tag{3}$$

where υ = mean occurrence rate of earthquakes or winds that may damage the structure and L = structure life.

The failure probability, P_f, is calculated in terms of the probability of occurrence of the hazard and the structure vulnerability.

Similarly, the loss due to business interruption is obtained from the loss of revenue as a consequence of the duration of the repairs or reconstruction works after the earthquake or hurricane, period assumed to be D years (Sthal, 1986).

$$C_{bi}(t) = \int_t^L R(\tau)e^{-q(\tau-t)}d\tau - \int_{t+D}^{L+D} R(\tau)e^{-q(\tau-t)}d\tau \tag{4}$$

where R = revenue obtained from the structure operation and q = net annual discount rate. Therefore,

$$E[C_{bi}] = \int_0^L P_F(C_{bi}(\tau))e^{-q\tau}d\tau \tag{5}$$

The expected cost of injuries is proposed to be:

$$E[C_{inj}] = C_{1I}(N_{in})P_f \tag{6}$$

where:
C_{1I} = average injury cost for an individual
N_{in} = average number of injuries on a typical structure.

For the expected cost related to loss of human lives, the cost corresponding to a life loss, C_{1L}, and the expected number of fatalities, N_D are considered. The details of this calculation are explained in previous works (De León, 2006 and Ang and De León, 2005).

$$E[C_L] = C_{1L}(N_D)P_f \tag{7}$$

3   BUILDING UNDER EARTHQUAKE HAZARD

A 7-storeys reinforced concrete building in Mexico is used to estimate the cost-benefit ratios above described. The floor plan area of the building is 6750 m². In the worst scenario case, it is assumed that there are no injuries but all people inside the building, at the collapse time, die.

The mean occurrence rate of significant earthquakes is $\nu = 0.142/yr$.

The following costs are all in US million.

$$C_{i1} = 0.4,\ C_{i2} = 0.55,\ C_r = 3.24,\ C_{1L} = 8.29$$

The expected number of fatalities if a failure occurs, $E[N_D]$, is estimated from a curve previously developed for typical buildings that collapsed due to earthquakes in Mexico, in terms of their plan areas, A, given an earthquake with a mean occurrence rate ν.

$$N_D = 45.48 + 5.53174(A/1000)^2 \tag{8}$$

In addition, the following data are used:

$$\gamma = 8\%,\ L = 50\ year,\ D = 2\ years,\ N_{in} = 0$$

Also, from the original and a stronger design of the building:

$$p_{f1} = 0.00875,\ p_{f2} = 0.003$$

The expected number of fatalities may be expressed:

$$E[N_D] = E\langle N_D | Failure \rangle P_f \tag{9}$$

where $E\langle N_D | Failure \rangle$ is the expected number of fatalities given the building failure. The failure probability P_f depends on the vulnerability of the structure and might be reduced through an increment on the structural design resistance. Therefore, the cost/benefit ratio of resistance investment increment versus fatalities avoided may also be assessed.

$$CB_1 = (C_{i2} - C_{i1})/(E[N_D]_1 - E[N_D]_2) \tag{10}$$

Another cost/benefit ratio is the resistance investment versus reduction on total losses.

$$CB_2 = (C_{i2} - C_{i1})/(E[C_L]_1 - E[C_L]_2) \tag{11}$$

These two ratios may be estimated by assuming alternative designs with additional resistances and by calculating the expected reductions on fatalities and losses as derived from the increased resistance of the structure.

With the above figures, the expected number of deaths and total loss are:

$$E[N_D]_1 = 297.5 * 0.00875 = 2.6$$
$$E[N_D]_2 = 297.5 * 0.003 = 0.89$$

And the cost-benefit ratio for fatality prevention is:

$$CB_1 = 0.081$$

Also,

$$E[C_L]_1 = 0.65$$
$$E[C_L]_2 = 0.45$$

231

And the cost-benefit ratio for losses prevention is:

$$CB_2 = 0.75$$

4 BRIDGE UNDER STRONG WINDS

In particular, the Tampico bridge, which was built over the Pánuco river in 1988 serves as a link between the Tamaulipas and Veracruz States and joins Tampico to the highway to Mexico City.

The structural system and bridge geometry are: a cable stayed steel orthotropic box sustained by concrete piers and piles with a total length of 1543 m and a central span of 360 m. The total width is 18.10 m for four lanes of traffic, two in each direction. See Fig. 1.

4.1 *Theoretical background*

In addition to the vehicle live loads, the bridge is exposed to strong winds because of its location on the coastal zone, its height and the openness of the area. Therefore, its reliability assessment requires a careful consideration of the failure probability associated with the aleatory uncertainty, and the range (or error bound) in the calculated probability as a result of the epistemic uncertainty (Ang and De León, 2005; De León, et al, 2004).

Once there is epistemic uncertainty on the mean wind velocity, the mean damage index becomes a random variable and, therefore, the failure probability and the expected life-cycle cost also become random variables.

$$C_i = C_1 - C_2 \ln(P_f) \tag{12}$$

The optimal failure probability is obtained (Sthal, 1986) from $\partial E[C_t]/\partial P_f = 0$:

$$P_f = C_2 /[PVF(C_d)] \tag{13}$$

The consequence of failure is expressed in terms of the cost of the actual design C_o. See Fig. 2.

4.2 *Uncertainty on failure probability and life-cycle cost*

$$E[P_f] = \int_0^\infty [P_f \mid V] f_V(v) dv \tag{14}$$

Figure 1. General view of Tampico bridge.

$$\sigma^2{}_{P_f} = \int\limits_0^\infty \{P_f - E[P_f]\}^2 f_V(v)dv \tag{15}$$

$f_V(v)$ is the *pdf* of the mean value of wind velocity. It is assumed that $E[v]$ is normal with COV $= 0.2$
Similarly,

$$E\{E[C_t]\} = C_i + E\{E[C_d]\} \tag{16}$$

and:

$$\sigma^2{}_{C_t} = (C_i \Delta C_i)^2 + \sigma^2 E(C_r) + \sigma^2 E(C_f) + \sigma^2 E(C_e) \tag{17}$$

where C_i is normal and ΔC_i is the COV (coefficient of variation) of C_i, which may be assumed to be 0.10 and $\sigma^2_{E(Cr)}$, $\sigma^2_{E(Cf)}$ and $\sigma^2_{E(Ce)}$ are the variances of the mean values of the economic consequences: repair, fatalities and business interruption which are assumed to be normals.

The failure probability is calculated as

$$P_f = E\{G\}/\sigma_G \tag{18}$$

where G is the limit state for the critical cross section:

$$G = 1 - \{P_a/P_r + M_a/M_r\} \tag{19}$$

And P_a, P_r, M_a and M_r are the applied axial force, axial force capacity, applied moment and moment capacity, respectively in the most critical column. Failure is conservatively defined as the event when the interaction ratio in the most critical column exceeds one.

Finally, the histograms of the failure probability and the expected life-cycle cost are estimated. Also, the 90 and 75 percentiles of these variables are calculated.

The epistemic uncertainty of the initial cost is assumed to be $\Delta C_i = 0.15$. The cost components and their respective epistemic uncertainties (COV) are summarized in Table 1.

Hence, the $E[C_i]$ are also random variables and are assumed to be normally distributed

$$N[E[E[C_i]], \sigma_{E[C_i]}] \tag{20}$$

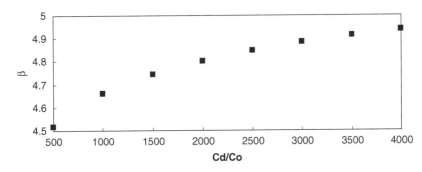

Figure 2.   Optimal reliability of a bridge for several costs of consequences C_d.

Table 1.   Cost components and epistemic uncertainties.

Cost component	COV
C_r (repair)	0.2
C_e (Economic loss)	0.4
C_f (Life loss)	0.8

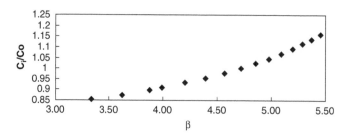

Figure 3.   Initial cost of Tampico bridge.

$$E[E[C_L]] = IC + E[E[C_D]] \tag{21}$$

and

$$\sigma^2{}_{E[C_L]} = (IC\Delta_{IC}{}^2) + \sum_1^k \sigma^2{}_{E[C_i]} \tag{22}$$

4.3   Estimation of cost items

The consequence of failure, C_d, is estimated to be 2000 Co $C_2 = 0.045C_i$, for a typical cable-stayed bridge, $r = 0.08$, for Mexico and $L = 200$ years.

The bridge is modeled and its response is obtained for several basic wind velocities and the original design is modified by reducing or increasing all cross sections according to the design requirements. Through this process, a set of alternative designs is obtained and their corresponding costs are estimated for typical local costs in the city of Tampico. This set serves to identify the specific design whose failure probability is closest to the optimal one. The following material properties were utilized: f'c = 250 Kg/cm^2 and fy = 4200 Kg/cm^2. The resulting curve for the set of alternative designs is shown in Fig. 3.

It is observed that, for the actual design, the reliability index of 4.72 is slightly below the optimal value of 4.8 for the case of $C_d = 2000 C_o$.

From a previous statistical analysis (Sánchez, 2003) a Type II extreme value distribution was found to be the best fit to the wind velocities based on past hurricanes recorded in the east cost of Mexico. The distribution is:

$$F_V(v) = \exp[-(21.82/v)^5] \tag{23}$$

4.4   Calculation of optimal failure probability and life-cycle cost

With the wind velocity distribution shown in Eq. (23), and, assuming a COV of 0.2 for the epistemic uncertainty in the estimation of the mean wind velocity, and the assumption that the mean wind has a lognormal distribution, a number of wind velocities were simulated. From these simulated velocities, the wind forces, dead load and live load effects were calculated for the optimal bridge design and the bridge responses were also obtained. From these responses, the limit state G was assessed and the distribution of the optimal reliability index was calculated. A sample of the calculations is shown in Tables 2 and 3. The cross section of a critical pile is a box of 240 cm × 640 cm dimensions and 50 mm thickness.

A histogram of the optimal expected reliability index is constructed. Similarly, the histogram for the optimal expected cost is obtained relative to the actual design cost. See Figs. 4 and 5.

Tables 4 and 5 show the mean values, the 75 and 90 percentile values for both results; i.e., for the optimal reliability and optimal life-cycle cost.

Table 2. Sample calculations of the reliability index for the optimal design (Rn = random number).

Rn	V (km/h)	P_{act} (ton)	M_{act} (ton-m)	E[G]	σ_G	β
0.3156	80.82	25711	30.11	0.007	0.00259	3.52
0.4703	74.24	25710	27.94	0.008	0.00238	3.83
0.6415	66.77	25709	25.66	0.008	0.00214	4.33
0.4198	76.35	25710	28.62	0.007	0.00245	3.72
0.2747	82.68	25711	30.76	0.007	0.00265	3.46
0.4156	76.53	25710	28.68	0.007	0.00245	3.71
0.7281	62.43	25709	24.42	0.008	0.00200	4.02
0.3500	79.31	25710	29.60	0.007	0.00254	3.58
0.9539	42.64	25707	19.49	0.009	0.00137	6.40
0.5679	70.09	25709	26.65	0.008	0.00225	4.08
0.2579	83.47	25711	31.04	0.007	0.00268	3.43
0.4286	75.98	25710	28.50	0.007	0.00244	3.73

Table 3. Critical member resistances for the calculation of the reliability index.

P_R (ton)	M_R Transv. (tn-m)
27000	8400

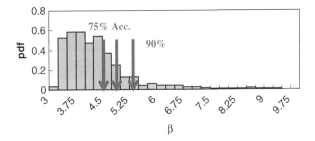

Figure 4. Histogram of optimal β index for Tampico bridge.

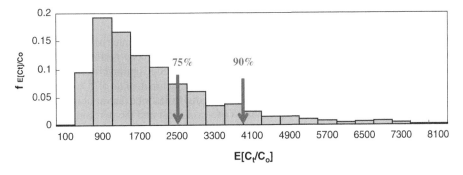

Figure 5. Histogram of optimal life-cycle costs for Tampico bridge.

Table 4. Mean and acceptable value, and 90 and 75 percentile values of optimal reliability.

Mean	4.17
90%	5.25
75%	4.42
Acceptable	4.80

Table 5. Mean value and the 90 and 75 percentile values of optimal cost.

Mean	1995
90%	3900
75%	2500

5 DISCUSSION

For the building under seismic loading, cost/benefit ratios have been calculated for the purpose of risk mitigation through optimal design of reinforced concrete buildings. The ratios may be further extended to improve the planning and funds allocation practices in Mexico.

Regarding the bridge under strong winds, histograms of the optimal failure probability and expected life-cycle costs have been calculated to be used a as a tool for conservative decisions. That is, instead of selecting the mean values, either the 75 or the 90 percentile value may be selected. By doing this, the epistemic uncertainty on the prediction model for the main design parameter, the wind velocity, is incorporated and taken into account in the design selection process. Recent changes on the meteorological and hurricane parameters and patterns, with stronger effects on urban and industrial developments on the shoreline, is becoming a concern for safety requirements and mitigation measures, specially for infrastructure located on the coast and exposed to very intense wind. Under these conditions, a more detailed assessment of the impact of the epistemic uncertainties on the load prediction and response models, such as the one formulated and applied here, appears to be justified. Authorities in charge of coastal development planning and industry managers may use the above mentioned results for conservative decision making depending on their risk-averseness.

6 CONCLUSIONS AND RECOMMENDATIONS

Cost/benefit ratios for buildings in Mexico City may be used for seismic risk management.

A structural reliability assessment of the Tampico bridge has been performed by considering the random features of the wind speed in the region and the bridge optimal failure probability was estimated. The actual design (with $\beta = 4.72$) is close to the optimal value of $\beta = 4.8$.

Aleatory and epistemic uncertainties have been included to describe the variability of the optimal reliability index and its corresponding expected life-cycle cost and the percentiles obtained may be used to set conservative decisions.

The proposed procedure may be further extended to set optimal retrofit plans and inspection and maintenance schedules. Also, the treatment may be applied to other important infrastructure facilities exposed to high speed winds in order to derive safety policies for coastal regions based on risk assessment and management.

ACKNOWLEDGEMENT

The authors thank SCT (Communications and Transportation Secretary in Mexico) for the information provided about the Tampico bridge.

REFERENCES

Ang, Alfredo H-S. and De León, D., (2005), "Modeling and analysis of uncertainties for risk-informed decisions in infrastructures engineering", *Journal of Structure and Infrastructure Engineering*, Vol. 1, 1, pp. 19–31.

De León, D., Ang, A. H-S. & Campos, D. 2004. "Effects of uncertainties on the reliability of a bridge connecting two offshore platforms". *Procs. IABMAS '04. Korea*. Rotterdam: Balkema.

De León, D. (2006). "Socio-economic factors on risk evaluation" *Internacional Colloquium*, ASRANET, Glasgow, UK.

Sánchez, Carlos O. (2003) "Regionalización Eólica para el Estado de Tamaulipas y aplicaciones prácticas en el Diseño Estructural de un edificio para la Ciudad de Tampico, Tamaulipas". *Thesis for Civil Engineering*. Unidad Académica Multidisciplinaria Zona Sur, CUTM, UAT.

Stahl, Bernhard, Edited by McClelland, B. and Reifel, M. D. 1986, *"Reliability Engineering and Risk Analysis"*, Chapter 5 from Planning and Design of Fixed Offshore Platforms. Van Nostrand Reinhold Co. New York.

Optimal fortification load decision-making and Life-Cycle Cost design for aseismic structures according to Chinese codes

Da-Gang Lu, Gang Li & Guang-Yuan Wang
School of Civil Engineering, Harbin Institute of Technology, China P.R.

ABSTRACT: A new model of life-cycle cost for aseismic structures is put forward in the present paper, which includes the minimum initial cost and the expected seismic loss under the future earthquakes during the design reference period. The function relationship between the minimum initial cost and the fortification load is derived through a series of minimum-cost seismic design subjected to the provisions of the Chinese codes by successively adjusting the fortification load. The expected seismic loss is the sum of the products of seismic risk probabilities with the corresponding economic losses. The probabilistic seismic risk analysis (PSRA) includes probabilistic seismic hazard analysis (PSHA) and probabilistic seismic fragility analysis (PSFA). The type IIIextreme distribution of the seismic intensity is adopted for the seismic hazard in the mainland of China. The simplified seismic fragility curves of four damage states are provided according to the three-level fortification principle in the Chinese seismic design codes of buildings, and then, the damage probability matrix can be developed by way of this practical seismic risk methodology. A two-stage life-cycle cost design methodology is presented for aseismic structures, in which the decision of the optimal fortification load (OFL) is made during the first stage, while the minimum-cost design under the optimal fortification load is taken in the second stage. The feasibility of the proposed procedure is demonstrated in a numerical example where the methodology is applied to the life-cycle cost design of a four-storey three-bay steel frame building.

1   INTRODUCTION

Minimization of the expected life cycle cost has received much attention for design of engineering structures under uncertain disaster hazard. In making decisions relative to the design of a structure that is situated in a seismically active region, consideration should be given to the expected cost of damage and other losses resulting from earthquakes occurring during the lifespan of the structure. Nowadays, life cycle cost has become one of the most important decision-making variables in the next-generation performance-based seismic design (PBSD).

Liu and Neghabat (1972) are among the first researchers who incorporated lifetime seismic damage cost into the initial design stage. Recent studies by Ang and De Leon (1997), Rackwitz (2000), Wen (2001), Wen and Kang (2001a, 2001b), Ang and Lee (2001), Esteva *et al*. (2002), Frangopol and Maute (2003), Liu *et al*. (2003, 2004), among others, have shown that the ultimate goal in PBSD of structures under uncertainty and risk is that an optimum is reached in terms of expected life cycle cost.

In the present paper, which is based on recent studies by Wang and Lu (2001), Lu *et al*. (2002), Wang *et al*. (2003), the optimal fortification load (OFL) for seismic intensity is chosen as the acceptable risk level, instead of the target reliability, then, a new model for performance-based seismic design optimization of structures considering life cycle cost is put forward. A new minimum expected life cycle cost optimization methodology for PBSD is advanced, which is demonstrated by a numerical example for a plane steel frame building.

2 A NEW MODEL FOR PERFORMANCE-BASED SEISMIC DESIGN OPTIMIZATION OF STRUCTURES CONSIDERING EXPECTED LIFE CYCLE COST

In 1972, Liu and Neghabat proposed to incorporate lifetime seismic damage cost into the initial design stage, i.e. the objective function of cost optimization model for seismic design of structures is

$$E[C_T(\mathbf{x})] = C_I(\mathbf{x}) + E[C_D(\mathbf{x})] \tag{1}$$

in which $E[\cdot] =$ expected value; $\mathbf{x} =$ the design variable vector of the building; $E[C_T(\mathbf{x})] =$ the expected total life cycle cost; $C_I(\mathbf{x}) =$ initial cost of building; and $E[C_D(\mathbf{x})] =$ the expected loss from seismic damage.

The optimization models adopted in most of the subsequent research are the extensions of Eq. (1). For example, the International Standard "General Principles on Reliability for Structures (ISO 2394; 1998)" proposed a minimum lifetime cost objective function:

$$C_{tot} = C_b + C_m + \sum P_f C_f \tag{2}$$

in which $C_b =$ the building cost, $C_m =$ the cost of maintenance and demolition, $C_f =$ the cost of failure, $P_f =$ the lifetime probability of failure.

Eqs. (1) and (2) consider neither the random occurrence and the intensity variation in time of the hazards nor the discounted factor of over time t. Under the assumption that hazard occurrences can be modeled by a simple Poisson process with occurrence rate of v/year and for resistance that is time-invariant, also considering discounting of cost over time, Wen and Kang (2001a) derived a closed analytical formulation of lifetime total expected cost model:

$$E[C(t,\mathbf{x})] = C_0 + \frac{v}{\lambda}(1 - e^{-\lambda t})\sum_i^k C_i P_i + \frac{(1 - e^{-\lambda t})}{\lambda} C_m \tag{3}$$

in which $C_0 =$ initial cost for new building; $C_k = k$th limit-state failure cost; $P_k = k$th limit-state probability; $C_m =$ operation and maintenance costs per year; $e^{-\lambda t} =$ discounted factor of over time t; $t =$ time period which is the design life of a new structure.

When used to decide the optimal target reliability for a new structure, Eqs. (1) to (3) can be transformed into the following formula:

$$E[C_T(p_f)] = C_I(p_f) + C_m(p_f) + E[C_D(p_f)] \tag{4}$$

Recently, the authors have taken the fortification load for seismic intensity I_d, instead of the target reliability, as the acceptable risk level of a structure. In other words, we choose the fortification intensity I_d as one key decision-making variable when we make the minimum total expected life cycle cost optimization for performance-based seismic design of structures. From our experience, the fortification intensity is more convenient in use than the target reliability.

During the variable design stage of a structure, the design scheme, that is, the design variable vector $\mathbf{x}$, can be denoted as the function $\mathbf{x}(I_d)$ of the fortification intensity I_d. Furthermore, the cost of maintenance and demolition can also be considered a kind of failure cost in some sense. Therefore, Eq. (3) can be transformed into the following form:

$$E[C(t,I_d)] = C_I[\mathbf{x}(I_d)] + \frac{v}{\lambda}(1 - e^{-\lambda t})L[\mathbf{x}(I_d)] \tag{5}$$

in which $L[\mathbf{x}(I_d)] =$ expected total damage cost considering all seismic performance leves.

With respect to the same fortification intensity I_d, many kinds of usable design schemes $\mathbf{x}_i(I_d)$ $(i = 1, 2, 3, \cdots)$ can be obtained, and accordingly, there are many cost functions. Therefore, structural initial cost $C_I[\mathbf{x}(I_d)]$ is a multi-value function of the intensity I_d, so it should not directly be used in Eq. (5) in a strict sense. On the other hand, given specific fortification intensity I_d, there should exit only one optimal design scheme in theory. Hence, from the viewpoint of more rational logical background, the function $C_I[\mathbf{x}(I_d)]$ should be replaced by its minimum counterpart $C_{min}[\mathbf{x}(I_d)]$, which is a single-value function.

Based on the above analysis, we herein propose a more scientific optimization objective function for performance-based seismic design considering expected life cycle cost as follows:

$$E[C(t, I_d)] = C_{min}[\mathbf{x}(I_d)] + \theta \frac{v}{\lambda}(1 - e^{-\lambda t})L[\mathbf{x}(I_d)] \tag{6}$$

in which $C_{min}[\mathbf{x}(I_d)] =$ the initial minimum cost under the fortification intensity I_d; $\theta =$ an adjusting parameter which considers the different importance of the initial minimum cost C_{min} and the loss expectation L, in general, we can let $\theta = 1$ if there are no specific requirements.

Because the minimum-cost design scheme $\mathbf{x}(I_d)$ of a structure is unique with respect to the given fortification intensity I_d, the objective function (6) can be expressed in the following simple formulation:

$$E[C(t, I_d)] = C_{min}(I_d) + \theta \frac{v}{\lambda}(1 - e^{-\lambda t})L(I_d) \tag{7}$$

Eq. (6) or (7) can be taken as a general framework for the next-generation performance-based seismic design. It includes five parts in general: (1) the minimum initial cost seismic design of structures; (2) the probabilistic seismic hazard analysis (PSHA) of the site; (3) the probabilistic seismic fragility analysis (PSFA) of structures; (4) the probabilistic seismic risk analysis (PSRA) of structures; and (5) the seismic loss assessment (SLA) of structures.

3 THE MINIMUM INITIAL COST SEISMIC DESIGN OF STRUCTURES SUBJECTED TO PROVISIONS OF CHINSES CODES

Under the given fortification intensity I_d, the minimum initial cost design problem of a structure can be conceptually stated as:

To find the design scheme $\mathbf{x}(I_d)$, so as to make the initial cost of the structure

$$C[\mathbf{x}(I_d)] \to min \tag{8}$$

subjected to seismic provisions of Chinese seismic design code of buildings (GB50011-2001) and other specific design codes of structures.

Since gradient information can greatly improve the optimization efficiency, we herein make use of the Polak-Ribiere conjugate gradient direction algorithm, which belongs to first-order optimization method. This algorithm performs the optimization loop according to the search direction as follows:

$$\mathbf{d}^{(j)} = -\nabla Q\left(\mathbf{x}^{(j)}\right) + r_{j-1}\mathbf{d}^{(j-1)} \tag{9}$$

in which $\mathbf{d}^{(j)} =$ search direction vector in the jth iteration; $\mathbf{x}^{(j)} =$ design variable vector in the jth iteration; $Q(\cdot) =$ the dimensionless, unconstrained objective function via penalty function method; $\nabla Q(\cdot) =$ the gradient vector of the function Q with respect to design variable vector; $r_{j-1} =$ the conjugate direction coefficient in the $(j-1)$th iteration, whose formula is

$$r_{j-1} = \frac{\left[\nabla Q\left(\mathbf{x}^{(j)}\right) - \nabla Q\left(\mathbf{x}^{(j-1)}\right)\right]^T \nabla Q\left(\mathbf{x}^{(j)}\right)}{\left\|\nabla Q\left(\mathbf{x}^{(j-1)}\right)\right\|^2} \tag{10}$$

where $\|\cdot\|$ represents l_2 norm.

Table 1. Relationship of three-level earthquake levels with the basic intensity.

Earthquake levels	Minor earthquake	Moderate earthquake	Major earthquake
Exceedance probability in 50 years	0.632	0.10	0.02 to 0.03
Relationships with the basic intensity	1.55 degree lower than the basic intensity	equals the basic intensity	about 1 degree higher than the basic intensity
Performance objectives	do not be damaged	be repaired	do not collapse

4 PROBABILISTIC SEISMIC HAZARD ANALYSIS IN MAINLAND OF CHINA

The seismic hazard at a building site is displayed through a cumulative distribution function (CDF) or its complimentary one (CCDF) of earthquake ground motion parameters, e.g., seismic intensity, peak ground acceleration, spectral acceleration, etc. The technology for performing probabilistic seismic hazard analysis (PSHA) for a site or a city has become mature since the Cornell's foundation work in 1968.

Gao and Bao (1986) analyzed 45 cities in the northern, northwestern and southwestern mainland of China by PSHA method, derived a conclusion that the cumulative distribution function of the seismic intensity during the design lifetime in mainland of China is type-three extreme value distribution, which takes the form of

$$F_I(i) = \exp\left[-\left(\frac{\omega - i}{\omega - \varepsilon}\right)^k\right] \tag{11}$$

in which $\exp(\cdot) =$ exponent distribution function; $\omega =$ the upper limit value of the random variable I, it takes 12 for seismic intensity; $\varepsilon =$ characteristic value of I, which equals to the basic intensity I_0 minus 1.55, i.e., $\varepsilon = I_0 - 1.55$; $k =$ shape parameter of the distribution function, whose value depends on the basic intensity I_0.

5 PROBABILISTIC SEISMIC FRAGILITY ANALYSIS OF STRUCTURES

The seismic fragility of a structural system is defined as the conditional probability of failure of the system for a given intensity of the ground motion. In a performance-based seismic design approach, the failure event is said to have occurred when the structure fails to satisfy the requirements of a prescribed performance level. If the intensity of the ground motion is expressed as a single variable (e.g., seismic intensity or the peak ground acceleration), the conditional probability of failure expressed as a function of the ground motion intensity is called a seismic fragility curve.

The assessment of seismic fragility ideally should employ as much objective information as possible. If there is lack of information on laboratory or field observations, then the current design code may provide valuable information for assessing the seismic fragility.

In the current Chinese seismic design code of building, a three-levels seismic design principle is used, which can be stated as for a structure not to be damaged under minor earthquake, to be repaired under moderate earthquake and not to collapse under major earthquake. The relationships of the three-level intensity with the basic intensity are summarized in Table 1. In China, five seismic damage states of engineering structures in general are specified: (1) nonstructural damage, (2) slight structural damage, (3) moderate structural damage, (4) severe structural damage, and (5) collapse.

Let the symbol B_j^* represents the seismic damage state larger than the state B_j, $P_f[B_j^*, \mathbf{x}(I_d)|i]$ be seismic fragility F_{Rj} for the above damage state $B_j^* (j = 1, \cdots, 4)$ of design scheme $\mathbf{x}(I_d)$ under the fortification intensity I_d when seismic intensity $I = i$.

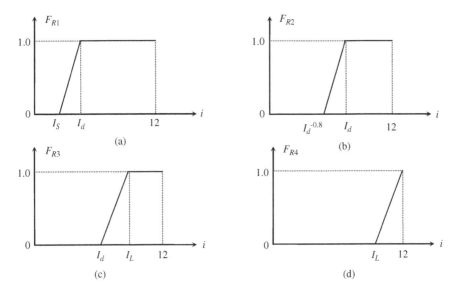

Figure 1.   Fragility curves for 4 larger than damage states.

Considering the regulations of the three-level performance objectives shown in Table 1, in this study, four simplified seismic fragility curves are proposed as shown in Figure 1, in which I_S = minor seismic intensity; I_L = major seismic intensity; and I_d = fortification seismic intensity. Fig.1 (a) to (d) represent four fragility curves for the "larger than slight damage state", "larger than moderate damage state", "larger than severe damage state" and "collapse damage state", respectively. Based on Figure 1, the seismic fragility functions for four above damage states can be easily derived.

6   PROBABILISTIC SEISMIC RISK ANALYSIS OF STRUCTURES

A point estimate of the failure probability for larger than damage state j can be obtained by convolving the fragility function $F_R(x)$ with the probability density function (PDF) of the seismic hazard CDF curve:

$$P_f[B_j^*, \mathbf{x}(I_d)] = \int_0^{12} F_{Rj} \cdot f_I(i) di \qquad (12)$$

in which $f_I(i)$ = PDF of seismic intensity I, which is the derivative of Eq. (11); F_{Rj} takes the form of one of Figs. (1a) to (1d).

The seismic risk probability for five damage states can be evaluated according to the following formula:

$$P_f[B_1, \mathbf{x}(I_d)] = 1 - P_f[B_1^*, \mathbf{x}(I_d)] \qquad (13a)$$

$$P_f[B_i, \mathbf{x}(I_d)] = P_f[B_{i-1}^*, \mathbf{x}(I_d)] - P_f[B_i^*, \mathbf{x}(I_d)] \quad (i = 2, 3, 4) \qquad (13b)$$

$$P_f[B_5, \mathbf{x}(I_d)] = P_f[B_4^*, \mathbf{x}(I_d)] \qquad (13c)$$

7 SEISMIC LOSS ASSESSMENT OF STRUCTURES

The seismic loss of a structure in damage state B_j can be expressed as follows:

$$D_j = D_j^{(1)} + D_j^{(2)} + D_j^{(2)} \qquad (j = 1, \cdots, 5) \tag{14}$$

in which $D_j^{(1)} =$ the direct loss of from both structural and non-structural damage as well as the cost of maintenance and demolition; $D_j^{(2)} =$ the indoor loss induced by the structural damage; $D_j^{(3)} =$ the indirect loss induced by the structural damage.

To simplify the seismic loss assessment approach, the three kinds of economic losses for five seismic damage states can be assessed according to the loss coefficients method which depends on the earthquake filed investigations and experts' judgment.

(1) For direct economic loss, the cost can be evaluated by

$$D_j^{(1)} = \beta(B_j)C_I(I_d) \tag{15}$$

in which $\beta(B_j) =$ the direct loss coefficient for damage state B_j.

(2) For indoor economic loss, the cost can be evaluated by

$$D_j^{(2)} = \gamma(B_j)C_{eq} \tag{16}$$

in which $\gamma(B_j) =$ the indoor loss coefficient for damage state B_j; $C_{eq} =$ the equivalent merit of the indoor asset.

(3) For indirect economic loss, the cost can be evaluated by

$$D_j^{(3)} = \delta(B_j)D_j^{(1)} \tag{17}$$

in which $\delta(B_j) =$ the indirect loss coefficient for damage state B_j.

The values of loss coefficients in Eqs. (17) to (19) depend on the types and importance of the buildings. The loss D_i must be evaluated according to the specific situation of a structure and the seismic damage states. The total loss expectation value with five seismic damage levels can be obtained by the following formula:

$$L[\mathbf{x}(I_d)] = \sum_{j=1}^{5} P_f[B_j, \mathbf{x}(I_d)] \cdot D_j \tag{18}$$

8 TWO-STAGE MINIMUM EXPECTED LIFE CYCLE COST OPTIMIZATION METHODOLOGY FOR PERFORMANCE-BASED SEISMCI DESIGN

In this study, the minimum expected life cycle cost optimization methodology for performance-based seismic design is proposed, which is divided into the following two design stages:

Stage 1: Decision-making for the optimal fortification load of aseismic structures considering expected life cycle cost.

In this stage, the optimal fortification load of aseismic structures is determined according to the following optimization model:

$$\min_{I_d} \quad E\left[C(t, I_d)\right] = C_{\min}[\mathbf{x}(I_d)] + \theta \frac{v}{\lambda}(1 - e^{-\lambda t})L[\mathbf{x}(I_d)] \tag{19}$$

in which the minimum-cost function $C_{\min}[\mathbf{x}(I_d)]$ can be obtained from the regression analysis of a series of minimum initial cost seismic design according to Eq. (3) by adjusting the fortification

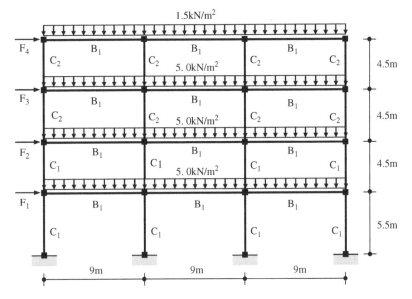

1.5kN/m²

| F₄ | B₁ | | B₁ | | B₁ | | 4.5m |

Figure 2. Four-storey plane steel frame.

intensity I_d, which is an increasing function of I_d; the total loss expectation function $L[\mathbf{x}(I_d)]$ can be obtained from the regression analysis of a series of seismic risk and loss assessment processes described in the above sections 4 to 7 by adjusting I_d, which is a decreasing curve of I_d.

The total expected life cycle cost curve $E[C(t, I_d)]$ by composing the above two curves must have the lowest point. The fortification intensity I_d corresponding to this point is called the optimal fortification intensity (OFI) I_d^*, which represents the minimum acceptable seismic risk level of a structure.

Stage2: Minimum initial cost seismic design under the optimal fortification load Once the optimal fortification load I_d^* has been obtained, the minimum initial cost seismic design can be made under this I_d^*. Therefore, the optimization model in this stage should be:

Find the design scheme $\mathbf{x}(I_d^*)$, to make the structural cost

$$C[\mathbf{x}(I_d^*)] \to \min \tag{20}$$

subjected to all constraints and requirements of codes.

The final solution is the optimal design scheme in consideration of the total loss expectation $L(\mathbf{x})$. Since the loss expectation has been taken into consideration when deciding I_d^* in the first design stage, it is only necessary to counteract the optimal resistance I_d^* by the minimum initial cost design scheme in the second design stage.

9 NUMERICAL EXAMPLE

A four-story and three-bay plane steel frame structure, as shown in Figure 2, is demonstrated by the method proposed in this paper. All the beams of the frame are made of Q235B steel, while all columns are made of Q345B steel. The soil type of the building site is type III, and the basic seismic intensity is 7 degree. The equivalent static horizontal seismic forces are calculated using base shear method according to the Chinese seismic design code.

The finite element model for this structure is built in ANSYS. All the beams and columns are modeled using Beam3 element during the elastic design stage under minor earthquake, while they

Table 2. Seismic risk probabilities for five damage states.

Intensity I_d	6	6.5	7	7.5	8	8.5	9
$P_f[B_1, \mathbf{x}(I_d)]$	0.29	0.49	0.68	0.82	0.91	0.96	0.99
$P_f[B_2, \mathbf{x}(I_d)]$	0.15	0.14	0.13	0.09	0.05	0.03	0.01
$P_f[B_3, \mathbf{x}(I_d)]$	0.37	0.26	0.14	0.07	0.03	0.01	0
$P_f[B_4, \mathbf{x}(I_d)]$	0.18	0.11	0.05	0.02	0.01	0	0
$P_f[B_5, \mathbf{x}(I_d)]$	0.01	0	0	0	0	0	0

Table 3. Total expected life cycle cost ($\times 10^5$ RMB).

Intensity I_d	6	6.5	7	7.5	8	8.5	9
C_0	27.812	35.213	45.897	73.911	87.296	97.553	115.27
C_{min}	21.032	23.875	30.846	36.336	40.374	55.867	61.254
L	19.859	13.215	8.506	4.542	2.443	0.615	0.107
$E[C_T]$	40.891	37.090	39.352	40.878	42.817	56.482	61.361

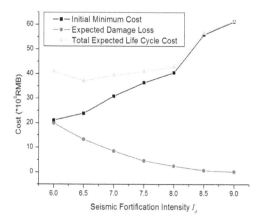

Figure 3. Decision-making of optimal fortification intensity.

are modeled using Beam24 element during the elastoplastic design stage under major earthquake. The minimum initial cost seismic design is performed using ANSYS design optimization tool. The global optimization strategy is used to treat with both elastic and elastoplastic inter-storey drift angle constraints simultaneously. The first-order optimization method is adopted, in which the gradients are calculated by the forward finite difference method.

The shape parameter $k = 8.3339$. The design lifetime is $t = 50$years. Since the earthquake occurrences are modeled as a Poisson process, according to Wen & Kang (2001b), the occurrence rate of ν per year is not required in Eq. (19). The discount rate $\lambda = 5\%$ is assumed. Let $\theta = 1$. For damage state B_2, $\beta = 0.10$, $\gamma = 0.05$, $\delta = 0.00$; For damage state B_3, $\beta = 0.30$, $\gamma = 0.15$, $\delta = 0.50$; For damage state B_4, $\beta = 0.90$, $\gamma = 0.50$, $\delta = 2.00$; For damage state B_5, $\beta = 1.00$, $\gamma = 0.95$, $\delta = 6.00$. Assume $C_{eq} = 1.5C_I$.

The seismic risk probabilities for five damage states are calculated using Eqs. (11) to (13), whose results are listed in Table 2. The original initial cost, the minimum initial cost, the expected damage cost as well as the total expected life cycle cost are summarized in Table 3 and Figure 3. From Figure 3, it can be readily seen that the optimal fortification intensity corresponding to the minimum total expected life cycle cost point is $I_d^* = 6.5$. The optimum solution corresponding to this optimal fortification intensity is the final seismic design scheme considering the expected life cycle cost.

10 CONCLUSIONS

The performance-based seismic design of civil engineering structures based on consideration of minimum total expected life cycle cost is studied in this paper. The conclusions can be summarized as follows:

 (i) Rational model of total expected life cycle cost should include the minimum initial cost plus the total expected damage cost.
 (ii) The optimal fortification load can decide the acceptable risk level more flexible and convenient than the target reliability.
(iii) The division of total optimum design process based on life cycle cost into the stage of the decision-making of the optimal fortification load and the stage of minimum initial cost design can greatly overcome some difficulties in the conventional design methods.

ACKNOWLEDGEMENT

The support of National Science Foundation of China through projects (Grant No. 50678057, 50108005, 59895410) is greatly appreciated.

REFERENCES

Ang A. H.-S. & De Leon D. (1997). "Development of target reliability for design and upgrading of structures." *Structural Safety*, 14(1): 91–103.
Ang A. H.-S. & Lee J.-C. (2001). "Cost optimal design of R/C buildings." *Reliability Engineering and System Safety*, 73: 233–238.
Esteva *et al.* (2002). "Life-cycle optimization in the establishment of performance-acceptance parameters for seismic design." *Structural Safety*, 24(2&4): 187–204.
Frangopol D. M. & Maute K. (2003). "Life-cycle reliability-based optimization of civil and aerospace structures." *Computers & Structures*, 81: 397–410.
Gao X. W. & Bao A. B. (1986). "Determination of Seismic design standards using probability methods." *China Building Structures Journal*, 7(2): 55–63.
Li G. (2005). "Life cycle seismic cost analysis and optimum design of steel frame structures." Master Thesis of Harbin Institute of Technology (Supervised by Prof. Lu D. G.).
Liu M. et al. (2003). "Optimal seismic design of steel frame buildings based on life cycle cost considerations." *Earthquake Engineering and Structural Dynamics*, 32: 1313–1332.
Liu M. et al. (2004). "Life cycle cost oriented seismic design optimization of steel moment frame structures with risk-taking preference." *Engineering Structures*, 26: 1407–1421.
Liu S. C. & Neghabat F. (1972). "A cost optimization model for seismic design of structures." *The Bell System Technical Journal*, 51(10): 2209-2225.
Lu D. G., et al. (2002). "Minimum total life cycle cost design of aseismic structures: principle and method." *The Second China-Japan-Korea Joint Symposium on Optimization of Structural and Mechanical Systems* (CJK-OSM2), Busan, Korea, Nov. 4–8, 2002, TS3-2, 129–134
National Standard of China P.R. (2001). "Seismic Design Code of Buildings (GB50011-2001)." Building Industry Press of China.
Rackwitz R. (2000). "Optimization - the basis of code-making and reliability verification." *Structural Safety*, 22(1): 27-60.
Wang G. Y. & Lu D. G. (2001). "Optimal fortification load and reliability of aseismic structures." *Earthquake Engineering Frontiers in the New Millennium*. Spencer and Hu, eds., Swets and Zeitlinger, Lisse, The Netherlands, 371-376.

Wang G. Y., et al (2003). "An optimal design for total lifetime cost of aseismic structures." *China Civil Engineering Journal*. 36(6): 1–6.

Wen Y. K. (2001). "Reliability and performance-based design." *Structural Safety*, 23(4): 407–428.

Wen Y. K. & Kang Y. J. (2001a). "Minimum building life-cycle cost design criteria. I: Methodology." *ASCE Journal of Structural Engineering*, 127(3): 330–337.

Wen Y. K. & Kang Y. J. (2001b). "Minimum building life-cycle cost design criteria. I: Applications." *ASCE Journal of Structural Engineering*, 127(3): 338–346.

Life-Cycle Cost and Performance of Civil Infrastructure Systems – Cho, Frangopol & Ang (eds)
© 2007 Taylor & Francis Group, London, ISBN 978-0-415-41356-5

Durability performance acceptance criteria for concrete structures

V. Malioka & M.H. Faber
Chair of Risk and Safety, Swiss Federal Institute of Technology, Institute of Structural Engineering, Zurich, Switzerland

A. Leemann & C. Hoffmann
EMPA (Swiss Federal Institute for Materials Testing and Research), Dübendorf, Switzerland

ABSTRACT: In this paper a novel probabilistic model framework for establishing durability performance acceptance criteria for newly built concrete structures is presented. The aim being to provide to owners of structures a consistent basis for assessing and specifying the acceptability of newly constructed structures. The spatial variability of the parameters which are decisive for the service life durability performance is modeled probabilistically in terms of sample statistics which may be assessed by testing of as built structures. This facilitates comparison of the performance of a given structure with specified requirements. The requirements to the service life performance are formulated in terms of the probability that the structure will be in a certain condition state after a given number of years. Requirement can thus be specified as e.g. maximum 40 percent of the surface of the structure exhibits visible corrosion after 50 years of service. Whereas the paper specifically addresses chloride induced deterioration the framework is general and may be applied equally well for the consideration of other phenomena as well. An example is provided illustrating the framework, the required information as well as the format of the acceptance criteria.

1 INTRODUCTION

The quality of newly constructed structures is an issue of special concern for the owners or operators of concrete structures. When it comes to the construction of a new structure the contractor is responsible for carrying out successfully the proposed design within the defined time schedule and construction cost. On the other end is the owner who has to be provided with a reliable forecast of the structural future performance, something that is not always, if not rarely, part of the obligations of the contractor. In principle the contractor will follow the, usually prescriptive design requirements but variability of the parameters governing the durability performance of the final structure cannot be avoided since procedures such as compaction and curing are highly influenced by the variable quality of the execution. Until now no criteria for the acceptability of newly constructed structures have been formulated which facilitates results of on-site measurements and tests of the properties of structures to be compared with acceptable values for a predefined desired overall service life durability performance of the structure. While the design procedures make allowance for uncertainties associated with e.g. documentation, interpretation and construction processes, the spatial variability of the achieved concrete properties in the realized structure is not so far consistently taken into account.

Based on results of recent years research on establishing models which allow for a representation of the spatial variability of the properties of concrete structures the present paper presents a probabilistic model framework, for the durability assessment of concrete structures subject to chloride induced corrosion, accounting for the relevant spatial variations. The ultimate aim is to enable owners to set quality control acceptance criteria for the spatial variability of the as built concrete material properties, for the assessment of the acceptability of newly built concrete structures. Hence the proposed framework is built according to that. The required models and utilization of information is outlined step by step. The approach facilitates to address important aspects such as the spatial representation of the structural surface by a subdivision into individual segments of semi-homogeneous conditions and the further discretization of these into smaller zones. The definition of the size of these zones is given through the so called correlation radius, a parameter which indicates the spatial variability within the considered surface segment. Finally an example is provided on how the presented framework may be utilized.

2 THE PROCEDURAL AND METHODICAL FRAMEWORK

After the completion of a structure the owner faces the decision problem of whether the structure may be accepted or not. The acceptance of the structure is based on the fulfilling of a number of requirements set by the owner. In any case these requirements cannot be based on individual preferences. They should rather have the form of acceptance criteria related to the quality control of the final structure. While available advances on research provide the tools for setting up such acceptance criteria the relevant spatial variations are not always consistently taken into account. Within this context the following questions, having all the same denominator, i.e. the spatial variability of the concrete material properties, need to be answered:

- Which steps/actions are necessary for setting up relevant quality control acceptance criteria?
- How can the acceptance criteria be expressed as requirements of the owner?
- How can these acceptance criteria be set up.

Figure 1 describes in the form of steps a framework within which the above questions can be answered. The framework is intended for individual newly built structures but it may be adapted for portfolios of structures and structures in service as illustrated in Faber et al. (2006).

2.1 *Representation of the structural surface*

The approach suggested in the following assumes that the considered structure exhibits homogeneous characteristics for what concerns exposure as well as execution and concrete characteristics. This implies that the exposure conditions as well as the material characteristics are homogeneous.

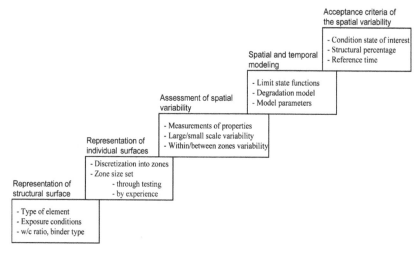

Figure 1. Framework for setting up quality control acceptance criteria.

It is clear that this will not be the case when a structure is considered as a whole. The requirement for exposure homogeneity can, however, be satisfied by subdividing the structure into smaller segments. Such segments can e.g. be the underside of a bridge deck, the lower parts of columns which are subjected into the splash of water and the upper part of these columns. Each identified segment can be associated with an exposure category based on its exposure conditions. Such categories or zones are distinguished and described in guidelines such as DuraCrete (1999). The requirement for homogeneity of the material characteristics can be satisfied by looking for differences on the surface resulting mainly from the compaction and curing processes. Furthermore batches produced with different w/c ratio and/or type of binder shall be identified and considered respectively.

2.2 *Representation of individuals surface segments*

The identified segments shall be further divided into smaller square zones the size of which shall be defined through statistical analysis of measurements of a material property and/or through experience from similar structural segments. Here "similar" refers to the exposure conditions, execution and material characteristics. In general, measurements of any concrete material property may be used for defining the size of the individual zones. However, it is more reasonable to carry out measurements of the concrete material property or properties that is/are more significant for the assessment of a future condition state of interest. For the case of corrosion initiation due to chloride ingress the main parameters are the concrete cover depth and the chloride diffusion coefficient. If measurements of more than one material property are made the smallest estimated zone size shall be used for the discretization of the structural segment, Malioka et al. (2006). If for any reason a concrete material property cannot be directly measured then measurements should be made of another relevant property. The last should have shown, in previous experimental results, to have a good correlation with the material property of interest, see for example Malioka et al. (2006).

2.3 *Assessment of the spatial variability*

Within the proposed framework the spatial variability is distinguished into within zones variability and between zones variability. Their assessment can be made through testing for measuring the material properties of interest. The locations of the measurements shall be chosen such that both the large and small scale variability is addressed. For this reason a testing screen is proposed with measurements in the horizontal and vertical direction and in a small scale, see example in Figure 2 left. The measurements do not need to be in regular intervals but the interval may vary. The number of the initial measurements may vary depending on the size of the structural segment. The test results from the measurements performed to assess the size of the individual square zones shall be used for the assessment of the within zones variability. The test results from measurements that fall within some of the individual square zones shall be used for the assessment of the between zones variability, Figure 2 right.

2.3.1 *Determination of the size of the individual zones*
A random field approach can provide the basis for assessing the dependency between material characteristics at different locations and for defining the size of the above mentioned zones. Detailed

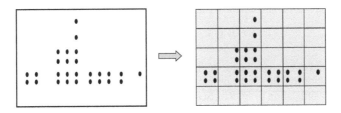

Figure 2.   Testing screen for the assessment of the spatial variability.

discussions on random fields may be found in e.g. Vanmarcke (1983). For illustrational purposes the main approach is outlined in the following; further detailed explanations may be found in Malioka and Faber (2004).

Consider a surface segment which satisfies the conditions of homogeneity. If s denotes a space coordinate such as length then $X(s)$ denotes a material property along this space coordinate with $X(s)$ considered a random field with $X(s_i)$ being random variables where s_i corresponds to a location along the space coordinate. If x_i represents the state of $X(s)$ at a location s_i then $f_m(x_1,s_1;x_2,s_2;\ldots x_m,s_m)$ is the joint probability density function of the random variables. In the following a two-dimensional random field of a property is considered i.e. variability across the area of a structural segment. Provided that the structural segment is homogeneous the assumption of a homogeneous random field is straight forward. Nevertheless if there is any doubt about the homogeneity of the considered sector then any trend that may exist should be examined with the help of testing, Fenton (1999a). Furthermore, the random field is assumed to be Gaussian in nature, which is the most common choice at the moment due to its simple mathematical properties and because it can be fully described by only its first two moments namely the expected value and the variance function. For such a field, with m random variables, the multivariate normal density function is then given by:

$$f_{\mathbf{X}}(\mathbf{x}) = (2\pi)^{-m/2} |\mathbf{C_{XX}}|^{-1/2} \exp\left[-\frac{1}{2}(\mathbf{x}-\mu_{\mathbf{XX}})^T \mathbf{C_{XX}}^{-1}(\mathbf{x}-\mu_{\mathbf{XX}})\right] \tag{1}$$

where $\mathbf{x}$ is the vector of realizations, $\mathbf{C_{XX}}$ is the m by m covariance matrix of the components in $\mathbf{x}$ and $\mu_{\mathbf{XX}}$ is the expected value of the random field. The covariance function of the random field is expressed as the product of the variance $\sigma_{\mathbf{XX}}^2$ and the correlation function $\rho_{\mathbf{XX}}$. The subscript in most characteristics indicates that a single random field is considered and in this case the correlation and covariance function maybe also referred to as auto-correlation and auto-covariance. Till now no specific correlation function has been proposed and one of the most common choices is given in Equation (2), Madsen et al. (1986).

$$\rho_{\mathbf{XX}} = e^{-\frac{\sqrt{x^2+y^2}}{h}}, \quad r = h\sqrt{2} \tag{2}$$

where r is the so called correlation radius, which is a measure of the distance within the random field where strong correlation can be assumed. Based on the correlation radius the structural segment under consideration may be divided into a number of zones based on its total area. As long as test results are available the maximum likelihood method (MLM) can be used to estimate the values of the mean and variance of the random field and the correlation radius that maximize Equation (1), see also Fenton (1999b).

2.4 Modeling the state of the structural segment

Following the descritization of a segment into N zones, which can be assumed to behave statistically independent due the descritization through the correlation radius, the probability that more than a percentage of it, is in a certain condition state Δ at some point in time t, can be assessed as:

$$P(\Delta_t \geq n/N) \quad 1 - B(n-1, N, \theta(t)) \tag{3}$$

where n is the number of zones being in the condition state Δ, $B(n-1, N, \theta(t))$ is the cumulative Binomial distribution and $\theta(t)$ is the probability of failure of the individual zones at time t and can be determined by simulation or FORM/SORM analysis on the limit state function.

The temporal modeling of the surface segment takes basis in the formulation of a proper limit state function that describes the condition state of interest. So for example if the condition state of interest is corrosion initiation then the time until corrosion initiation can be described by the following limit state function:

$$g(T_I \leq t) = X_I T_I - t \qquad (4)$$

where T_I is the time till corrosion initiation, t is a reference time and X_I is the model uncertainty, for which typical values can be found in e.g. Faber and Sorensen (2002). Assuming that corrosion is due to chloride ingress, the time till corrosion initiates at the depth of the reinforcement d may be written as, DuraCrete (2000):

$$T_I = \left(\frac{d^2}{4k_e k_t k_c D_o (t_o)^n} \left(erf^{-1}\left(1 - \frac{C_{CR}}{C_S} \right) \right)^{-2} \right)^{\frac{1}{1-n}} \qquad (5)$$

What is represented by each parameter in the model and typical values for them can be seen in Table 1. Similarly a limit state function can be formulated for the case of e.g. visual corrosion and/or corrosion due to other agents than chloride ingress, e.g. carbonation. The parameters of the limit state function are modeled as random variables, hence described by a certain distribution with mean μ_v and standard deviation σ_v (the subscript v stands for variable). The parameters which are considered spatially varying are modeled similarly but with the mean being itself a random variable with mean value $\mu_{\mu v}$ and standard deviation $\sigma_{\mu v}$. For these parameters σ_v represents the within zones variability, while $\sigma_{\mu v}$ represents the between zones variability.

2.5 Acceptance criteria

The framework described so far is built up in such a way that it can enable owners to set up acceptance criteria for the spatial variability of the material properties of a considered structural segment. Following the formulation in Section 2.4 it is straight forward to formulate the relevant acceptance criteria which can be expressed as: "*The probability that more than a percentage of a segment of a structure is in a certain condition state, at a reference future time after its construction, is less than a certain (acceptable) value*". Based on this statement it is clear that the owner needs to define the following:

- the condition state of interest
- the reference time
- the percentage of the structural segment that may be accepted when it is in a certain condition state
- the acceptable probability of the considered condition state at the reference time.

Based on the results outlined in Section 2.4 the combinations of values of the within zones variability and between zones variability of a spatially variable material property, for which the criteria are satisfied, can be estimated. When measurements are made available the within zones variability is calculated directly. From measurements that fall within the same squares the between zone variability can be calculated and it can then be checked whether this combination is acceptable or not. Nonetheless the preferences of the owner shall not be set arbitrary but rather with due consideration of life-cycle cost where the percentage of the structural segment being in a condition state and the probability of this event should be treated as optimization variables. In such an approach the associated costs can be taken into account and the option with the smaller associated life-cycle cost can be used to identify the cost optimal decision on the acceptance criteria.

3 ILLUSTRATION

In the following a surface segment of an ordinary Portland cement bridge deck is considered that satisfies the requirements for homogeneity as those described in Section 2.1. It is assumed that the

Table 1. Values of the parameters in Equations (4) and (5) – DuraCrete (2000).

		Units	Density function	μ_v	σ_v
Concrete cover depth	d	Mm	Log-normal	μ_d	σ_d
Mean value of d	μ_d	Mm	Normal	μ_{μ_d}	σ_{μ_d}
Surface concentration	C_S	% weight of concrete	Log-normal	3.103	1.232
Critical concentration	C_{CR}	% weight of concrete	Log-normal	0.48	0.15
Diffusion coefficient	D_o	mm^2/year	Log-normal	220.92	25.41
Environmental variable	k_e	–	Gamma	0.265	0.045
Test variable	k_t	–	Deterministic	1	0
Curing variable	k_c	–	Beta	0.793	0.102
Age factor	n	–	Normal	0.362	0.245
Reference (test) time	t_o	years	Normal	0.362	0.245

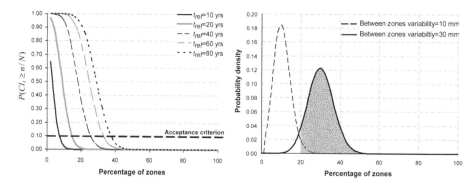

Figure 3. Probability of corrosion initiation over a structural percentage at different points in time (left) and probability density function of the zones with corrosion initiation (right).

w/c ratio is equal to 0.4 and that the segment belongs to the esposure category "splash zone". Based on experience from similar structural segments the correlation radius r is taken equal to 50 cm and it is assumed that this results in a descritization of the segment into 50 zones. For an analytical estimation of the correlation radius based on measurements the reader may refer to e.g. Malioka et al. (2006). The condition state of interest is the initiation of corrosion due to chloride ingress and the parameters in the limit state function, Equation (4), are given in Table 1. The concrete cover depth is considered as spatially variable.

For the segment considered, prior to any measurements, the statistical characteristics of the concrete cover depth can be $\mu_{\mu d} = 60$ mm, $\sigma_d = 18$ mm, DuraCrete (2000). A few references are available where the uncertainty of the mean value is accounted for and for the purposes of this illustration the standard deviation of the uncertain mean of the concrete cover depth is assumed equal to $\sigma_{\mu d} = 10$ mm. Using Equation (3) the probability that corrosion initiation is present at more than a certain percentage of the segment is estimated, Figure 3 – left. Provided that an acceptable probability has been set, $P(\Delta_t \geq n/N) \leq P_{f\ accceptable}$, one may check whether this criterion is satisfied for a certain percentage of the segment at a certain point in time. For a reference time t of e.g. 20 years and an acceptable percentage of 20% it is seen that the probability of corrosion initiation, CI, is much smaller than the set acceptance criterion of 0.1. Figure 3 – right shows the probability density function of the percentage of zones that exhibit corrosion initiation after 20 years. As it is seen for the case where $\sigma_{\mu d} = 10$ mm, the probability of corrosion initiation over 20% of the zones is clearly small (clear gray shaded area). However if the between zones variability is $\sigma_{\mu d} = 30$ mm the probability increases significantly (clear gray and dotted gray shaded area together). Similar

Table 2. Scenarios examined based on the assumed criteria set by the owner.

	Case 1	Case 2
Reference time	20 years	20 years
Critical percentage	40%	40%
$P(\Delta_t \geq n/N)$	0.1	$5 \cdot 10^{-4}$

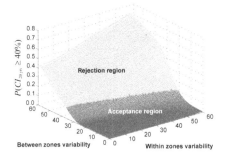 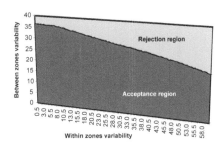

Figure 4. Acceptance and rejection regions for the within and between zones spatial variability of the concrete cover depth (units in mm)-Case 1.

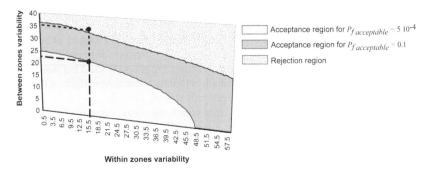

Figure 5. Acceptance and rejection regions for the within and between zones spatial variability of the concrete cover depth (units in mm)-Case 1 and Case 2.

illustrations can be provided for different values of the within zones variability of the concrete cover depth, σ_d.

The owner, however, needs to know which combinations of the within and between zones variability can be accepted. Measurements are performed on the considered segment using a screen such as the one shown in Figure 2. It is assumed that from the measurements the mean value of the uncertain mean is calculated equal to $\mu_{\mu d} = 60$ mm. Table 2 shows the cases examined based on the criteria assumed to have been set by the owner, as described in Section 2.5.

Figure 4 illustrates the results for Case 1. In the left hand side the probability of corrosion initiation, after 20 years, in more than 40% of the segment is shown for different combinations of the within and between zones variability. The dark shaded area, in both the left and right hand side, represents the acceptance region i.e. the combinations of the two considered types of variability for which the probability is less than 0.1. Figure 5 illustrates the acceptance regions for Cases 1 and 2 together. If from measurements the within zones variability is estimated equal to 17.5 mm then

255

Table 3. Values of the parameters in Equations (4) and (5) – DuraCrete (2000).

		Units	Density function	μ_v	σ_v
Diffusion coefficient	D_o	mm²/year	Log-normal	μ_D	σ_D
Concrete cover depth	d	mm	Log-normal	μ_{μ_D}	σ_{μ_D}
Concrete cover depth	d	mm	Normal	60	18

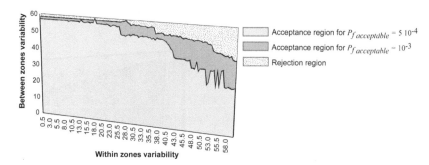

Figure 6. Acceptance and rejection regions for the within and between zones spatial variability of the diffusion coefficient (units in mm²/year).

the maximum acceptable variability between the zones is 33.2 mm for Case 1 while for Case 2 is equal to 22 mm. Of course any value within the designated as acceptance regions can be accepted.

In the previous considered cases the concrete cover depth has been assumed as being the dominating variable. Nevertheless the spatial variability of other concrete material properties should not be ignored. In the following the previously considered segment is used but now the chloride diffusion coefficient is considered as being spatially variable. For this investigation the parameters shown in Table 1 are reused with the few changes provided in, Table 3.

It is assumed measurements are available from which the mean value of the uncertain mean is calculated equal to $\mu_{\mu_D} = 220.92$ mm²/year. Figure 6 illustrates the combinations of the within and between zone variability of the diffusion coefficient for which the probability of corrosion initiation over 40% of the segment, after 20 years, is less than the acceptable probability of this condition state. From measurements performed in the segment the two types of variability can be estimated and whether they are within the acceptance regions can be checked.

4 DISCUSSION

The presented framework aims to provide a tool for assessing and setting up acceptance criteria for the spatial variability of material properties in concrete structures. Aspects such as the representation of the structural surface by segments and then zones, the assessment of the spatial variability and the definition of the size of the aforementioned zones as well as the spatial and temporal modeling are treated. Within the proposed framework the spatial variability is distinguished into within zones variability and between zones variability. The representation of the structural surface as suggested enables and simplifies the modeling of the spatial variability of a material property as well as the modeling of future degradation. An illustration is provided considering one surface segment. An application of the framework with real measurements may be found in Malioka et al. (2006). While the framework is promising as a tool for identifying the acceptable spatial variation of a concrete material property and thereby to support acceptance or rejection of newly built structures, it should be noted that to avoid arbitrary preferences of individuals the setting of acceptance criteria should be established based on a life-cycle costs minimization where future inspections and maintenance actions are taken into account. Furthermore, the case of non-compliance with given acceptance criteria should be treated and actions such as those suggested in CEB-fib (2006) should be considered.

ACKNOWLEDGMENTS

The authors would like to thank the Swiss Federal Roads Authority (ASTRA) for the financial support of the project "Spatial variability of concrete properties within a building component", project number: AGB2002/027.

REFERENCES

CEB-fib (2006). Model code for service life design. *Model Code, Bulletin 34. ISSN 1562-3610*, Sprint-Digital-Druck, Stuttgart.

DuraCrete (1999). Probabilistic performance based durability design of concrete structures. BRITE EU-RAM Project no. 1347.

DuraCrete (2000). Statistical quantification of the variables in the limit state functions. BRITE EU-RAM Project no. 1347, Document BE95-1347/R9.

Faber, M.H. & Sorensen, J.D. (2002). Indicators for inspection and maintenance planning of concrete structures. *Structural Safety,* 24(4), pp 377–396.

Faber, M.H., Straub, D. & Maes, M. (2006). A computational framework for risk assessment of RC structures using indicators. *Accepted for publication in the journal of Computer-Aided Civil and Infrastructure Engineering.*

Fenton G.A. (1999a). Random field modeling of CPT data. *Journal of Geotechnical and Geoenvironmental Engineering,* 125(6), pp. 486–498.

Fenton G.A. (1999b). Estimation for stochastic soil models. *Journal of Geotechnical and Geoenvironmental Engineering,* 125(6), pp. 470–485.

Madsen, H.O. et al. (1986). *Methods of structural safety.* Prentice-Hall, Inc., New Jersey.

Malioka, V. & Faber, M.H. (2004). Modeling of the Spatial variability for Concrete Structures. *Proceedings of the 2nd International Conference on Bridge Maintenance, Safety and Management, IABMAS'04*, Kyoto, Japan, October 18–22, 2004. pp. 825–826.

Malioka, V., Leeman, A., Hoffmann, C. & Faber, M.H. (2006). Spatial variability of concrete properties within a building component. Final report of the ASTRA project with number AGB2002/027: To be published as a report of the Swiss Federal Roads Authority (ASTRA), Switzerland.

Straub, D., Malioka, V. & Faber, M.H. (2006). A framework for the asset integrity management of large deteriorating concrete structures. *Accepted for publication in Structure and Infrastructure Engineering.*

Vanmarcke, E. (1983). *Random fields: analysis and synthesis.* MIT Press, Cambridge, Massachussets.

Optimum repairing level of concrete received chloride induced damage considering earthquake in life span

M. Matsushima
Kagawa University, Japan

M. Yokota
Shikoku Research Institute, Japan

ABSTRACT: Deterioration of concrete structure has been drawing a greater social attention and severe chloride damage has been observed in seaside area. In this paper the method to seek the optimum repairing level of concrete structure is proposed by evaluating the seismic capacity of concrete structure deteriorated by chloride induced damage. Since deterioration and seismic parameters involve various uncertain factors, the proposed method is defined the parameters based on reliability theory. The optimum repairing level is obtained by minimizing the total expected cost in life span.

1 INTRODUCTION

Deterioration of concrete structure has been drawing a greater social attention and severe chloride damage has been observed in seaside area. The maintenance technology and scheme are widely recognized in order to maintain the structure in life span. Figure 1[1],[4] shows the relationship between elapsed year and the numbers of bridge concerning over 50 year or not. Half of bridge numbers are over the life span after 2030 year. It is widely known that the maintenance cost in life span may be not small in comparison with construction cost. Life cycle cost model is important to maintain the structure. However, to estimate the deterioration and to compute the life cycle cost of structure in life span can not be established at present.

In this paper, the optimum repairing damage level can be obtained based on maximizing the total expected cost in life span of structure. The total expected cost is obtained by the sum of repairing cost and failure expected cost reduced the target earthquake in life span. The failure probability

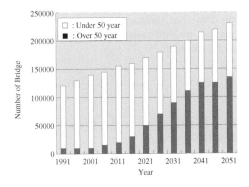

Figure 1. Aging of bridge and elapsed year[1],[4].

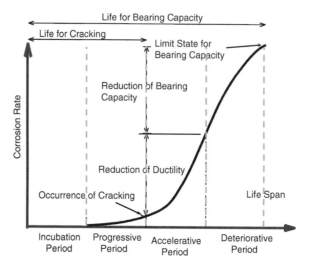

Figure 2.   Deterioration model[3].

is computed by assuming that the structure with corrosion rate over 20% is collapsed by expected earthquake.

2   MODEL OF CHLORIDE INDUCED DAMAGE

The process of chloride induced damage may divided into four stages, (1) incubation period, (2) progressive period, (3) accelerative period and (4) deteriorated period as shown in Figure 2. The chloride ion from the surface of concrete penetrates into concrete and the beginning of corrosion of steel bar is induced after the density of chloride ion near steel bar exceeds the critical value. The cracking of surface concrete is occurred by the pressure of corrosion. After cracking, corrosion speed of steel bar is accelerated and come to the end of bearing capacity of structure. To estimate the process of deterioration is to compute the period of each process and draw the each deterioration period as shown in Figure 2.

2.1   *Incubation period*

The incubation period is determined by the penetration of chloride ion and initial contains of chloride ion density. The chloride ion density near steel bar is computed by the equivalent diffusion coefficient and chloride ion density at surface concrete supplied from seawater. Eq. (1) can be obtained by using Fick's second low and assuming that the value of chloride ion per unit time supplied from seawater may be constant.

$$C(X,t) = C' + W \cdot \left[2\sqrt{\frac{t}{\pi D}} \cdot \exp\left(-\frac{X^2}{4Dt}\right) - \frac{X}{D}\left\{1 - \mathrm{erf}\left(\frac{X}{2\sqrt{Dt}}\right)\right\}\right] \qquad (1)$$

Where, $C(X,t)$: chloride ion density at time t and depth X from surface of concrete (kg/m^3)
C': Initial chloride ion density (kg/m^3)
D: Equivalent diffusion coefficient of chloride ion density ($cm^2/year$)
    Critical chloride ion density indicates $1.2\,kg/m^3$ according to the standard Specification of JSCE. The corrosion of steel bar starts when chloride ion density near the steel bar exceeds above critical value.

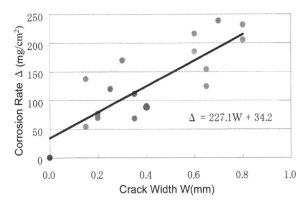

Figure 3.   Relationship of crack width W and corrosion rate Δ.

Table 1.   Relationship of damage level L and crack width W.

Damage Level L	Description	Remarks
1.0	Peeling off of Cover, Severe Damage	Deteriorative Period
2.0	Progress cracking. Maximum Crack width over 0.8 mm	Accelerative Period
3.0	Many cracking. Crack width over 0.4 mm	Progressive Period
4.0	Miner cracks. Crack width under 0.2 mm	Progressive Period
5.0	No damage	Incubation Period

2.2   Progressive period

The progressive period is governed by the volume of the corrosion rate. The corrosion speed can be computed by concerning with the increasing of chloride ion density. The cracking of surface concrete assumes to be occurred when the corrosion rate may be over 50 mg/cm^2.

2.3   Accelerative period

The accelerative period is governed by the volume of corrosion rate in same as the progressive period. The oxygen, water and chloride ion come easily into the inside of concrete and the corrosion speed is rapidly increased. The corrosion speed of steel bar assumes to be three times as one of progressive period.

3   DAMAGE LEVEL

Damage level of concrete structure can be judged by visual inspection and the crack width induced by the pressure of corrosion products is important index among them as shown in Table 1. The relationship between the volume of corrosion rate obtained using the laboratory experiments and crack width shows in Figure 3[3]. As shown in Figure, the relationship of crack width and corrosion rate is linear. Therefore, the damage level can be obtained by computing the volume of corrosion rate from crack width as shown in Table 1.

Deteriorations are occurred also after the repairing of structures because of the difficulty to eliminate completely the chloride ion penetrated into concrete body at present. Therefore, the structure after repairing is not same situation as the newly-built structure but the deterioration of structure is accelerated. The deterioration of structure after repairing is assumed to be described by the deterioration coefficient in this paper.

The repairing interval assumes to be obtained by being multiplied by the deterioration coefficient α as shown in Eq. (2).

$$T_k = T_1 \cdot \alpha^{k-1} \qquad (2)$$

Where, T_k: repairing interval at k-th, T_1: repairing interval at 1-th, α: deterioration coefficient, α assumes to be 0.5 according to previous study[2].

4 DAMAGE LEVEL

Target structure assumes to be failed by earthquake when corrosion produces of steel bar may exceed 20%. This reason is due to that the knot of deformation bar may be eliminated and the basic requirement of concrete structure is disappeared. The variation of corrosion rate assumes to be normal distribution and 0.4 assumes as the coefficient of variation according to previous studies such as laboratory experiments, observations in situ.

The return period of earthquake assumes to be T year. The annual occurrence probability P_i is described as Eq. (3).

$$P_i = \frac{1}{T} \qquad (3)$$

The occurrence probability of earthquake for n years is described in Eq. (4).

$$P_0 = 1 - (1 - P_i)^n \qquad (4)$$

The return period is obtained using Eq. (4) by assuming the annual occurrence probability. The annual failure probability of structure deteriorated by chloride induced damage induced by earthquake is obtained by multiplying the annual occurrence probability of earthquake P_i and failure probability of structure failed by earthquake at t year $P_f(t)$. The structure with the loss of cross section area over 20% assumes to be failed by earthquake. The failure probability P_{fm} at the return period m is described as Eq. (5).

$$P_{fm} = 1 - \prod_{i=1}^{m}(1 - P_f(t) \cdot P_i) \qquad (5)$$

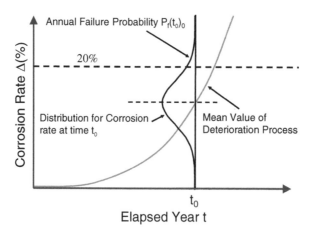

Figure 4.   Conception of annual failure probability.

Where, $P_f(t)$: The probability of that loss of cross section area of steel bar occurred by chloride induced damage exceeds 20%, P_i: annual occurrence probability of earthquake, m: life span of concrete structure.

5 EXPECTED COST MODEL[5)]

The total expected cost C_T is described as shown in Eq. (6) by computing the sum of repairing cost in life span $n \cdot C_R$ and expected loss cost induced by earthquake $P_{fm} \cdot C_f$.

$$C_T = n \cdot C_R + P_{fm} \cdot C_f \qquad (6)$$

Where, n: repairing numbers in life span, C_R: repairing cost, P_{fm}: failure probability in life span, C_f: loss cost in earthquake.

The expected loss cost C_D is described by the sum of the newly-build cost and the cost to remove the failed structure. The expected loss cost C_D assumes to describe twice times as initial construction cost Cc. However, the social loss cost at failure Cs is varied to estimate by the difficulty to estimate the rating of structure and the difference for user of structure. For example, when the target bridge is failed by earthquake, the huge loss cost occurs in case of cutting off the main roadway for long time. By the way, the only few loss cost is occurred in case of local roadway. The social loss cost is widely affected by the variation of the social conditions.

Therefore, the loss cost at failure C_f is described by construction cost Cc and important coefficient γ as shown in Eq. (7).

$$C_f = \gamma \cdot C_C \qquad (7)$$

The non dimension total cost C_{0T} is obtained by assuming $C_R = \eta \cdot Cc$ and $C_f = \gamma \cdot Cc$. C_{0T} is obtained as shown in Eq. (8) by substituting above assumption for Eq. (6).

$$C_{0T} = \frac{C_T}{C_C} = n \cdot \eta + P_{fm} \cdot \gamma \qquad (8)$$

Where, α is repairing coefficient. The repairing cost is divided two segments. One is the cost for temporary stage, another one is the direct cost for repairing.

Especially, in case of low damage level, not repairing but reinforcement is carried out. The repairing cost model assumes that the repairing coefficient η be increased with the decreasing of damage level L as shown in Figure 5. The repairing coefficient η is exponentially increased in order to proceed not repairing but reinforcing below damage level L = 2.0.

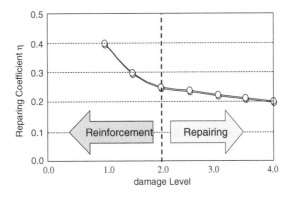

Figure 5.   Repairing coefficient α and damage level L.

263

Table 2. Computation condition.

Equivalent Diffusion Coefficient D	4.0×10^{-8} cm²/sec
Adherent chloride ion density W	4.23×10^{-8} mg/cm²/sec
Cover of thickness C	4.0 cm
Initial chloride ion density C_0	0.3 kg/m³

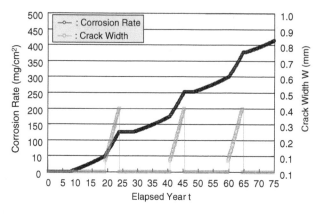

Figure 6(a).   Corrosion rate Δ and elapsed year t.

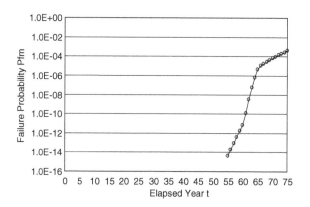

Figure 6(b).   Failure probability P_f and elapsed year t (Repairing Level L = 3.0).

6   OPTIMUM REPAIRING SCHEME USING LIFE CYCLE COST

The process of chloride induced damage using proposed deteriorated model in order to draw up the optimum repairing scheme using life cycle cost. Target structure is the bridge located at coast side of sea such as SETO Inland Sea. The environment of target place is splash zone. The computation condition for target structures describe in Table 2. The life span of target structure is 75 year. The relationship between the mean value of corrosion rate obtained in computation and elapsed year t shows in Figure 6(a) for damage level L = 3.0 in repairing and Figure 7(a) for damage level L = 4.0 in repairing. The crack width induced by the pressure of corrosion is eliminated by the repairing and increases with the increasing of deterioration. The repetition of repairing and the deterioration is carried out in life span. Besides, repairing interval is decreased with time by the repetition of repairing. The corrosion rate increase with time as shown in Figure. The repairing is three times in life span for damage level L = 3.0 in repairing and four times for damage level

264

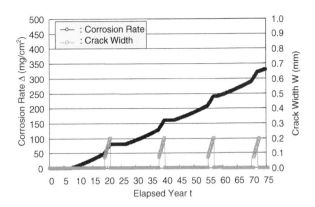

Figure 7(a).   Corrosion rate Δ and elapsed year t.

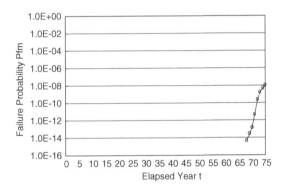

Figure 7(b).   Failure probability P_f and elapsed year t (Repairing Level L = 4.0).

L = 4.0 in repairing. The relationship between the failure probability P_{fm} and elapsed year t shows in Figure 6(b) for damage level L = 3.0 in repairing and Figure 7(b) for damage level L = 4.0 in repairing. The corrosion rate in life span indicates 420 mg/cm^2 for the damage level L=3.0 in repairing and 330 mg/cm^2 for damage level L = 4.0 in repairing. The relationship between the failure probability P_{fm} and elapsed year shows in Figure 6(b) for damage level L = 3.0 in repairing and Figure 7(b) for damage level L = 4.0 in repairing. For the damage level L = 3.0 in repairing, the failure probability is exponentially increased with elapsed year after passing away 50 year. For the damage level L = 4.0 in repairing, increasing after passing away 65 year. The relationship between the number of repairing and the damage level L in repairing shows in Figure 8. The number of repairing is twice times bellow the damage level L = 2.5 in repairing and increases exponentially with the increasing of damage level L = 2.5 in repairing.

The relationship between the failure probability P_{fm} and the damage level L in repairing shows in Figure 9. The failure probability increases exponentially with the decreasing of damage level in repairing. The relationship between the expected cost and damage level in repairing shows in Figure 10(a) and (b). (a) indicates the case of important coefficient $\gamma = 2.0$, and (b) indicates $\gamma = 50.0$. The expected loss cost is exponentially decreased with the increasing of damage level in repairing. The important coefficient doesn't affect the shape of relationship of expected loss cost and damage level in repairing. The repairing cost increases at low damage level in repairing because not repairing but reinforcement is carried out and increases at high damage level in repairing because the number of repairing increases. Total expected cost is obtained by the sum of the expected loss cost and the repairing cost in life span. The optimum repairing level is chosen as the damage level

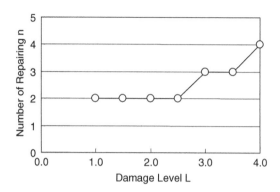

Figure 8.   Repairing numbers and repairing level L.

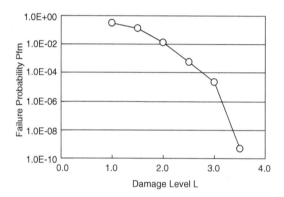

Figure 9.   Failure probability P_{fm} and repairing level L.

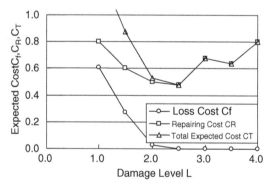

Figure 10(a).   Optimum repairing level ($\gamma = 2.0$).

in repairing when the total expected cost may take a minimum. The optimum damage level in repairing is not affected by the variation of important coefficient and choose $L_{opt} = 2.5$.

7   CONCLUSION

In this paper, the method to seek the damage level in repairing of concrete structures is proposed by evaluating the seismic capacity of concrete structures deteriorated by chloride induced damage.

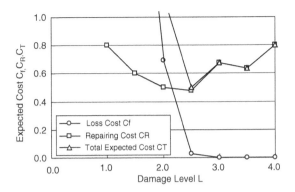

Figure 10(b).   Optimum repairing level ($\gamma = 50.0$).

(1) The method to obtain the optimum repairing level is proposed using the total expected cost computed by the sum of the expected loss cost induced the failure due to earthquake and repairing cost in life span.

(2) The optimum damage level in repairing is obtained using the total expected cost. The life span assumes to be 75 year in computation. The optimum damage level in repairing is not affected by the variation of important coefficient and chooses $L_{opt} = 2.5$.

REFERENCES

Diagnosis Engineering of Concrete '06 (Basic), JCI, pp.186–187.

Iba et al: Basic Study of Evaluation on LCC of RC Structure received Chloride induced Attack, Proceeding of JSCE, No.704/V-55, pp.1–11, 2002.5.

M. Matsushima et al: Design Thickness of Cover of Concrete Structures received Chloride induced Damage, No.490/V-23, pp.41–49, 1994.5.

Standard Specification of Design and Construction of Concrete Structure (Durability Design), JSEC, p.185. 2001.1.

Y. Miyagawa: Early Chloride Corrosion of Reinforced Steel in Concrete, Doctoral thesis, 1985.2.

Life-Cycle Cost and Performance of Civil Infrastructure Systems – Cho, Frangopol & Ang (eds)
© 2007 Taylor & Francis Group, London, ISBN 978-0-415-41356-5

Updating the time-dependent reliability using load monitoring data and the statistics of extremes

Thomas B. Messervey
Department of Mathematical Sciences, United States Military Academy, West Point, NY, USA

Dan M. Frangopol
Fazlur R. Khan Endowed Chair of Structural Engineering and Architecture, Department of Civil and Environmental Engineering, ATLSS Center, Lehigh University, Bethlehem, PA, USA

ABSTRACT: This paper investigates how to incorporate in-service live load data obtained via structural health monitoring (SHM) into the calculation of the reliability index β over time. The idea is that a structure could either be overly conservative in its design, in which case one would want to take advantage of this fact in maintenance scheduling, or a structure could be subjected to greater than anticipated loading, wear, or decay, also demanding maintenance adjustments. In this analysis a very important question is uncovered. What live load is most appropriate in calculating the reliability of a structure, the initial code-driven design load, a projected anticipated future load, or a load distribution created from an on-site historical demand history? To answer this question, a model is proposed that incorporates all three elements as they change over time. An innovative approach using the statistics of extremes is introduced to update, over time, a prior assumed load distribution to reflect the actual in-service loads on site. The results are then incorporated into the calculation of the reliability index through Bayesian updating and projected forward in time.

1 INTRODUCTION

1.1 *Background*

The reliability index, β, provides a probabilistic measure of a structure's ability to resist anticipated demand. In order to achieve a holistic, life-cycle treatment of cost, performance, or safety, this index must be projected forward in a probabilistic manner. Typically, corrosion effects and deterioration decrease a structure's resistance capacity, while increased uncertainty about loads increases the demand. This results in the progressive decrease of the reliability index over time. In any such analysis, one of the main contributors to the decrement of the reliability index is the uncertainty itself (Messervey et al., 2006b). Hence, the ability to reduce uncertainty is of benefit. Structural health monitoring offers a potentially powerful mechanism to not only reduce uncertainty, but to improve random variable input parameters by making them more accurate through site-specific data. Improved accuracy could result in a higher or lower reliability rating depending upon the data collected. As such, one of the most important parts of this study is to begin investigating how to incorporate updated information into a reliability analysis. For example, if a bridge is observed for a day and only light vehicle traffic is recorded, is the bridge determined to be safe? Obviously, no. However, if a bridge is observed for 50 years and the load demand is always significantly less than anticipated, should that be considered in the calculation of bridge reliability and maintenance scheduling? Obviously, yes. This paper begins to explore how to reasonably combine design, current, and historical data over time.

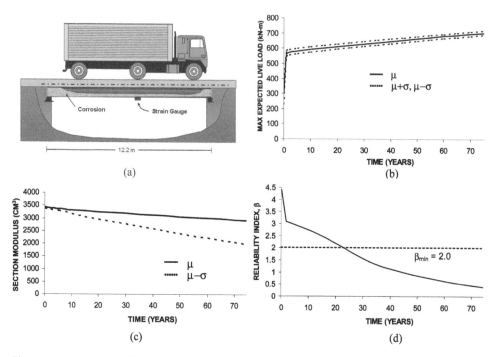

(a)

(b)

(c)

(d)

Figure 1. Flexure reliability analysis of a short span steel beam highway bridge designed under the HS20 truck, subjected to live load effects, and corrosion over time.

1.2 Scenario of interest

This study builds upon a simple example reported in (Messervey & Frangopol, 2006a) and (Messervey et al., 2006b) which investigates the reliability of a short span, simply supported, W690 x125 steel beam bridge as shown in Figure 1a.

The bridge is subjected to the HS-20 truck static live load and live load effects are calculated in accordance with the live load model reported in Nowak (1993). This results in greater anticipated maximum expected moment demand over time as predicted by the statistics of extremes (Messervey & Frangopol, 2006c). The increase in the expected maximum live load moment is shown in Figure 1b. Concurrently, the beam undergoes corrosive effects. Albrecht and Naemmi's (1984) study is utilized to predict the depth of corrosion over time as $C(t) = At^B$, and Estes (1997) corrosion pattern where corrosion extends one quarter of the way up the web at center span is followed. The resulting decrease in the elastic section modulus is shown in Figure 1c. A reliability analysis with respect to flexure is conducted over a 75 year time period. The random variable input parameters are the same as those in (Messervey et al., 2006b) and the result of this analysis is shown in Figure 3d. It is important to note that this is an elastic analysis as strain gauges are envisioned for data collection. The impact of this decision (not considering the plastic section modulus) will be a subject of future research. Here, employing a $\beta_{min} = 2.0$, maintenance action would be required at about 23 years.

1.3 Proposed monitoring solution

The idea is to leverage simplicity and to propose a low cost, low oversight solution that provides a bridge manager with actionable information. For this, several strain gauges are placed at or near midspan of a simply supported member to capture daily peak strain demands. This would require a fairly simple but continuous onboard monitoring system. Assuming linear operational behavior, strain recordings would be transformed to moment via Hooke's law. Once peak moment

270

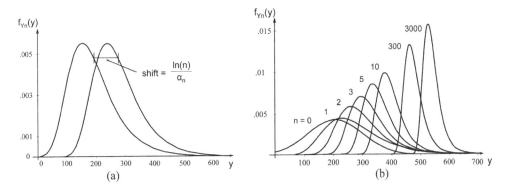

Figure 2. Transformation of a Gumbel distribution which results in no change of shape and transformation of a normal distribution for $n = 1, 2, 3, 5, 10, 300$, and 3000 respectively.

demands are recorded, the statistics of extremes can be utilized to determine the characteristics of the underlying moment demand distribution. This distribution can then be compared to the one reported in Nowak (1993) and be introduced into the 75 year life-cycle analysis. Strain gauges and peak picking offer some benefits of interest. Strain gauges are reliable, inexpensive, and record actual load demand. As such, they are not limited to what a designer remembers to include in a performance function and over time, the recorded data would include seasonal, environmental, and any other stressors introduced. Peak picking also minimizes the amount of data collected. Managing this data should take low computational effort and demand little power. Aside from the goal in this study to conduct a reliability analysis, such a system could readily be used for early warning once a maximum allowable level of strain is specified.

2 STATISTICS OF EXTREMES: DETERMINING A SITE SPECIFIC MOMENT DEMAND DISTRIBUTION

2.1 Truck traffic model and distribution selection

In a recent study of 10 years of weigh in motion (WIM) data, Gindy and Nassif (2006) make some interesting observations about the characteristics of truck gross weights across 33 sites in New Jersey, USA. These observations include population statistical descriptors, variability between site locations, and variability with respect to the length of the time window in which data is observed. It is noted that although the results reported in Nowak (1993) have formed the basis for calibration factors of the AASHTO LRFD Bridge Design Code (1994), it should be considered that the study was based off of 9,250 trucks and that this data is now over 30 years old. As such, the idea here is to utilize the Nowak study as an initial estimate for the moment demand distribution and for the live load effects of this demand as the distribution is projected forward in time with respect to a particular volume of traffic, and then to update it with observed information simulated from the Gindy & Nassif study. In Messervey & Frangopol (2006c), a method is developed to update an initial estimate of a load distribution assuming the initial distribution is a Gumbel distribution. This transformation is unique in extreme statistics as it only causes a shift of the average maximum expected value based only on the size of the observed sample (number of trucks) as shown in Figure 2a. As such, the shape parameter α_n, and the characteristic value μ_n are independent. However, since there is no change in shape the amount of shift is greater than when there is a change in shape. If a change of shape does occur, the shape parameter α_n and the characteristic value μ_n are dependent. In a practical sense, modeling an initial traffic distribution with a Gumbel distribution may be overly conservative as the distribution itself begins with a characteristically large upper tail and the shifts of the average expected maximum value can be very large. As such,

271

Table 1. Simulation and reliability analysis parameters.

Scenario	Mean moment demand μ_x(kN-m)	Std Dev of moment demand σ_x(kN-m)	Number of trucks	Utility
HS20 Design Truck	305	78.7		Used to calculated the initial reliability index basis for the theoretical average truck.
Nowak (1993)	228.75	73.2	500	Initial moment distribution based on the average expected truck. Used to calculate live load effects.
Simulated Data	167.5	77.2	300	Based on the Gindy and Nassif (2006) study. Replicates SHM data. Used to update the intial estimated moment distribution and volume of traffic.

it may be desired to model the transformation of non-Gumbel distribution. Figure 2b shows the transformation of a normal distribution that has been fit to the population truck data obtained in the Gindy and Nassif study N(204.6, 94)kN. The distribution is observed in its original form (no trucks) and then as the number of trucks observed ranges from $n = 1, 2, 3, 5, 10, 300, 3000$. One notices that the transformation to the characteristic Gumbel shape is rapid and would likely occur in the first few days of an analysis.

Exact solutions that describe the above transformations are readily available in Ang & Tang (1984). Here, the goal is to develop a method to reverse-engineer the transformation process. Given peak strain recordings in the transformed (extreme) space, can the initial (untransformed) distribution and volume of traffic be validated or updated?

2.2 Model and Simulation to update an initial moment demand distribution

Building upon the scenario presented in Figure 1, the average truck from the Gindy and Nassif (2006) study produces a maximum static live load moment $M_{LL} = 167.5$ kN-m as one wheel line is placed upon the beam at the position of maximum moment. The same coefficient of variation of COV = .46 that is listed in the study is applied resulting in a standard deviation $\sigma_{MLL} = 77.18$ kN-m. Microsoft Excel is then utilized to simulate a normal distribution N(167.5, 77.18)kN. A volume of 300 trucks/day is created each "day" of the simulation which intentionally does not match the 500 trucks/day utilized to develop the live load effects in Figure 1b. Maximum peak values are recorded and logged for a period of two years to coincide with a bi-annual inspection requirement. It should be noted that any uncertainty involved with sensor accuracy or in the random variables associated with Hooke's law and the transformation from strain to moment are not treated in this analysis to facilitate model development. In a practical sense, this would be applicable to a newer structure where deterioration has not yet begun or to a structure where a known load could be placed on it periodically to calibrate the resistance capacity. Table 1 summarizes the input parameters and the intent of each as it relates to the experiment.

From the simulated data, a reduction of the initial assumed moment distribution and volume of traffic is anticipated.

Let Y_n be the maximum daily observed moment demand. After 180 days of simulation time, an average daily maximum moment demand, $\mu_{Yn} = 385.7$ kN-m and standard deviation of $\sigma_{Yn} = 27.7$ kN-m are recorded. The following equations can be used to compare these results with their theoretical counterparts (Ang & Tang, 1984)

$$\mu_{Yn} = \sigma_X \mu_n + \mu_X + \frac{\gamma \sigma_X}{\alpha_n} \tag{1}$$

272

$$\sigma_{Y_n} = \left(\frac{\pi}{6}\right)\left(\frac{\sigma_X}{\alpha_n}\right) \tag{2}$$

where $\gamma = .577216$ (Euler's number), σ and μ are the descriptors of the initial normal distribution, and

$$\alpha_n = \sqrt{2\ln(n)} \tag{3}$$

$$\mu_n = \alpha_n - \frac{\ln[\ln(n)] + \ln(4\pi)}{2\alpha_n} \tag{4}$$

where n is the number of observed trucks and μ_n is the characteristic value. These equations conduct the transformation shown in Figure 2b. Conducting these calculations for $t = 1$ day ($n = 500$ trucks) result in, $\mu_{Y_n,} = 453.5$ kN-m and $\sigma_{Y_n,} = 26.6$ kN-m. The disparity between these values and those observed by the simulation indicate that updating is necessary.

Working with the observed data, the shape factor of a Gumbel distribution can be calculated as (Menun, 2003).

$$\alpha'_n = \frac{\pi}{\sqrt{6}\sigma_{Y_n}} \tag{5}$$

It should be noted that (3) and (5) do not produce the same result, but can be related through the relationship

$$\alpha'_n = \frac{\alpha_n}{\sigma_X} \tag{6}$$

but only after convergence is achieved between the theoretical and monitoring information. (6) can be arranged to estimate the number of trucks in the transformed space as

$$\alpha'_n = \frac{\sqrt{2\ln(n)}}{\sigma_X} \tag{7}$$

which is then inverted to solve for n. Equations 1–7 fully define each moment distribution in the transformed space. These two Gumbel distributions can then be combined in a probabilistic manner using Bayesian updating as (Casella & Berger, 2002)

$$E(\theta \mid x) = \frac{\lambda^2}{\lambda^2 + \sigma^2}x + \frac{\sigma^2}{\sigma^2 + \lambda^2}\mu \tag{8}$$

where θ represents the random variable of interest, μ its theoretical mean, λ^2 the theoretical variance, x the monitoring-based mean and σ^2 the monitoring-based variance. (8) provides weighting factors to the theoretical data and observed data based upon their respective variances. This equation is then used to combine $\mu_{Y_n}, \sigma_{Y_n}, \alpha_n$, and n. With these values combined, the updated Gumbel distribution can be mapped back to the original space ($n = 0$) by reevaluating (3) and (4) successively and then by inverting (2) and (1) successively to solve for the initial mean and standard deviation of the original distribution, μ_X and σ_X. Table 2 shows the results of this process over a two year period.

The process shows a reasonable convergence after two years to the simulated data. In this scenario, updating occurred every three months. More frequent updating (or more total updates) would likely have led to a more exact result. Also of interest, the possibility exists to include different sample spaces (number of trucks) in the analysis. For example, if peak values are picked every week in addition to every day, one can observe the transformed space $n = 2100$ trucks and $n = 300$ trucks concurrently utilize information from both in the analysis. Important to this approach is

Table 2. Simulation experiment results.

Parameters	Mean moment demand μ_x(kN-m)	Number of trucks	Std Dev of moment demand σ_x(kN-m)
Nowak (1993)	228.75	500	73.2
Simulated Data	167.5	300	77.2
Updated Parameters			
6 mo.	200.0	410	73.4
12 mo.	182.0	296	73.5
18 mo.	176.9	311	73.7
24 mo.	176.0	303	73.7

that each space must be well defined before the information is utilized. These topics will be the subject of future research. However, the information obtained shows an almost surprising ability to determine what is acting upon the structure of interest using very little prior information. On an actual structure, the number of trucks and true load distribution would likely vary over time. Additionally, environmental, seasonal, and other unforeseen stressors would become part of the distribution, which is exactly what is desired, a true representation of the load demand over time that is able to change and impact the reliability analysis of the structure.

3 INCORPORATING AN UPDATED MOMENT LIVE LOAD DISTRIBUTION INTO THE 75 YEAR RELIABILITY ANALYSIS

How to incorporate updated information into the life-cycle reliability analysis is a non trivial question that will only find its beginning in this work. What is an appropriate interval for updating? The more one updates, the sooner the observed data is approached. Is the answer to this question different if the monitoring data determines that the structure is being subjected to loads greater than the design load? Through updating, are we losing a level of performance or safety intended by the code if the structure is routinely not subjected to demands associated with the HS20 truck? Conversely, if it is identified that a structure is not subject to HS20 type demands, should one continue to allocate the maintenance and inspection dollars to maintain HS20 type performance?

In this analysis, Bayesian updating is proposed as an appropriate method to combine design and monitoring information. In this context, there are two likely choices: 1) Make the appropriate adjustments to the initial model parameters, and restart the analysis from $t = 0$, thus effectively recreating the past; 2) Restart the clock and begin a new life-cycle analysis at the time the information becomes mature by combining the last known reliability index with the reliability index calculated using the distribution of the observed demand. Then, the live load effects would again be projected forward for the updated distribution and updated volume of traffic. Figure 3 shows the results of restarting the analysis with updated parameters. In this example, the HS20 design moment of 305 kN-m is combined with the average observed moment of 176 kN-m using (8). This results in an increase of β_o from 4.41 to 4.86 as seen in Figure 3b. The updated moment demand and updated volume of traffic change the live load effects as shown in Figure 3a which results in an additional increase of the reliability index. In this scenario, the β_{min} threshold is crossed approximately eight years later than before.

Of interest may be the second approach to including updated information where the reliability analysis is essentially interrupted and updated in-stride. In such an analysis, it would make sense to update the last known value of the reliability index but not the reliability index associated with the live load effects as actual data has replaced what was predicted. Hence, at two years, β_o would be combined with the reliability index from the observed moment demand distribution creating β_2. If a successive update occurred at $t = 10$ years, β_2 would again be combined with the reliability index

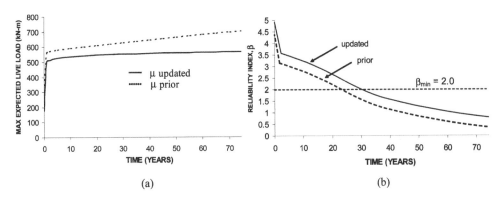

(a)   (b)

Figure 3.   Recalculating the reliability from t = 0 years with updated input parameters.

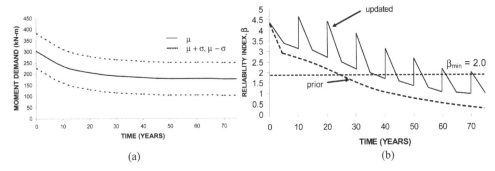

(a)   (b)

Figure 4.   Periodic Bayesian updating of the design load and reliability index.

calculated from the monitoring data to create β_{10}. It will be left for future investigation but it will likely be reasonable to apply Bayesian updating to the monitoring demand profile as well. The two choices would be to average all load data collected over 75 years, or to incrementally update new load information in discrete time segments such that recent load data has the ability to shift more quickly to reflect a change in load demand. The latter is likely the better choice. Another area for investigation is the potential loss of visibility on the wear and tear of the structure. In this analysis, it will be assumed that an inspection occurs or that a calibration load is placed upon the bridge to identify wear and tear in the resistance portion of the performance function.

In the following example updating is conducted at 10 year intervals. At each update, it will be assumed that calibration confirms the mean value of all other variables to include section loss. As such, the structure is essentially performing exactly as predicted and the changes due to updating of the moment distribution can be isolated. At each update interval, live load effects are restarted as the analysis again projects into the future. Hence, even though the curve is a 75 year analysis, it is an analysis that shapes itself over time. Additionally, although live load effects are restarted, the associated moment demands are continually combined with different values for the section modulus as the structure is assumed to corrode as predicted.

Figure 4a shows the decrement of the original HS20 moment live load as it is combined with the observed live load distribution over time. Here, after 40 years (or four updates) the design load has been essentially replaced with the observed loads. The sawtooth appearance of the reliability profile shows that the live load effects dominate the reliability curve. The updated curve trends higher than the prior curve due to the decremented moment demand shown in Figure 4a and due to resetting the clock on live load effects. Here, the updated profile crosses the β_{min} threshold at about 37 years or 14 years after the initial model. Improving this curve further would likely require investigation of the resistance part of the performance function to observe if the corrosive effects are as predicted.

275

4 SUMMARY AND CONCLUSIONS

In this paper, the idea of utilizing Bayesian updating and the statistics of extremes to analyze SHM data is explored with the intent of efficiently obtaining a site-specific live-load moment demand distribution for inclusion into a life-cycle reliability analysis. A method centered upon the "peak picking" of maximum daily values is developed using moment demands simulated from a recent study of WIM data. Using this method, an initial normally distributed moment demand distribution and volume of traffic are updated to match observed information using only maximum strain recordings. Using the updated moment demand distribution, two methods are discussed and presented for incorporating this information into a 75 year reliability analysis. In doing so, the analysis raises some important questions about how to appropriately combine design values, recorded historical data, and anticipated future demands which will undoubtedly be the subject of future research.

ACKNOWLEDGMENTS

The support of the National Science Foundation through grants CMS-0217290 and CMS-0509772 to the University of Colorado, Boulder, and grants CMS-0638728 and CMS-0639428 to Lehigh University are gratefully acknowledged. Also, the support provided by the Collaborative Research Center SFB 477 "Monitoring of Structures" funded at the Technical University of Braunschweig by the German Research Foundation (DFG) is appreciated. The opinions and conclusions presented in this paper are those of the authors and do not necessarily reflect the views of the sponsoring organizations.

REFERENCES

AASHTO, 1994 *AASHTO LRFD Bridge Design Specifications*, American Association of State Highway and Transportation Officials, First Edition, Washington DC.

Albrecht, P. & Naeemi, A.H. (1984). Performance of weathering steel in bridges. National Cooperative Highway Research Program, Report 272.

Ang, A.H. and Tang, W.H. 1984. Probability Concepts in Engineering Planning and Design Volume II, Wiley, New York.

Casella, G. and Berger, R.L. 2002. Statistical inference. Duxbury, Thompson Learning, Pacific Grove USA.

Estes, A.C. 1997. A System Reliability Approach to the Lifetime Optimization of Inspection and Repair of Highway Bridges. Ph.D. Dissertation, Department of Civil, Environmental and Architectural Engineering, University of Colorado at Boulder.

Gindy, M. and Nassif, H. 2006. Effect of bridge live load based on 10 years of WIM data, Proceedings of the Third International Conference on Bridge Maintenance and Safety, Porto, Potugal, July 16-19, 2006, IABMAS'06, Taylor & Francis, 9 pages on CD-ROM.

Kong, J.S. & Frangopol, D.F. 2005. Sensitivity Analysis in Reliability-Based Lifetime Performance Prediction Using Simulation, *Journal of Materials in Civil Engineering*, Vol. 17 pp. 296–306.

Menun, C., 2003. CEE203 Probability Models in Civil Engineering, Course Notes, Stanford University, CA, pp.102–119.

Messervey, T.B. and Frangopol, D.M. 2006a. A Framework to Incorporate Structural Health Monitoring into Reliability-Based Life-Cycle Bridge Management Models, Proceedings of the Fifth International Conference on Computational Stochastic Mechanics, Rhodes, Greece, June 21–23, 2006.

Messervey, T.B., Frangopol, D.M., and Estes, A.C. 2006b. Reliability-based life-cycle bridge management using structural health monitoring, Proceedings of the Third International Conference on Bridge Maintenance and Safety, Porto, Portugal, July 16–19, 2006, IABMAS'06, Taylor & Francis, 8 pages on CD-ROM.

Messervey, T. B., and Frangopol, D. M. 2006c. Bridge live load effects based on statistics of extremes using on-site load monitoring, Proceedings of the 13th WG 7.5 Working Conference on Reliability and Optimization of Structural Systems, IFIP 2006, Kobe, Japan. Proceedings to appear.

Nowak, A.S. 1993. Live load model for highway bridges. *Structural Safety*, Elsevier, 13, 53–66.

Life-Cycle Cost and Performance of Civil Infrastructure Systems – Cho, Frangopol & Ang (eds)
© 2007 Taylor & Francis Group, London, ISBN 978-0-415-41356-5

JSCE's guidelines for assessment of bridge structures by using monitoring data

A. Nakajima
Utsunomiya University, Utsunomiya, Japan

K. Maeda
Tokyo Metropolitan University, Tokyo, Japan

T. Obata
Hokkaido University, Sapporo, Japan

ABSTRACT: It is important to assess the structural performance of existing bridges from the viewpoint of their adequate maintenance and extending their life cycle. Then, the guideline for evaluating the structural performance of the bridge based on the monitoring was published by one of the scientific committee of Japan Society of Civil Engineers (JSCE). The guideline prescribes the basic issue for assessment of the bridge through the monitoring, the method of the structural monitoring and assessment, and the intervention according to each life cycle stage of bridges such as during construction and commissioning trials, and in service. The guideline also establishes the harmonization of three international standards, that is, ISO 13822, ISO 14963 and ISO 18649.

1 INTRODUCTION

In recent years, the maintenance and the long-life service of existing infrastructure has become a worldwide issue. The life extension of infrastructures makes the large economic profit for the society from the viewpoint of Life Cycle Cost (LCC) of structures. ISO codes for assessment, evaluation and monitoring of structures, that is, ISO 13822, ISO 14963 and ISO 18649 have been also established.

In Japan, many bridges were constructed to support the spreading transportation network from the 1950's to the 1970's and they have become older. As a result, the number of bridges that have been in service for more than 50 years is increasing dramatically. Then, there is fear that these will be more than 50,000 such bridges in 2021. On the other hand, the public investment for the infrastructure development, particularly, new transportation network construction is very difficult in Japan due to the economic conditions and environmental impact.

It is important to assess the structural performance of existing bridges from the viewpoint of their adequate maintenance and extending their life cycle. Various monitoring is useful in assessing the structural performance of the bridges. So far, there are many monitoring examples for assessment of the structural performance, damage identification, and evaluation of the effect of repair and rehabilitation of the bridges. However, the results of these monitoring examples are summarized individually under the client and engineer, and are not generally disclosed to the public. Therefore, it is required to establish the guideline for evaluating the structural performance based on the defined performance requirement by using the systematic monitoring technique and summarizing the results of the assessment.

Then, this paper introduces the guideline entitled "Guidelines for Assessment of Bridge Structures by Using Monitoring Data" and published by one of the scientific committee of Japan Society of Civil Engineers.

2 PURPOSE OF GUIDELINE

The guideline entitled "Guidelines for Assessment of Bridge Structures by Using Monitoring Data" was published in March 2006. This guideline aims to assess the structural performance of the bridge during construction and commissioning trials as well as in service based on the static and dynamic monitoring according to each life cycle stage of bridges. The guideline is also based on three international standards and other relevant publications. In evaluating the structural performance of the bridge, this guideline considers the performance requirement such as the safety, serviceability and environmental compatibility. The application of the guideline to the assessment of the bridge structures is also effective from the viewpoint of LCC.

In the past, the term "Monitoring" is only measure the variation of the structural behavior with time. Recently, however, the term "Monitoring" is used as the term with more extensive meaning. The term "Monitoring" is to obtain the useful findings about the design, the construction practice and the maintenance of the structures as well as to measure their behavior. Then, the guideline uses the monitoring in evaluating the structural performance of the bridge. However, if the assessment of the performance is difficult based on the monitoring data, the reliability assessment of the bridge is adopted referring to ISO 13822.

3 CONTENT OF GUIDELINE

This guideline aims to standardize the procedure for assessing the structural performance of the bridge during the construction and commissioning trials as well as in service by using the monitoring data as mentioned above. The guideline consists of nine chapters and two annexes. Chapter 1 presents the scope of the guideline, normative references and technical terms. Chapter 2 presents the bases of the assessment of the structural performance of the bridge based on the monitoring data. Chapters 3 to 6 present the data for assessment of the performance, the method of monitoring, the structural analysis for assessment of the performance and the method of the assessment of the performance. Chapters 7 to 9 present the check of the intervention based on the assessment, report of the assessment and decision based on the results of the assessment. Furthermore, the informative articles related to the detailed classification of bridges and the verification based on the reliability assessment are indicated in the annexes.

3.1 *General rules (Chapter 1)*

This chapter explains the scope of the guideline, the normative documents and relevant references, and defines the technical terms employed in the guideline. The outline of the content of the referenced normative documents, that is, ISO 13822, ISO 14963 and ISO 18649 is explained here briefly.

The purpose of ISO 14963 entitled "Guidelines for dynamic tests and investigations on bridges and viaducts"(ISO, 2003) is to provide general criteria for dynamic tests of bridges and viaducts. This code can supply the information on the dynamic behavior of bridges that can serve as a basis for condition monitoring or system identification. The scope of this international standard is as follows:

- classification of dynamic test for construction and service condition;
- indication of investigation and control of whole structures and their parts;
- list of the equipment for excitation and measurement;
- classification of techniques of investigation with reference to suitable methods for signal processing, data presentation and reporting.

However, the evaluation of the data obtained from the dynamic test is not included in this document. Then ISO 18649 which is associated with the evaluation of the results from dynamic tests is also proposed.

The purpose of ISO 18649 entitled "Evaluation of results from dynamic tests and investigations on bridges"(ISO, 2004) is to provide the method for evaluating the results from dynamic tests and investigations on bridges and viaducts. It complements the procedure for conducting the tests as given in ISO 14963 and describes the techniques for data analysis and system identification, the modeling of the bridge and surrounding environment, and the evaluation of the measured data. This international standard gives the guidance of the assessment of measurements carried out over the life cycle of the bridge, such as the one during construction and commissioning trials and in service. The stage of the life cycles and the overview of the bridge vibration monitoring are quoted from those in "Guidelines for Bridge Vibration Monitoring" (JSCE, 2000) published by one of the scientific committee of JSCE. The relationship between the life cycle of the bridge and the objective of the vibration monitoring, the required monitoring and evaluation, the measuring technique/testing, the analysis/theory and the evaluation method are shown as the overview of the bridge vibration monitoring.

On the other hand, the purpose of ISO 13822 entitled "Bases for design of structures – Assessment of existing structures"(ISO, 2001) is to provide the general requirements and procedures for the assessment of existing structures based on the principles of structural reliability and consequences of failure. The assessment is required under the following circumstances:

- anticipated change in use or extension of design working life;
- reliability check as required by authorities, insurance companies, owners, etc.;
- structural deterioration due to time-dependent actions;
- structural damage by accidental actions.

This guideline establishes the harmonization of above three international standards and then indicates the purpose and framework of the assessment of the structural performance of bridges based on the monitoring.

3.2 *Bases of assessment of performance (Chapter 2)*

This chapter explains the general framework of the assessment and overviews the purpose of the assessment. First, the general framework of the assessment of the performance of the bridge and the corresponding flowchart are specified in service and during the commissioning trials. The procedure consists of the steps such as specification of assessment objective, scenarios, preliminary assessment, detailed assessment and reporting results of assessment and so on. The assessment of the performance of the bridge is conducted according to the flowchart shown in Figure 1.

Although the general framework of the assessment indicated here is quoted from the one in ISO 13822, some items are revised. Namely, the assessment of the performance is also applied to the bridge during construction and commissioning trials, and the verification of the performance of the bridge after the implementation of intervention and the feedback to the future design are added to the procedure and flowchart.

When the assessment of the structural performance of the bridge is conducted, the objective of the assessment must be definitely specified to secure the performance requirement according to each stage of the life cycle of the bridge. Here, the safety, serviceability and environmental compatibility of the bridge during construction, commissioning trials and in service are considered as the performance requirement. However, the safety also includes the durability of the bridge. These required structural performances must be indicated clearly in the plan of construction, utilization, safety and environment conservation.

In conducting the assessment of the performance of the bridge, the identification of the scenario presents the basis for the assessment and design of intervention to be taken to ensure the structural safety and serviceability. Then, the importance of the scenario for the assessment is given referring to ISO 13822.

In addition, the purpose of the assessment of the structural performance by using the monitoring is explained depending on the stage of the life cycle of the bridge, which corresponds to the one of the bridge during construction, commissioning trials and in service. Moreover, the purpose

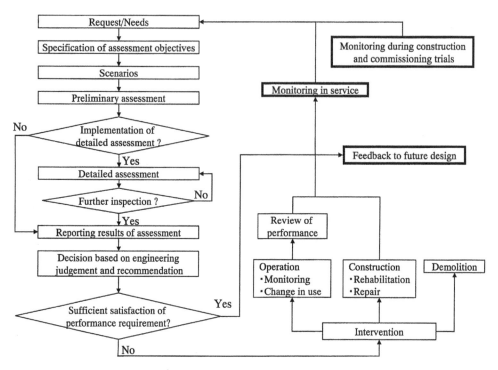

Figure 1. General flowchart for assessment of performance.

of the monitoring for evaluating and securing the performance requirement such as the safety, serviceability and compatibility to the surrounding environment is explained. In this case, the related monitoring is conducted when the change in use, the fatigue damage under the cyclic loading and the damage due to the accidental action occur in the bridge. This description is mainly quoted from the publication entitled "Guidelines for Bridge Vibration Monitoring" by one of the past scientific committee of JSCE. Although the publication does not deal with the static monitoring, the guideline deals with the static monitoring as well as the dynamic one.

3.3 Data for assessment (Chapter 3)

This chapter explains the data for conducting the assessment. The data of the material properties, structural properties, dimensions and other conditions based on the actual structural condition and related to the past and/or the future actions should be employed for the assessment. The data specified in the current codes is not employed directly for the assessment, since the current codes are normally employed for the design ones. The content of this chapter is based on that of ISO 13822.

3.4 Method of monitoring (Chapter 4)

This chapter explains the procedure and plan of the monitoring of the bridge required for the assessment. It is required to specify the objective of the assessment with the monitoring in the general framework shown in Figure 1, and to determine the period of monitoring. For example, there is a short term monitoring to secure the safety during construction of the bridge and measure the vibration and noise of the bridge in environmental problem, while there is a long term monitoring to ensure the time-dependent behavior of the bridge under successive load.

It is preferable to select the adequate method which fits the monitoring objective from many methods for monitoring the static behavior and dynamic behavior. It is also preferable to predict the property of the data according to the monitoring one by the analysis of the numerical model, to select the adequate monitoring data and to refer to the past monitoring example in advance. Furthermore, it is required to select the loading method, sensors and data analyzing system to obtain the data that suits the purpose of the monitoring. This guideline indicates the detailed information of the dynamic excitation method, sensors and the data analyzing method such as the structural identification in time domain and frequency domain.

The monitoring during construction is conducted to secure the safety, to control the profile of the bridge and to evaluate the structural performance. The monitoring during the commissioning trials is also conducted to ensure the assumed design value and to collect the initial data for the future maintenance work. On the contrary, the monitoring in service such as the one of the traveling load and the environmental compatibility is conducted to maintain the bridge adequately in the working life.

3.5 *Structural analysis for assessment of performance (Chapter 5)*

This chapter explains the structural analysis method employed for the assessment of the performance of the bridge. It is preferable to use the adequate structural model based on the actual condition of the bridge such as the applied actions, the structural behavior and the resistance of its member. It is also required to use the adequate proven analysis method which takes into account the bridge modeling, surrounding ground and the modeling of the traffic loads, human walk, earthquake and wind load.

3.6 *Method of assessment of performance (Chapter 6)*

This chapter explains the method of the assessment of the structural performance by employing the monitoring data. The objective, reliable and reasonable assessment is required to evaluate the monitoring data. Then, it is necessary to specify the basic evaluation criteria. The values of limit states, the defined values and ranges, and standard values are employed as the evaluation criteria for the assessment of the structural performance according to the purpose of the assessment, the condition of the bridge and the characteristics of the monitoring data.

Since the standard values for the assessment will be different according to the condition of the assessment items, the standard values are divided into the following categories;

Category 1: When the values of limit states exist, the basis of the evaluation is to check whether the monitoring data exceeds the one of the limit state.

Category 2: When the defined values or defined ranges of the condition exist, the basis of the evaluation is to compare the monitoring data with those defined values or ranges.

Category 3: If the evaluation values of limit states or defined states are not clear, the values which are estimated numerically for the case of the initial completed condition or similar condition of the bridge are useful. When the estimated values for the initial completed condition or similar condition of the bridge exist, the basis of the evaluation is to check whether the monitoring data exceeds those estimated values or whether they are within the range of estimated values with an acceptable error.

Category 4: When the monitoring data for the previous condition or the numerical values for the similar condition exist, the current measured data can be compared with those data to evaluate how much the bridge is improved or deteriorated.

In evaluating the performance of the bridge, the measurement accuracy in the monitoring data, the variety of the structural characteristics and the difference between the measured state and the evaluated state must be considered.

Furthermore, this chapter explains the general consideration about the evaluation of the safety during construction and in service, the evaluation of the serviceability of the various types of bridges and the evaluation of the environmental compatibility.

However, if the assessment of the performance is difficult based on the monitoring data, the reliability assessment of the bridge is adopted. If the reliability assessment of the bridge is required, new bridge or the existing bridge must be distinguished definitely. It is preferable to refer to the Annex 2 in conducting the reliability assessment. The content of this part is based on that of ISO 13822.

3.7 *Interventions (Chapter 7)*

This chapter explains how does the intervention recommend or propose based on the results of the assessment of the performance and indicates the items to be considered in conducting the recommended or proposed intervention. The possible interventions such as the repair, rehabilitation, upgrading, monitoring of the performance, enhancement of the maintenance and demolition are recommended or proposed according to the results of the assessment of the bridge during construction, commissioning trials and in service. It is also required to check the cost and risk accompanied with the proposed intervention, and the effect of the intervention on the recycling and surrounding environment.

Furthermore, this chapter indicates the concrete items to be checked such as the purpose, procedure, design, construction process and performance of the interventions. The content of this chapter is also quoted from the one of ISO 13822.

3.8 *Report (Chapter 8)*

This chapter mainly explains the format of the final report related to the planning step to the step of the final results, when the assessment of the performance is conducted through the monitoring of the bridge. The assessment of the bridge is generally conducted in a manner involving a number of phases of the work, and some form of the report is required at the end of each phase of the work. The required items related to the assessment of the performance of the bridge and its monitoring is as follows;

- bridge to be investigated;
- objective of investigations;
- methods of investigations;
- equipments for investigations;
- relevant results;
- customer, manager and engineer of work;
- data of investigations;
- references.

3.9 *Decision based on results of assessment (Chapter 9)*

This chapter explains the items to be noticed in deciding finally the intervention recommended in Chapter 7 and reported in Chapter 8. While the engineer related to the assessment only recommends the adequate intervention, the client of the assessment work need to decide the implementation of the intervention. If the client does not respond to the intervention during a reasonable period, the engineer related to the assessment might have the legal duty to inform the relevant authority.

Furthermore, this chapter indicates that the results of the assessment and the investigated intervention should be fed back to the future design of the similar bridge in order to avoid the occurrence of similar defects due to the design work and contribute to the development and succession of the technique.

3.10 *Annex (Informative)*

The detailed classification of bridges is presented in Annex 1 and the verification based on the reliability assessment is also presented in Annex 2.

The behavior of the bridge is strongly affected by the type of the superstructure and substructure, and the construction work. It is also required to consider these effects sufficiently in conducting the assessment. Then, the detailed classification of bridges such as the type of superstructure, appendage and substructure, the function of the bridges and methods of the construction are summarized in Annex 1.

This guideline aims to conduct the assessment of the performance of the bridge by using the monitoring in principle. However, if the monitoring data can not be obtained, the reliability assessment is required. Then, in Annex 2, the method of verification based on the reliability assessment is presented referring to the related articles of ISO 13822.

4 SUMMARY

This paper briefly introduces the guideline entitled "Guidelines for Assessment of Bridge Structures by Using Monitoring Data" and published by one of the scientific committee of Japan Society of Civil Engineers in March 2006. This guideline aims to standardize the procedure for evaluating the structural performance of the bridge during the construction, commissioning trials as well as in service by using the monitoring data. In evaluating the structural performance of the bridge, this guideline considers the performance requirement such as the safety, serviceability and environmental compatibility. This guideline indicates the basic evaluation criteria which consist of four categories for the assessment of the structural performance according to the purpose of the assessment, the condition of the bridge and the characteristics of the monitoring data.

Furthermore, this guideline establishes the harmonization of three international standards, that is, ISO 13822, ISO 14963 and ISO 18649 and then indicates the purpose and framework of the assessment of the structural performance of bridges based on the monitoring.

Then, the guideline will be very useful for the assessment of the structural performance of the bridges and its application to the assessment of the bridge structures is also effective from the viewpoint of LCC.

ACKNOWLEDGEMENTS

The authors are grateful to the member of Task Committee on Monitoring of Bridge Vibration and its Standardization, Committee of Structural Engineering, Japan Society of Civil Engineers, especially, the member of Working Group for drafting the guideline.

REFERENCES

ISO. 2001. Bases for design of structures – Assessment of existing structures. ISO 13822.
ISO. 2003. Guidelines for dynamic tests and investigations on bridges and viaducts. ISO 14963.
ISO. 2004. Evaluation of results from dynamic tests and investigations on bridges. ISO 18649.
Task committee on Bridge Vibration Monitoring. Committee of Structural Engineering. 2000. Guidelines for Bridge Vibration Monitoring. Tokyo. Japan Society of Civil Engineers(in Japanese).

Life-Cycle Cost and Performance of Civil Infrastructure Systems – Cho, Frangopol & Ang (eds)
© 2007 Taylor & Francis Group, London, ISBN 978-0-415-41356-5

Rehabilitations of Finnish road bridges

S. Noponen
Ramboll Ltd, Finland

A. Jutila
Helsinki University of Technology, Laboratory of Bridge Engineering, Finland

ABSTRACT: This paper is part of a Nordic research project entitled Bridge Life Cycle Optimisation (ETSI). The research work carried out in the project is based on the effective inspection and management systems developed in the Nordic countries. The main aim of the project is to create appropriate tools for decision-making and for resolving practical problems related to bridges. For example, the project will provide basis for choosing the proper bridge type or material, taking into account long-term effects on a detailed level. The paper focuses on the rehabilitations of Finnish road bridges in the years 1995–2005.

The main task is to determine the average number of rehabilitation thus far for various bridge types. This study prevents estimates of the future development of rehabilitations and their average costs. As far as the repair of old bridges is concerned, little information is available, for instance, in the registers of the Finnish Road Administration (Finnra), but nowadays the amount of information stored is rapidly increasing due to recent developments in the bridge management system.

Concrete, timber, steel, and stone bridge rehabilitations information used in this current study has been gathered from Finnra's register. The average costs of the rehabilitations are estimated. Future research will add this information on rehabilitations to the life cycle optimisation of different bridge types and materials.

1 INTRODUCTION

As of the beginning of 2006, Finnra administers 14 282 bridges. The most common bridge superstructure materials are concrete and steel: 67% of Finnish bridges are constructed of concrete and 27% of steel. Other materials include timber and stone: 5% of Finnish bridges are constructed of timber, and 1% are stone bridges. The age distribution of bridges appears in Fig. 1.

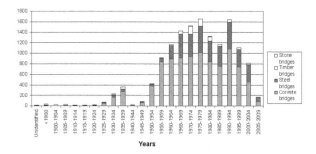

Figure 1. Age distribution of Finnra's bridges by different superstructure materials [1].

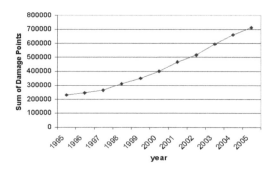

Figure 2.   The development of the Sum of Damage points for Finnra's bridges [1].

Finnra's Bridge Register contains plenty of information on it's bridges. All bridges have their own names and signs. Signs include one letter, which indicates the road district and number of the bridge. Using this information, the locations of the bridges are easy to find from special maps. Other information found in the register is the year of erection, bridge type, materials, and the bridge's most important dimensions. The register also contains information about inspections, deteriorations and rehabilitations of the bridges. The rehabilitation information includes facts about actions, times, and costs of rehabilitations.

The Finnra's Bridge Management System is mainly based on general inspections. The average time interval for general inspections is about five years. An inspector evaluates the damages and deterioration visually and records them in the register. The ranking system of bridge rehabilitation is based on a Sum of Damage points. This sum consists of the addition of single Damage points (D_{point}) to the overall sum concerning the whole bridge. Damage points are calculated from the following equation

$$D_{po\,int} = S_{tructure} \cdot C_{ondition} \cdot D_{class} \cdot D_{urgency} \tag{1}$$

where $S_{tructure}$ is a factor of the structural main component. A bridge is divided into ten main structural components, and the factor indicates just how important the condition of this component is to the whole bridge. $C_{ondition}$ (Inspector's evaluation points 0 to 4) is the inspectors' evaluation of the condition of the main components. D_{class} is the deterioration class (1 to 4) which describes how serious the deteriorations are. The deterioration classification appears in Finnra's Bridge Inspection Guide [2]. $D_{urgency}$ is the urgency of the rehabilitation (1 to 5). The Sum of Damage points of all Finnra's bridges are continuously increasing (Figure 2), because the rehabilitation budgets have not yet met rehabilitation needs [3]. The Sum of Damage points prior to the year 2000 are estimates, because the system came into use that year.

From 1996 to 2005, Finnra's bridge rehabilitation budget grew from 8 M€ to 31 M€. In the next few years, this budget may reach Finnra's estimated needs level [3].

2   REHABILITATION OF FINNISH ROAD BRIDGES IN 1996–2005

This study gathers the information on bridge rehabilitation from Finnra's Bridge Register. The task was to determine, which parts of the bridges are the most vulnerable and how costly they are to renovate. This study also examines and compares the rehabilitation of different superstructure materials as well as the growth in the number of rehabilitations.

Unfortunately, not all cost information can be found in the Bridge Register, which is why the average cost of certain rehabilitations was calculated and multiplied by the number of rehabilitation actions. This was how the total rehabilitation costs were determined. The portions of various actions were calculated from all the rehabilitations (Figures 3, 4 and 5).

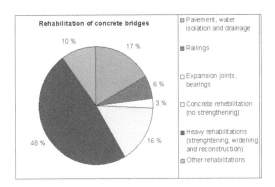

Figure 3. Cost contribution of concrete bridge rehabilitations carried out during the years 1996–2005.

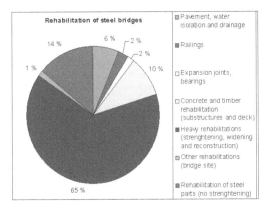

Figure 4. Cost contribution of steel bridge rehabilitations carried out during years 1996–2005.

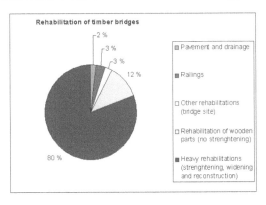

Figure 5. Cost contribution of timber bridge rehabilitations carried out during years 1996–2005.

Of course, the total costs of certain rehabilitations are estimates and therefore conclusions must be drawn carefully. In all superstructure materials, the heavy actions, including strengthening, widening, and reconstruction of the bridge, represent the bulk of the costs. Although for reinforced concrete bridges, for example heavy activities comprised only 5.4% of the rehabilitations, but represented 45.9% of the costs. For all superstructure materials, heavy actions dominate the cost distributions. The portions of heavy activities are greater for timber and for steel bridges than for concrete. Timber and steel rehabilitation needs, may be underestimated compared to those of concrete, although less than all heavy rehabilitations are performed because of deterioration of the

Table 1.   Quantity, age, rehabilitation probability and cost information of Finnra's bridges in 1996–2005. The costs are on year 2001 level.

Bridge type	Number of bridges in year 2005	Average age of bridges at the end of 2005	Probability of heavy rehabilitation activities in 10 years [%]	Average cost of rehabilitation contract [€/m²]	Money used for rehabilitation per bridges decks during 10 years [€/m²]
Concrete bridges	**9596**	**32.8 years**	**4.9**	**139**	**34**
Cast in-situ reinforced concrete bridges	6876	35.8 years	5.4	160	40
Precast reinforced concrete bridges	1681	28.0 years	4.2	217	46
Cast in-situ prestressed concrete bridges	639	17.6 years	2.2	63	10
Precast prestressed concrete bridges	400	25.4 years	1.5	100	62
Steel bridges and composite bridges	**3840**	**27.4 years**	**11.2**	**299**	**34**
Steel bridges	3615	28.4 years	11.9	336	38
Composite steel bridges	225	11.9 years	2.6	89	11
Timber bridges	**648**	**33.0 years**	**24.8**	**357**	**382**
Stone bridges	**190**	**90.3 years**	**3.2**	**157**	**32**
All Finnra's bridges	**14274**	**32.1 years**	**7.4**	**203**	**38**

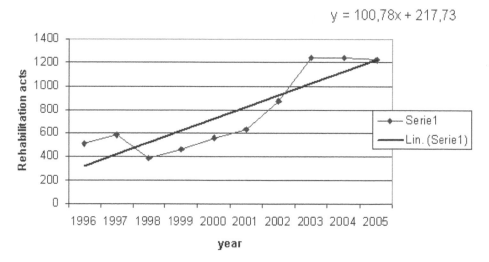

$$y = 100{,}78x + 217{,}73$$

Figure 6.   The development of all rehabilitation actions for reinforced concrete bridges during the years 1996–2005.

superstructure material. Also, the lifecycle costs could be less, if more money had been available for rehabilitation and if the work had been carried out earlier.

Comparison of the superstructure material information shown in Table 1 shows that the average ages of Finnish concrete, steel, and timber bridges are very much the same. In this comparison the likelihood of heavy rehabilitation and the rehabilitation costs themselves are much higher for timber bridges than for concrete and steel bridges.

The development of all reinforced concrete bridge rehabilitation and heavy rehabilitation activities are represented in Figures 6 and 7. The set of points is illustrated and simplified by linear

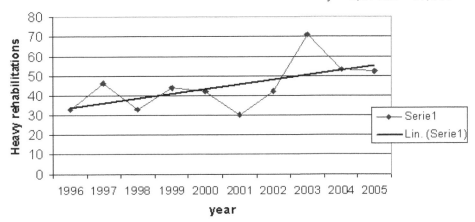

Figure 7. The development of heavy rehabilitation activity of reinforced concrete bridges during the years 1996–2005.

regression. Figure 6 indicates that the slope of the line, which describes the growth in rehabilitations, is 100.8. The rehabilitations slopes indicating the number of rehabilitations for steel bridges are 33.4, and for timber bridges, 8.0. For all materials, the number of rehabilitations is growing because of growing rehabilitation budgets.

The slope of the straight line indicating the development of heavy rehabilitation activities is of course less than the previous ones. For reinforced concrete bridges, the slope indicating heavy activities is 2.4 (Figure 7), for steel bridges, 5.5, and for timber bridges, −0.6. Rehabilitation of timber bridges are declining because the number of Finnra's wooden bridges is also decreasing.

3   PREDICTION OF BRIDGE SERVICE LIFE INCLUDING OBSOLESCENCE FACTORS

In this previous chapter, one particular example demonstrates the estimation of bridge service life including obsolescence factors. Obsolescence means here loss of ability of the structure to perform satisfactorily due changes in human based fuctional or safety requirements. This kind of phenomena are widening and strengthening of bridges. In the example of reinforced concrete bridges takes into account the known age distribution (Figure 1) and the development of heavy rehabilitation activities. By extrapolating a line from the linear regression method (Figure 7), one can predict the service life of bridges. The area below the curves corresponds to the number of bridges and bridges forced into heavy rehabilitations. In this study, the need to perform heavy activities could be considered the end of the service life of a bridge. In the example, the average age of the bridges is in 2005 is 63.3 years, and the number of the heavy rehabilitations is 50% of the numbers of reinforced concrete bridges (Figure 8). This indicates that the heavy rehabilitations increases at the same speed for 25 years, after which it remains constant, while the number of bridges increases in the same way as nowadays. The annual number of bridge demolitions is estimated to be the same as in the year 2005.

4   CONCLUSIONS

The estimated average cost of a Finnish bridge rehabilitation contract has been quite high: 203 €/m², over a fifth of the cost of a new bridge (Table 1). This calculation is based on the information available in the Bridge Register, and is only trend-setting. Hopefully, future research will clarify whether

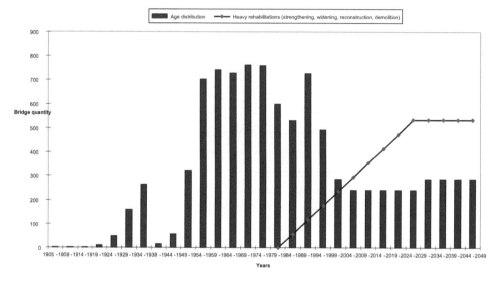

Figure 8.   Prediction of the service life including obsolescence factors of reinforced concrete bridges.

shortening rehabilitation intervals and making contracts smaller save costs. The high portions of heavy rehabilitation, compared to those of other repair activities, suggest that this could be possible.

When comparing the rehabilitation of bridges made by different superstructure materials, one must keep in mind that not all rehabilitation work results from deterioration and that one must be careful with such conclusions. Timber bridges have been less competitive than bridges built from other materials, whereas stone bridges have been quite serviceable, with an average age 90.3 years. Also, big bridges, built of pre-stressed concrete cast in-situ and composite bridges do well in rehabilitation cost comparisons, but are generally not as old as Finnish bridges.

A simple model was developed for predicting bridge service life including obsolescence factors. The given example indicates that by 2050, half of Finnish reinforced concrete bridges will not reach their expected one hundred years age.

This study will continue to analyse the life cycle costs of different bridge types and materials, to compare these costs, and to ascertain the most effective rehabilitation intervals.

REFERENCES

Finnra. Sillat 1.1.2006 (Bridges 1.1.2006), Internal publications of the Finnish Road Administration (Finnra). 2006. 74 p.

Finnra. Sillantarkastuskäsikirja (Bridge Inspection Manual, The Directives for Bridge Inspection Procedures. English 1st edition 1989) 6th renewed edition in Finnish, ISBN 951-726-873-4, TIEH 2000003-04 Helsinki, 2006, Edita. 98 p.

Söderqvist, M-K. Experience in the Finnish Bridge Management System Development, IABMAS Congress Report, IABMAS'04, Kyoto. 2004. 8 p.

Life-Cycle Cost and Performance of Civil Infrastructure Systems – Cho, Frangopol & Ang (eds)
© 2007 Taylor & Francis Group, London, ISBN 978-0-415-41356-5

Study on application of reliability analysis for bridge management system on steel bridge structures

T. Obata
Graduate school of Hokkaido University, Sapporo, Japan

A. Nakajima
Graduate school of Utunomiya University, Utunomiya, Japan

K. Maeda
Graduate school of Tokyo Metropolitan University, Tokyo, Japan

ABSTRACT: The purpose of this study is investigated the applicability, usability and practicality of reliability analysis for the bridge management system (BMS). In this study, the bridge management system is carried out by using the event-tree analysis. The parameters of the BMS are damage identify level, future life expectancy and life cycle cost (LCC). The optimum maintenance which minimum LCC of inspection and repair case is find out from analytical result. The possibility of effects and application for real structures of the methods in this study is discussed from these results.

1 INSTRUCTIONS

In recent years, the long-life of active use of infrastructure installations has become a worldwide issue also in Japan. The developmental spending or investment for preparation of infrastructures such as construction of new transportation network is very difficult in Japan from the problem of economic condition and environmental impact. Also the all over the World, the life extension of the infrastructures makes the large economic profit, reduce the industrial waste for the society and effective decrease in the global warming and other environmental pollutions in life cycle of structures (K. Nishikawa, 1994).

In Japan, much infrastructure was constructed in improvement of civic service and to support the spreading transportation network from the 1950's to the 1970's. In recently, much infrastructures in Japan are become old, and the numbers of bridges that have been in active service more than 50 years are increasing dramatically and it is expected that these bridges will be more than 50,000 in 2021. From reasons, the inspection, maintenance and rehabilitation planning are very important problem for long-lived active use of bridge and infrastructures. For realization of bridges life extension, it is necessary to establish an effective bridge management system which considering life cycle cost, damage level, future life expectancy and others.

The purpose of this study is to investigate the bridge management system (BMS) in consideration of damage identify level by inspection, future life expectancy and life cycle cost (LCC). The analytical methods in this study is applied the second moment method of reliability theory and event tree analysis (Mori, et al. 1993. Yoshikawa, 1995, Frangopol et al. 1997). The analytical structural model in this study, five steel girders with RC deck superstructure bridge model was used. Too many factors deterioration of steel bridges are existence, the deterioration factor in this study is used as steel corrosion. The damage of steel superstructure is occured by the cross section loss of the steel girder and steel bar in the RC deck. The damage inspection accuracy level and steel corrosion rate are set three cases for each parameter of the analyses, and the limitation of target reliability index β is set to 3.72 and life cycle span of bridge is 100 years in this study. The results are considered minimum LCC strategy of each corrosion rate case. And the possibility of application of analytical processes in this study for BMS is discussed from these results.

2 RELIABILITY EVALUATION OF STRUCTURES

The load function S and limit state strength R is abides by some probability functions, and each probability function S and R are independent random variable. In the case of exceed the limit state strength of a structure is become a load S larger than strength R. The probability of this relation is given by Equation 1.

$$p_f \equiv P(S > R) = P(S - R > 0) \tag{1}$$

where p_f probability of failure; $S =$ load function; $R =$ limit state strength.

The normal distribution functions are defined for the $R:N(R_0, \sigma_r)$ and $S:N(S_0, \sigma_s)$. And the Z is defined as $Z = R - S$, also $Z : N(Z_0, S_z)$ to be the normal distribution function.

$$Z : N(Z_0, \sigma_Z) = N\left(R_0 - S_0, \sqrt{\sigma_r^2 + \sigma_S^2}\right) \tag{2}$$

where $Z =$ performance function; $N =$ the normal distribution function; $Z_0 =$ mean of the performance function; $\sigma_z =$ standard deviation of Z; $R_0 =$ mean of the limit state strength R; $\sigma_r =$ standard deviation of R; $S_0 =$ mean of load function S; $\sigma_s =$ standard deviation of S.

The performance function Z meaning to afford of the design performance, and it is a very important of the probability value of performance function Z. This value express probability of failure $P (Z < 0)$ is given by Equation 3.

$$p_f = p(Z < 0) = 1 - \Phi\frac{Z_0}{\sigma_Z} \tag{3}$$

where $p_f = P(Z < 0) =$ probability of failure; $\Phi =$ standard normal probability distribution function (mean $= 0$, standard deviation $= 1$); $Z_0 =$ mean of performance function Z; $\sigma_z =$ standard deviation of Z.

The safety index β is defined by performance function Z. The equation 4 is given by several times deference of Z_0 and σ_z.

$$\beta \equiv \frac{Z_0}{\sigma_Z} = \frac{R_0 - S_0}{\sqrt{\sigma_r^2 + \sigma_S^2}} \tag{4}$$

where $\beta =$ safety index of performance function Z.

When safety index b defined by the Equation 4, also probability of failure pf is given simply by Equation 5.

$$p_f = 1 - \Phi(\beta) \tag{5}$$

The method of obtaining the probability of failure p_f from mean and standard deviation of the load function S and limit state strength R is called second moment method. This method is not obtained p_f directly, calculate safety index b correspondence to each limit state condition. The second moment method is possible to investigate of structural safety, and this analytical method was adopted in this study.

3 ANALYTICAL PROCEDURE

3.1 Relation of deterioration model and steel corrosion

Too many factors deterioration of steel structures are existence by passage of active use years. The well-known factors of deterioration of stiffness of structures are corrosion, fatigue, damage of external condition and others. In Japan, the reason of re-construction of steel bridges in 1986 to 1996 as followed; Most of reasons are improvement of road width, alinement and others, and 12% of bridges re-constructed by superstructure damage (Nishikawa et al. 1997). The reasons of detail in damages of the re-construction of steel superstructures, 26% of bridges are damaged by corrosion of steel members. It is considered that the damage of corrosion of member is one of the most important problems of the maintenance and rehabilitation of steel infrastructures.

From above reasons, it is thought that the damage of corrosion in steel member is main item of deterioration by passage of the time of structural limit state resistance. When the structural limit state resistance is decreased just only deterioration by passage of the time, the structural limit state resistance $r(t)$ in time t is obtained by Equation 6 (Mori et al. 1993).

$$r(t) = r_0 \cdot g(t) \tag{6}$$

where $r(t) =$ structural limit state resistance; $t =$ passage of the time; $r_0 =$ initial resistance of the member; $g(t) =$ deterioration function. The deterioration function means deterioration by environmental factors like a corrosion and others by passage time. In usually, the deterioration function modeling is expressed by Equation 7 (Clifton et al. 1989).

$$g(t) = 1 - a \cdot (t - T_I)^b \tag{7}$$

where $a =$ parameter of progress velocity of deterioration; $b =$ dependence parameter of environmental factor for the cause of deterioration; $T_I =$ start time of deterioration. In this study, the parameter of T_I and b set to $T_I = 0$ and $b = 1$. When the tests of expose a steel material to air, corrosion rate is 0.005 mm/year in good environment of the air. In the terrible air condition included salt like a seaside, the corrosion rate is over 0.2 mm/year (Ohoi, 2002). In usually, structural damage by corrosion is passable to repairing. The structural limit state resistance after repeating also considered in estimation of deterioration by passage time on reliability analysis in this study.

3.2 Inspection, repair and event tree analysis

Infrastructures are gradually deteriorating by damage of long passage using time. The maintenance and rehabilitation on infrastructures, it is very important that to identifying accurately of damage level. The performance inspection of infrastructures is necessary to keep structural performance, maintenance and rehabilitation. Too many inspection methods are already established, non-destructive inspection methods are used in this study. Also many methods are known in non-destructive inspection of the radiological testing, ultrasonic testing, magnetic particle testing, liquid penetrant testing and others, the effective of inspections are different in each method. In general, the higher performance testing is also expensive of the cost. The optimum inspection method is selected by conform to environmental condition, progress velocity of deterioration and life cycle cost of strictures.

When something damages are founded by inspection, it is necessary to repair and rehabilitation for prevents of the deterioration and secures the safety of infrastructures. But, to find the all damage by non-destructive inspection testing methods is difficult. From reason, physically damage level η and damage identification probability of each non-destructive inspection testing method are considerate in the analysis. In this study, physically damage level and damage identification probability are given by Equation 8 and Equation 9.

$$\eta(t) = \frac{D_0 - D(t)}{D_0} \tag{8}$$

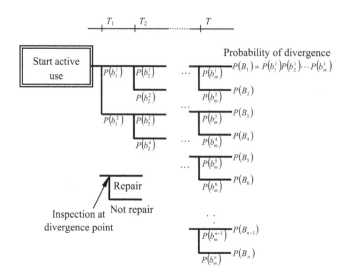

Figure 1. Event tree analysis.

$$d(\eta) = \begin{cases} 0 & (0 \leq \eta \leq \eta_{\min}) \\ \Phi\left(\dfrac{\eta - \eta_{0.5}}{\sigma}\right) & (\eta_{\min} \leq \eta \leq \eta_{\max}) \\ 1 & (\eta > \eta_{\max}) \end{cases} \qquad (9)$$

where D_0 = main girder plate thickness at the completion; $D(t)$ = main girder plate thickness after t years; $\eta_{0.5}$ = 50% average of damage identification probability; $\eta_{\min}$ = minimum damage level to found by inspection; $\eta_{\max}$ = maximum damage level to found by inspection. In this study, set $\eta_{0.5}$ to three kind of value of 0.05, 0.10 and 0.15 (5%, 10%, 15%). The too match combination pattern occurred by damage level of by passage time, inspection, damage identification probability and repairing. For the analytical method in this study is applied event tree analysis.

The event tree analysis is set the divergence probability of each event for the event tree divergence figure. Each event is analyzed in time domain according to divergence probability, and results are investigated by each finally event probabilities. This method also can be obtaining the social risk analysis result from occurrence probability of the events. The event tree is shown in Figure 1. In this study, set the inspection carried out in divergence point. The repairing divergence probability is decided by Equation 9, and occurrence probability of each event is using the probability of failure pf from reliability analysis. When the numbers of inspection set to m, the numbers of divergence of event tree are 2^m patterns. The reliability analyses are performed in the all patterns of event tree divergence points.

3.3 Optimization of life cycle cost

In general, life cycle cost (LCC) is total cost of maintenance, inspection, repair, rehabilitation, user cost and others. In this study, investigate the social risk cost when excess of ultimate limit state condition of structures. The calculation of social risk cost is investigated by the product of user cost by probability of failure. The total cost of LCC in this study is obtained by Equation 10.

$$C_T = C_I + C_{PM} + C_{INS} + C_{REP} + C_F \qquad (10)$$

where C_T = Total cost of LCC; C_I = initial cost; C_{PM} = periodical maintenance cost; C_{INS} = inspection cost; C_{REF} = repairing cost; C_F = social risk cost. The inspection cost C_I is set 100 as standard price in this study. The other costs are calculated simply by Equation 11 to Equation 15(Frangopol et al. 1997).

$$C_{PM} = \sum_{t=2}^{98} C_{pm} \cdot t \tag{11}$$

$$C_{INS} = \sum_{i=1}^{m} (1 - \eta_{\min})^{20} \tag{12}$$

$$C_{REP} = \sum_{i=1}^{2^m} C_{rep,i} P(B_i) \tag{13}$$

$$C_{rep,i} = \sum_{i=1}^{m} \alpha_{rep} e_{rep,i}^{0.5} \tag{14}$$

$$C_F = \sum_{i=1}^{2^m} p_f P(B_i) C_f \tag{15}$$

where the periodical inspection of Equation 11 is done every tow years in active using term. The inspection cost C_{pm} is supposed product of 0.001 by C_I, and total inspection cost express the C_{pm} and the number of inspection in passage time t. The passage time of active use is set to 100 years. The standard price of inspection α_{ins} is supposed product of 0.001 by C_I, to be related quality of inspection and inspection cost by using parameter α_{ins} and η_{min}. The C_{REP} of Equation 14, at first the repairing cost of i order of event $C_{rpe,i}$ is obtained by Equation 14. The next process of calculation of C_{REP} is product of $C_{rpe,i}$ by $P(B_i)$. The $P(B_i)$ is divergence probability of i order event. Equation 14 is express the re-construction cost α_{rep} (this cost is supposed as same as C_I) and repairing level parameter e_{rep}. The e_{rep} is obtained by comparison with limit state strength of before and after repairing. The social risk cost in Equation 15 is possible to expressed by using maximum value of probability of failure p_f in i order of event, divergence probability $P(B_i)$ in i event and social risk cost C_f. The C_f is defined by over the value of ultimate limit state while active service. In this study, value of C_f is supposed product of 500 by C_I as used.

It is necessary to minimized life cycle cost for the effective maintenance of infrastructures. The Equation 16 and Equation 17 are objective function and restriction conditions for minimize the LCC to keep the reliability of safety of infrastructures.

$$\min C_T \tag{16}$$

$$\beta_{life} \geq \beta_{life}^* \tag{17}$$

where C_T = Total cost of LCC β_{life} = safety index of passage time of active use of structures; β_{life}^* = serviceability limit state safety index.

The optimum maintenance planning support system is investigated to keep the structural performance and minimize the LCC by using these methods.

4 RESULTS AND DISCUSSTIONS

The analytical structural model in this study is shown in Figure 2 and Table 1. This bridge model is considering to one of the most universally steel girder and concrete deck bridge system. The design of bridge model in this study is based on Japanese standard of bridges (Japan society of load, 2002), span length 350,000 mm and load width 11,800 mm of non-composite steel girder simple beam bridge.

The analytical conditions in this study, active service passage time is set 100 years, initial cost is 100 of non-dimensional number and the numbers of inspection is 49 times in 100 years active service (the inspection are done every tow years). The corrosion ratio of steel members are set to 0.05 mm/year, 0.1 mm/year and 0.2 mm/year, and inspection precision $\eta_{0.5}$ are set to 0.05, 0.10 and

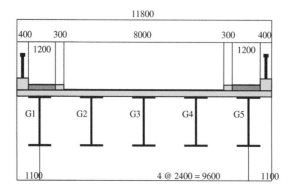

Figure 2. Section of bridge model.

Table 1. The parameters bridge model.

Type		Non-composite steel girder
Span		350,000 mm
Width		11,800 mm
Depth of deck		220 mm
Main	Upper flange	520 × 25 mm
girder	Lower flange	470 × 25 mm
	Web	1850 × 9 mm

Table 2. The analytical cases.

	Corrosion ratio	Inspection precision
Case 1	0.05 mm/year	$\eta_{0.5} = 0.05$ $\eta_{0.5} = 0.10$ $\eta_{0.5} = 0.15$
Case 2	0.10 mm/year	$\eta_{0.5} = 0.05$ $\eta_{0.5} = 0.10$ $\eta_{0.5} = 0.15$
Case 3	0.20 mm/year	$\eta_{0.5} = 0.05$ $\eta_{0.5} = 0.10$ $\eta_{0.5} = 0.15$

0.15. The total analyses are performed nine patterns in this study. The allowable safety index is set to $\beta^*_{life} = 3.72$ ($p_f = 10^{-4}$). The analytical cases in this study are shown in Table 2.

The analytical results in this study are obtained total cost of LCC based on number of inspection times and deterioration function of the structure. The relation of number of inspection times and total costs are shown in Figure 3, and the deterioration functions minimum cost result in each case are shown in Figure 4.

The results of optimum inspection precision of non-destructive inspection are 0.10 in Case 1 and Case 2, 0.15 in Case 3. The Case 1 and Case 2 means that enough of middle class level of inspection precision when deterioration velocity is not so large. In the Case 3, it is easy to find out damages by using low precision type of inspection in terrible environment of high corrosion ratio. From this reason, the low cost inspection method of $\eta_{0.5} = 0.15$ was chosen in optimum maintenance planning.

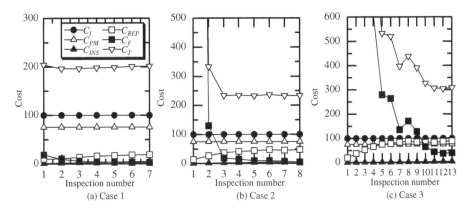

(a) Case 1   (b) Case 2   (c) Case 3

Figure 3. Inspection number and total cose (LCC).

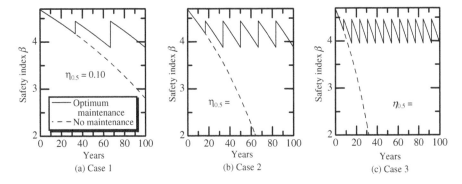

(a) Case 1   (b) Case 2   (c) Case 3

Figure 4. Deterioration function of optiumum maintenance planning.

The number of times of inspection and repairing to keep allowable safety index are explained. In the Case 1, the number of minimum inspection to keep allowable limit are tow times and also in the case of repeating are tow times is identified of the minimum LCC maintenance strategy. It is considered that only lowest rehabilitation times to keep allowable limit in Case 2 and Case 3 are four times and ten times, but these cases are not minimum of LCC. The LCC minimum strategy in Case 2 and Case 3 are very important to do the preventive maintenance for control the social risk cost. The LCC minimum strategy in Case 2 is five times of inspection and repeating was performed, and Case 3 is twelve times of inspection and repeating. It is considered that the preventive maintenance is very effective than large-scale rehabilitation for reduce the LCC. The analytical method in this study is obtained the deterioration function and life cycle cost and also can be identified minimum LCC maintenance strategy. From results, this method has applicability for the bridge management system to long life active service strategy of infrastructures.

5 CONCLUSIONS

This study is to investigate the bridge management system (BMS) in consideration of damage identify level, future life expectancy and life cycle cost (LCC). The analytical methods in this study is applied the second-moment method of reliability theory and event tree analysis. This analytical bridge model is considering to one of the most universally steel girder and concrete deck bridge system, and the parameters of the study are damage identify level of inspection, future life expectancy and life cycle cost. The optimum maintenance which minimum LCC of inspection and repair case is find out from analytical result. The major conclusions obtained in this study are summarized as follows:

(1) The LCC minimum strategy in bad environmental condition are very important to do the preventive maintenance for control the social risk cost.
(2) It is possible to find out of optimum maintenance strategy of minimum LCC, and this method can be chosen inspection level to appropriate for environment.
(3) This method has applicability for the bridge management system to long life active service strategy of infrastructures.
(4) The other parameters (example: fatigue) considered in this analysis, more higher applicability is obtained for the analyze of optimum maintenance strategy.

REFERENCES

Clifton, J. R. and Knab, L. I. 1989. *Service life of concrete*, National Bureau of Standards. NUREG/CR5466.
Frangopol, D.M., Lin, K.-Y., and Estes, A. C. 1997. Life-Cycle Cost Design of Deteriorating Structures, *Journal of Structural Engineering*, ASCE, Vol.123, No.10: pp.1390–1401.
Japan load association. 2002. *Specification of highway bridges part I and part II*. Tpkyo: Maruzen. (In Japanese)
Mori, Y. and Ellingwood, B. 1993. Reliability-Based Service-Life Assessment of Aging Concrete Structures, *Journal of Structural Engineering, ASCE*, Vol.119, No.5, pp.1600–1621, 1993.
Nishikawa, K. 1994. Life Time and Maintenance of Highway Bridges. *Journal of structural mechanics and earthquake engineering, JSCE*, No.501/I-29: pp.1–10. (In Japanese)
Nishikawa, K., Murakoshi, J. and Kamizen, Y. 1997. The results of re-construction of bridges (III). The data on public works research institute, No.3512. (In Japanese)
Ohoi, K. 2002. The limit state of buildings 3 steel strictures. *Journal of Architecture and Building Engineering*, No.1494, Architectural Institute of Japan: pp.18–19.
Yoshikawa, H. 1995. *Design and analysis of reinforced concrete*, Tokyo: Maruzen. (In Japanese)

Life-Cycle Cost and Performance of Civil Infrastructure Systems – Cho, Frangopol & Ang (eds)
© 2007 Taylor & Francis Group, London, ISBN 978-0-415-41356-5

Multi-performance based optimum maintenance system framework for existing bridges

Kyung-Hoon Park, Y.K. Hwang & S.Y. Lee
Structure Research Department, Korea Institute of Construction Technology, Korea

Jung S. Kong
Department of Civil and Environmental Engineering, Korea University, Korea

Hyo-Nam Cho
Department of Civil and Environmental Engineering, Hanyang University, Korea

Jong-Kwon Lim
Infra Asset Management Corporation, Korea

ABSTRACT: For the performance- and cost-effective bridge maintenance, many researchers have focused on the optimum maintenance method and system of deteriorating bridges. To establish a lifetime optimum maintenance strategy of a deteriorating bridge considering the lifetime performance as well as the life-cycle cost, a more practical and realistic method is proposed and the prototype bridge maintenance system is developed. The proposed method and development system can produce the set of optimum tradeoff maintenance scenarios among other conflicting objectives such as cost minimization and performance maximization. The multi-performance indexes and the systematic integrated performance assessment model are described in this paper. The optimum maintenance scenario could be generated at the system level of the bridge through applying system reliability and weight factors by taking account of the subordinate relation due to the replacement of member as well as the individual member level. Applied examples using the developed system are also presented to verify the proposed model and method.

1 INTRODUCTION

Recently, the demand for the practical application of life-cycle cost (LCC) for maintenance of bridges is rapidly growing in the bridge engineering practice. Most of bridge maintenance systems try to implement such an LCC analysis technique (Thompson et al. 1998 & KICT 1999). Various techniques and systems have been developed for maintaining a bridge which requires more attention in the maintenance due to it's social and economical importance and structural feature. Although the study of the optimum maintenance strategy has been in progress to overcome the limit of the existing bridge maintenance systems and balance the conflicting objectives such as the life-cycle performance and cost, the systemization is not significant for its practical application. To establish the maintenance strategy through a prediction of the bridge performance and cost, it is necessary to make not only the theoretical approach but also the practical application for the bridge maintenance appropriate to the domestic situation, and a computational system should be developed in a simple and comprehensive form for its utilization. For this purpose, this study suggests a method to select the optimum maintenance scenario considering the life-cycle performance and cost and develops a systematic integrated system for the bridge maintenance. This paper briefly describes the proposed method and a general model for LCC-effective optimum maintenance of bridges with an emphasis on considering multiple and often conflicting objectives such as condition, safety, and cost. And then, the framework of the developed optimum maintenance system for next-generation is presented. Finally, by using the developed system, it proposes a process of generating optimum maintenance scenario of the steel-girder bridge in the Korean National Road.

2 OPTIMUM MAINTENANCE STRATEGY

The optimum maintenance scenario should minimize the maintenance cost and maximize the target performance of the structure required by a bridge manager during the bridge lifetime. In order to establish a performance-based optimum maintenance strategy considering the life-cycle cost, it is required to evaluate the life-cycle performance and cost and generate the optimum maintenance scenario (KICT 2006).

2.1 *Lifetime performance evaluation*

At first, it is required to select a reasonable performance index to evaluate the bridge performance, and in the second place, to establish a performance evaluation model and a quantitative analysis model of the repair/reinforcement effects appropriate to the life-cycle analysis. To develop a performance evaluation model, the multi performance indexes – reliability index and condition index are introduced in this study. For evaluating the reliability index, a response surface method (RSM) which is one of the probabilistic safety evaluation methods is applied to the time-variant reliability analysis and the quantification of maintenance effects (Kong et al. 2006). The condition index used in this study sequentializes and generalizes the discontinuous condition states based on the sight inspection applied to the current bridge maintenance in Korea (MOCT & KISTEC 2003). The information required in the condition index evaluation is obtained from the statistical analysis using the existing data or the professional research data (KICT 2006). These two performance indexes can be integrated into a system level through the system reliability and the weight factors for the bridge safety guidelines in Korea respectively. A profile of the life-cycle performance and cost is executed by using MLTR (Kong & Frangopol 2001) program which is an analysis tool for the lifetime performance and cost profile considering the effect of maintenance interventions.

2.2 *Life-cycle cost estimation*

For the life-cycle cost analysis, the important problems are not only the direct maintenance cost paid by the bridge agency but also the indirect cost and the failure cost. The expected value of the total cumulative maintenance cost obtained from the sum of all expenditure by the applied maintenance interventions for each failure mode until the considered time. This study tries to evaluate the failure cost as a function of the reliability reduction with the time passage. The expected value of the total cumulative failure cost can be obtained by the sum of all costs generated by each failure mode applied up to a considered time. A new road user cost model has been proposed which is consist of the vehicle operating cost and the user delay cost extending the driver delay cost considered in the exiting study to all users riding the vehicle. To make the reasonable estimation of the user cost, the traffic analysis and regression analysis are used for developing a regression model (KICT 2006).

2.3 *Optimum maintenance scenario generation*

The optimum maintenance scenario could mainly depend on the budget of the bridge agency and the importance level of the bridge performance. And the satisfied maintenance scenario should be decided so as to balance the different objectives such as the improvement of the structural performance and the reduction of the maintenance cost. The single optimum maintenance plan computed by a single objective cannot satisfy the specific requirement of the optimum tradeoff between the life-cycle maintenance cost and the structural performance during the lifetime. As an alternative approach for this maintenance plan, the combination of the available maintenance interventions can be applied to generate the optimum maintenance scenario during the period of time considered (Liu & Frangopol 2005, Furuta et al. 2004). For the life-cycle maintenance of the deteriorating bridge, this study generates a set of optimum maintenance scenario minimizing LCC and maximizing the reliability and condition as the separated objective functions based on the Pareto optimum concept. Multi-objective genetic algorithm (GA) is used for generating a set of the various maintenance scenarios.

3 OPTIMUM MAINTENANCE SYSTEM FOR BRIDGES

3.1 *Multi-objective optimization*

The problem generating maintenance scenario considered in this paper can be formulated in the multi-objective combinatorial optimization problem as follows:
Objective:

$$F_1 = C_T^p = \sum_{t=t_p}^{T} C_t^p \quad \rightarrow \text{ min.} \tag{1}$$

$$F_2 = \rho_t = \min[\rho_{t_p}, \rho_{t_p+1}, \rho_{t_p+2}, \cdots, \rho_T] \quad \rightarrow \text{ max.} \tag{2}$$

$$F_3 = \beta_t = \min[\beta_{t_p}, \beta_{t_p+1}, \beta_{t_p+2}, \cdots, \beta_T] \quad \rightarrow \text{ max.} \tag{3}$$

Subject to:

$$g_i(\cdot) \leq 0 \qquad i = 1, 2, \cdots, N_s \tag{4}$$

where C_T^p = current value of total life-cycle cost; ρ_t = lifetime condition index; β_t = lifetime reliability index; t_p = present time; T = total period of lifetime; $g_i(\cdot)$ = i-th constraint related to the budget, performance or specific requirements of bridge manager; N_s = number of constraints.

Bridge managers will face the problem selecting the alternative scenario among many suggested scenarios and the final maintenance scenario will rely on the decision of a bridge manager. This study applies a fitness function to provide the optimum tradeoff solution, which is possible not only for a case of the optimum tradeoff solution among the conflicting objective functions but also for a case of bridge manager's specific requirements and constraints. The maintenance problem of a bridge that systematically consists of the various members has considered a single member until now. But this study proposes an advanced technique to select the optimum tradeoff maintenance scenario over the entire bridge system considering the subordinate relation due to the replacement of the bridge members for securing a realistic application.

3.2 *Optimum maintenance system framework*

With an evaluation model and a cost model related to the performance including the condition index and the reliability index, this study introduces a systematic procedure selecting the optimum tradeoff solution among the conflicting objectives based on the multi-objective GA. A MCS method is used so as to consider the probabilistic structural performance because of the uncertainty related to the deteriorating process of each bridge member and the uncertainty of the life-cycle maintenance cost. This process applies to all parts considered in members of a bridge, and in case that the replacement of a member influences on the other members, a maintenance scenario of the bridge system is proposed considering the subordinate relation between members, that is, the influenced members can be replaced at the same time. Also, the cost items applied to generate the optimum maintenance scenario can be selectively considered and the suitable optimum maintenance scenario for the considered cost items can be proposed. This study developed a GA_MLTR with the GA-based program using MLTR by the suggested methods. Figure 1 shows the proposed optimum maintenance system framework.

3.3 *Optimum maintenance scenario generation system*

The GA_MLTR-based optimum maintenance scenario generation system is developed for the general steel-girder bridges due to the limit of bridge performance evaluation, and it is named LCMSTEB (Performance-based Life-Cycle Maintenance System for Steel Bridges). Visual C++ is used as a development language for LCMSTEB, and MS-SQL is used for the internal database. GA_MLTR is embedded as a solver and the result of analysis is expressed as a graphic form in the program and stores in an Excel file. For executing the analysis with the minimum input from the user, LCMSTEB has the internal database including bridge information, traffic information, characteristics of performance and cost data, etc. By considering the convergence of solutions (scenarios) analyzed by GA_MLTR, the number of individuals and generations for GA is applied with 1,000 and 30 respectively, and the probability of crossover and mutation is applied with 50% and 5% respectively. Figure 2 shows the typical input and output screens of LCMSTEB.

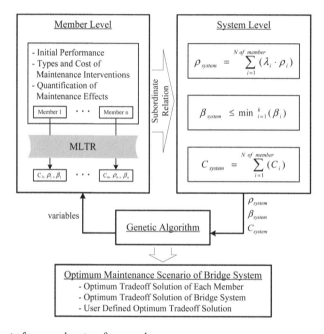

Figure 1.   Concept of proposed system framework.

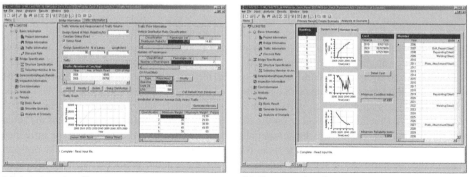

(a) An example of input screen   (b) An example of output screen

Figure 2.   Typical screens of LCMSTEB.

4 ANALYSIS EXAMPLES

In order to verify the applicability and rationality of the optimum maintenance method and the developed system, a sample bridge is selected for the typical and simple case among steel bridges in Korea. The bridge is built in 2001 with 45 m length, 19.5 m width and 3-steel boxes in a simple span and is used in service without special maintenance interventions. This paper shows two analysis examples: (1) girder at the member level, and (2) slab and pavement at the system level.

The analysis example of girder considers the condition and reliability index. And the condition index of slab and pavement is considered in the analysis process to generate a system maintenance scenario. For the convenience of analysis, the direct maintenance cost is only considered in the system level analysis and the failure cost is additionally considered in the member level analysis. The maintenance interventions considered in this paper for each member are shown in Table 1. For the probabilistic analysis, the type of distribution and coefficient of variation of each maintenance intervention are assumed as normal distribution and 0.3 respectively. And then the quantification of deterioration model and the repair/reinforcement effect of the members are considered. The reliability index evaluation is executed by using the developed response surface model (Kong et al. 2006), and the condition index profile is applied to the deterioration model which is assumed and decided by the statistical analysis. For example, the initial deterioration models of the girder can be presented as the following Figure 3 and Figure 4. The change of the performance index profile can be evaluated by superposing the initial performance index profile and additional performance index profiles associated with all subsequent maintenance interventions.

4.1 Life-cycle analysis at member level

Among the maintenance scenario solutions derived by the system, elite solutions in 3 dimensional area are shown as the following Figure 5. Pareto front solutions obtain close to the ideal optimal point in which the cost gets minimized and condition index and reliability index get maximized, and it is located on the right side and lower part of Figure 5. Each point, as shown in Figure 5, means each maintenance scenario and contains the profile of cost, condition index and performance index in a set of optimum solution. The maintenance scenario provides the annual maintenance intervention during the considered analysis period. Table 2 presents the ranking of optimum tradeoff maintenance scenario established in the solution area of Figure 5. As for the degree of importance which is the ratio of each single objective against the minimum and maximum value respectively among solutions, the optimum tradeoff maintenance scenarios among 3 objectives locate on the upper part.

Table 1. Maintenance Interventions.

Member	Maintenance Intervention	Unit Cost (Thousand Won)	Application Rate (%)	Improved Condition Index (%)
Steel Girder	Painting (PG)	$30.0/m^2$	100	30
	Welding (WE)	$20.0/m$	10	40
	Bolting(BO)	$3.0/ea.$	10	30
	Steel Attachment(SA)	$750.0/m^2$	10	70
	Replacement (RG)	$2,000.0/m^2$	100	100
Slab	Epoxy Injection (EI)	$3.0/m$	30	24
	Waterproofing (WP)	$27.6/m^2$	80	24
	FRP Attaching (FA)	$78.9/m^2$	50	72
	Replacement (RS)	$150.0/m^2$	100	100
Pavement	Surface Treatment (ST)	$10.0/m^2$	10	30
	Cutting-overlay (CO)	$20.0/m^2$	30	80
	Patching (PA)	$10.0/m^2$	5	40
	Re-pavement (RP)	$25.0/m^2$	100	100

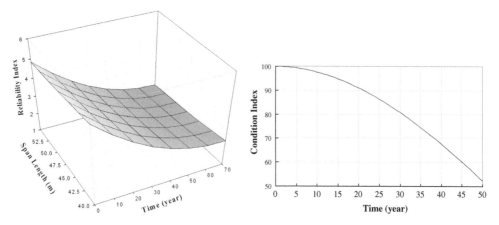

Figure 3. Initial reliability index profile for girder.   Figure 4. Initial condition index profile for girder.

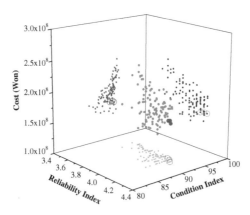

Figure 5.   Solution set for the elite.

Table 2.   Scenario ranking of optimum tradeoff maintenance.

Rank	Cost (Thousand Won)	Condition Index	Reliability Index	Degree of Importance		
				Cost	Condition	Reliability
1	161,000	92.8	4.08	0.859	0.890	0.893
2	163,600	92.5	4.12	0.840	0.858	0.946
3	162,300	91.9	4.16	0.850	0.799	1.000
4	156,200	92.8	4.05	0.895	0.884	0.842
5	142,000	92.6	3.97	1.000	0.865	0.738
6	182,200	93.0	4.12	0.702	0.898	0.946
7	147,900	89.8	4.12	0.956	0.599	0.946
8	148,000	93.3	3.87	0.956	0.932	0.599
9	151,600	91.9	4.01	0.929	0.799	0.791
10	194,500	93.3	4.08	0.611	0.940	0.890

304

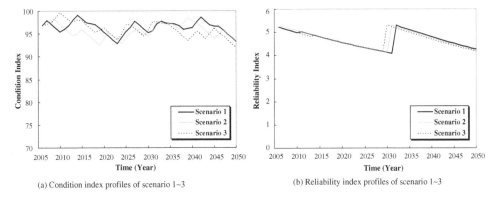

(a) Condition index profiles of scenario 1~3

(b) Reliability index profiles of scenario 1~3

Figure 6.   Condition index and reliability index profiles of scenario from 1 to 3.

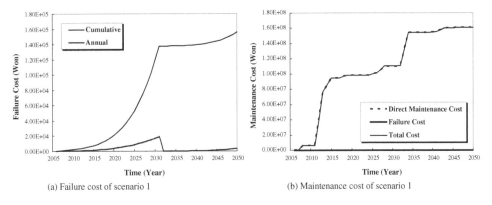

(a) Failure cost of scenario 1

(b) Maintenance cost of scenario 1

Figure 7.   Cost profiles of scenario 1.

Up to ranking 3, the condition index and reliability index profiles for each scenario are shown in Figure 6. During the considered analysis time, the change of condition index appears to be sensitive with the influence of maintenance interventions. On the contrary, due to the high safety of steel-box girder, the reliability index profile does not show the significant change except by the retrofiting. Figure 7 shows the cost profiles. The failure cost estimated by multiplying the failure probability with retrofit cost takes small part of total maintenance cost because of its high reliability index.

4.2 Life-cycle analysis at system level

The developed system can generate the maintenance scenarios for each individual member of the bridge such as the slab, pavement and girder, etc. But the bridge as a system structure has the sub-ordinate relationship among members that the replacement of a member causes the replacement of other members so that the simple combination of the optimum maintenance scenario for each member does not mean the optimum maintenance scenario of the entire bridge system. Therefore, the maintenance scenario of the system level simultaneously considering the maintenance interventions in each member can become a realistic maintenance plan. This section presents the generation of the system level maintenance scenario considering slab and pavement.

The elite solutions obtained from the analysis are distributed with respect to two objectives of the condition index and cost as the following Figure 8. Table 2 shows the annual maintenance scenarios of the slab and pavement in priority order up to ranking 3 from Figure 8. The maintenance interventions uniformly distributed in a 1-year time interval are applied to each member, and

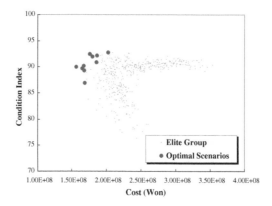

Figure 8.   Solution set for elite.

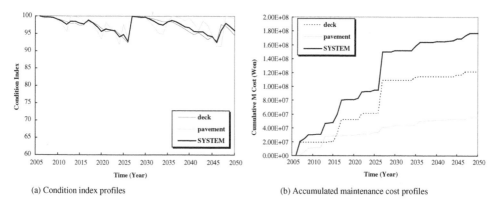

(a) Condition index profiles   (b) Accumulated maintenance cost profiles

Figure 9.   Condition index and cost profiles of Scenario 1.

more detailed time interval can be applied. The system level analysis gives a significant result that is possible to replace the pavement followed by the slab replacement by considering the subordinate relationship along with the replacement of a specific member, that is, it can generate more reasonable maintenance scenario.

During the period of analysis, the maintenance cost and condition index of Scenario 1 ranked first in Table 3 are shown in Figure 9. According to Figure 9(a) showing the condition index profile of each member and system, it can be known that the bridge condition depends on the condition of slab, because the slab is governed to calculate the system condition index in accordance with the bridge safety guidelines in Korea. According to Figure 9(b) showing the cumulative maintenance cost of the member and the system, the annual maintenance cost generating a member replacement increases rapidly and the expected cost converted into the current value increases slowly with an effect of the discount rate.

5   CONCLUSIONS

To establish a lifetime optimum maintenance strategy of a deteriorating bridge considering the lifetime performance and cost, a more practical method is proposed and the prototype bridge maintenance system is developed. The study results in the following conclusion.

Table 3. Annual maintenance scenarios.

No.	Year	Scenario 1 Slab*	Scenario 1 Pave.**	Scenario 2 Slab	Scenario 2 Pave.	Scenario 3 Slab	Scenario 3 Pave.
1	2006	EI	ST	EI		WP	CO
2	2007	WP		FA	CO		
3	2008		CO			EI	ST
4	2009			WP	ST		
5	2010		PA		PA		
6	2011		ST				
7	2012					WP	
8	2013		RP		PA		CO
9	2014	EI			ST		
10	2015		ST	EI	CO		
11	2016	WP			PA		PA
12	2017	FA	PA		ST	EI	RP
13	2018		ST			WP	
14	2019				PA		
15	2020				ST	FA	ST
16	2021		CO				CO
17	2022	WP	ST	WP	PA		
18	2023	EI	PA		ST		ST
19	2024						PA
20	2025		CO		RP		
21	2026				ST		
22	2027	**RS**	**RP**			EI	
23	2028			EI	CO		PA
24	2029		ST	WP		**RS**	**RP**
25	2030	EI	CO		ST		
26	2031			FA			
27	2032					WP	PA
28	2033				ST		CO
29	2034		PA	EI	CO		
30	2035	WP					
31	2036	EI	RP				
32	2037		ST	**RS**	**RP**		
33	2038		PA		ST	WP	RP
34	2039				CO		CO
35	2040		CO	WP			
36	2041						
37	2042	EI		EI	PA		
38	2043						ST
39	2044		ST				
40	2045	WP			CO	WP	CO
41	2046			WP	ST		PA
42	2047	FA	PA		RP	EI	
43	2048	EI	RP			FA	ST
44	2049			EI	ST		
45	2050						PA
Condition Index		92.4	91.2	87.7	86.1	91.6	90.2
LCC (Million Won)		121.3	55.4	112.6	44.8	123.4	57.4

*Note: The abbreviations refer to Table 1.

(1) The proposed method and the developed system can generate the optimum tradeoff maintenance scenario among the conflicting objectives, lifetime safety, condition, and cost.
(2) The optimum tradeoff maintenance scenario is generated not only in the level of bridge member but also in the level of bridge system by considering the subordinate relationship in accordance with the replacement between members.
(3) The proposed method and developed system can improve the current bridge maintenance methods only minimizing the life-cycle cost and can be utilized as an efficient tool to provide an optimum bridge maintenance scenario appropriate to the various limit and specifications required by a bridge agency.

ACKNOWLEDGEMENTS

This study was performed in appreciation of support of the Ministry of Construction and Transportation (MOCT) and Korea Institute of Construction and Transportation Technology Evaluation and Planning (KICTTEP) with the research and development work of construction technology (Technical Innovation 10).

REFERENCES

Furuta, H., Kameda, T., Fkuda, Y. & Fangopol, D.M. 2004. Life-cycle cost analysis for infrastructure systems: Life Cycle cost vs. safety level vs. service life, *Life-cycle performance of deteriorating structures*, *ASCE*, Reston, Vergina, 19–25.
Kong, J.S. & Frangopol, D.M. 2001. *MLTR user's manual*, Department of Civil, Environmental, and Architectural Engineering, University of Colorado, Boulder, Colorado, USA.
Kong, J.S., Park, S.H., Kim, S.H., Park, K.H. & Fangopol, D.M. 2006. Novel management system for steel bridges in Korea, *IABMAS; Proc. intern. conf.*, Porto, Portugal.
Korea Institute of Construction Technology (KICT) 1999. Study on the improvement of bridge management system (BMS) in year 1998, *Research Report*, MOCT, Korea.
Korea Institute of Construction Technology (KICT) 2006. Development of life-cycle cost analysis method and system for the life-cycle cost optimum design and the life-time management of steel bridges, *Research Report*, MOCT & KICTTEP, Korea.
Liu, M. & Frangopol, D.M. 2005. Multiobjective maintenance planning optimization for deteriorating bridges considering condition, safety, and life-cycle cost, *Journal of Structural Engineering*, *ASCE*, 131 (5), 833–842.
MOCT & KISTEC 2003. Safety inspection and precise safety diagnosis guide (bridge), MOCT, Korea.
Thompson, P.D., Small, E.P., Johnson, M. & Marshall, A.R. 1998. The Pontis bridge management system, Structural Engineering Institute, *IABSE*; *Proc. intern. conf.*, Zurich, Switzerland, 8 (4), 303–308.

Life-Cycle Cost and Performance of Civil Infrastructure Systems – Cho, Frangopol & Ang (eds)
© 2007 Taylor & Francis Group, London, ISBN 978-0-415-41356-5

An approach to identify and localize damages with lamb-waves excited by piezoelectric members

U. Peil & S. Loppe

Institute for Steel Structures, Technical University Braunschweig, Germany

ABSTRACT: The procedure presented serves for an automatic large-area monitoring of steel structures by means of non-destructive testing. Piezoceramic members are applied as a cluster to the structure under observation. These elements are used both as sensors and as actuators and initiate guided wave propagation in the structure. Damages are detected by the changed structural response. Beside other methods of signal processing the principle of beamforming is used with the sensor signals. The example of a 6 mm steel plate shows the sensitivity of the developed procedure for damage detection.

1 INTRODUCTION

The Collaborative Research Center 477 "Life cycle assessment of structures via innovative monitoring" at the Technical University Braunschweig started in 1998 and works on different methods, strategies und procedures for the monitoring of structures. The project "Crack detection by means of piezo arrays" as a part of the Collaborative Research Center focuses on the detection and localization of structural changes e.g. cracks.

2 MOTIVATION AND OBJECTIVE OF THE PROJECT

The monitoring of large areas of steel structures is still essentially accomplished by visual inspection. Regions that are considered as critical are usually examined additionally by non-destructive testing methods after having been checked visually. With structures not loaded statically, e.g. railway bridges, cranes and crane tracks, reactors with changing pressure, towers and masts, one has to pay attention especially to fatigue cracks, since these can limit the load-carrying capacity directly. Sudden crack initiation may also appear in conjunction with buildings exposed to chemical attacks (stress corrosion in chemical reactors, blast furnaces, etc.).

The visual detection as well as the detection by means of non-destructive testing methods are not easy to perform everywhere and sometimes even impossible, e.g. hardly accessible structures at huge heights, the inner regions of reactors or welded box girders or insulated coated members. Sometimes structures have to be monitored continuously to prevent damages. In such cases the above-mentioned visual inspection at certain intervals is not possible. Then a fully automatic equipment is required that checks the state of the structure in short intervals and that is able to detect and localize the damage at an unknown point in time.

If the position of the crack can be anticipated relatively exactly, there are a number of established methods for automated crack monitoring. The situation becomes more difficult if the crack position cannot be predicted precisely, for example, if in a plane structure the crack occurs at an arbitrary point due to local micro notches caused by stress corrosion. It is the objective of this project to develop a robust, affordable and permanently installed measuring equipment that is capable of monitoring the critical structures reliably.

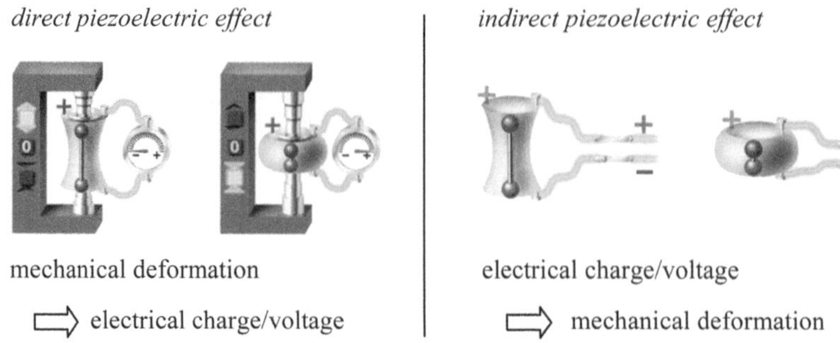

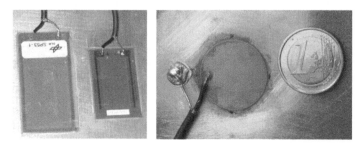

direct piezoelectric effect   *indirect piezoelectric effect*

mechanical deformation   electrical charge/voltage

⟹ electrical charge/voltage   ⟹ mechanical deformation

Figure 1.   Direct and indirect piezoelectric effect [CeramTec 2004].

Figure 2.   Encapsulated modules and circular piezoceramics electroded on both sides.

3   GENERAL APPROACH, USE OF PIEZOCERAMIC ELEMENTS

By means of piezoceramics and the use of the indirect piezoelectric effect (Fig. 1), symmetrical and antisymmetrical Lamb waves are initiated that spread over the whole structure. By an elaborate arrangement of those piezo elements and the choice of an optimized stimulation, the responding wave-modes can be controlled purposefully.

At the same time identical piezo members are used as sensors for the detection of the propagating waves (use of the direct piezoelectric effect, see figure 1). A change between both functions (actuation and sensing) is easily possible and also intended by this method.

Two different procedures of stimulation have to be distinguished. A continuous stimulation focuses on the vibration of the structure whereas with a short-time excitation the wave propagation and their reflections as well as their attenuations are of interest. The presented paper works on the wave propagation phenomena.

3.1   *Selection of piezoceramic elements and measurement hardware*

The application of diverse piezoelectric ceramics, both simple double-sided electroded discs and encapsulated modules [Keats Wilkie 2000], was tested, see Figure 2. The use of simple circular discs [PI Ceramic 2004] with a diameter of 10 to 30 mm and a thickness of less than 0.5 mm was found to be most successful. A 2-component-epoxy was used to bond those discs under pressure to the steel structure. The epoxy was made electrically conductive by adding copper powder with a particle size <100 μm. In contrast to mobile equipments that are used e.g. for the investigation of pipelines [Guided Ultrasonics], the presented system is permanently installed.

The piezos working as actuators are driven by a so-called hybrid amplifier. This amplifier provides both requirements for an improved control of the piezos compared to a regular voltage-amplifier: on the one hand a quasi steady control of the actuator voltage for bidirectional working space and on the other hand the highly dynamic charge control to provide nearly hysteresis-free and linear actuator behavior [Doerlemann et al.], see figure 3.

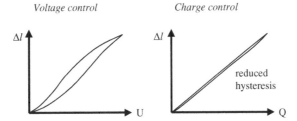

Voltage control   Charge control

Figure 3.   Piezo control.

With this equipment the chosen actuators with a relative small capacitance can be driven with signal frequencies greater than 100 kHz. The advantage of the compensation of the nonlinear relation between piezo-voltage and piezo-elongation is the direct comparability to the results of FE calculations performed in parallel to the experiments. The used FE software ANSYS implies a linear model for the piezoelectric properties as shown in equation 1 [Ansys 2005].

$$\begin{bmatrix} \{T\} \\ \{D\} \end{bmatrix} = \begin{bmatrix} [c] & [e] \\ [e]^T & -[\varepsilon] \end{bmatrix} \cdot \begin{bmatrix} \{S\} \\ -\{E\} \end{bmatrix}$$

(1)

T: mechanical stress
D: electric displacement
S: strain
E: electric field

4   EXCITATION AND DETECTION OF LAMB-WAVES

The powering of the piezoceramics with an electrical charge causes strain in the piezos in the direction of their polarization and orthogonal in the plane of the disc due to the transversal contraction. FE simulations show that the thin epoxy-layer can be regarded as a stiff bonding. As a result of the free boundary conditions on its backside, the thickness mode of vibration of the piezo discs does not affect the steel plate significantly. In fact the radial component as a result of the lateral strain is important. This distinguishes the presented procedure from the typical ultrasonic application where normally a fluid is used for the coupling of the transceivers.

The transmission of the piezo deformations to the plate initiates guided waves in the plane structure, namely Lamb waves [Viktorov 1967] that are to be examined (Fig. 4). Guided waves can propagate over large distances so that relative big areas can be monitored.

Two types of Lamb waves exist, symmetrical and antisymmetrical Lamb wave modes, see Figure 4. In contrast to other wave types, these Lamb waves exhibit a dispersive behavior, i.e. their phase velocity and thus their group velocity are not only dependent on the material but also on the frequency. The result of this dispersive property is the widening of the shape of the time-limited signal with the ongoing wave-propagation. This phenomenon worsens the resolution and makes experimental data hard to interpret because of signal overlap. The Rayleigh-Lamb-frequency equation (eq. 2) describes this property [Achenbach 1975].

$$\frac{\tan\left[\frac{1}{2}\pi \cdot \sqrt{\Omega^2 - \xi^2}\right]}{\tan\left[\frac{1}{2}\pi \cdot \sqrt{\frac{\Omega^2}{\kappa^2} - \xi^2}\right]} = -\frac{4\xi^2 \cdot \sqrt{\frac{\Omega^2}{\kappa^2} - \xi^2} \cdot \sqrt{\Omega^2 - \xi^2}}{\left(\Omega^2 - 2\xi^2\right)^2} \qquad (Symmetrical)$$

(2a)

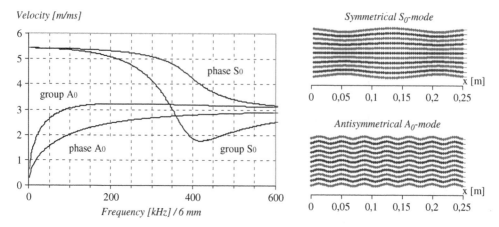

Figure 4. Dispersion-curves 6 mm steel, deformations f = 35 kHz (S$_0$ and A$_0$-wave).

$$\frac{\tan\left[\frac{1}{2}\pi\cdot\sqrt{\Omega^2-\xi^2}\right]}{\tan\left[\frac{1}{2}\pi\cdot\sqrt{\frac{\Omega^2}{\kappa^2}-\xi^2}\right]} = -\frac{\left(\Omega^2-2\xi^2\right)^2}{4\xi^2\cdot\sqrt{\frac{\Omega^2}{\kappa^2}-\xi^2}\cdot\sqrt{\Omega^2-\xi^2}} \qquad (Antisymmetrical) \qquad (2b)$$

$$\text{where}\quad \Omega=\frac{2h\cdot\omega}{\pi\cdot c_T};\ \ \xi=\frac{2kh}{\pi};\ \ \kappa=\frac{c_L}{c_T} \qquad (2c)$$

Figure 4 shows the solution of equation 2 for a 6 mm steel plate. Only the two zero modes are depicted since higher modes belonging to higher frequencies cannot be excited or detected with the measuring equipment available for this work so far.

Especially for the sensor-piezos the stiff bonding is a significant advantage because the generated charge is nearly proportional to the strain of the plate surface. Therefore also wave modes with particle movements mainly in the plane and not orthogonal to the plane can be detected very well.

If the excitation is not symmetrical both lamb-modes A$_0$ and S$_0$ develop. By means of elaborate signal processing techniques (e.g. the dispersion-compensation [Wilcox 2003]) the measured signal can be separated into both modes. The dispersion-compensation plays an important role within the processing of the signals. This procedure is based on the transformation from the time domain into the frequency domain, a further transformation into the wavenumber domain taking into account the dispersion relations (i.e. Rayleigh-Lamb-Equation) and finally a last transformation into the spatial domain. Depending on the accounted wave mode the results are signals with a shape compressed back to its original shape. To reduce the characteristic part of the signal to a single peak the Hilbert-envelope [Oppenheim 1999] has to be calculated. Figure 5 explains the dispersion-compensation with simulated signals of piezos applied on an unlimited plate. The distance between actuator and sensor-piezo is 2500 mm, the excitation signal is a 3-cycle Hanning windowed toneburst with a center frequency of 40 kHz.

It can be noticed that with both dispersion compensated signals (A$_0$ and S$_0$ wave) the starting point of the wave-train is at 2500 mm, that means that the real distance is precisely detected. Furthermore the compression of the widened signals (especially A$_0$-wave) to its original compact shape in the spatial domain is shown. Thus the main benefits of the dispersion compensation are the possibility not only to use the S$_0$-mode for damage detection but also the A$_0$-mode and the ability to compress the widened Signal to its original shape.

312

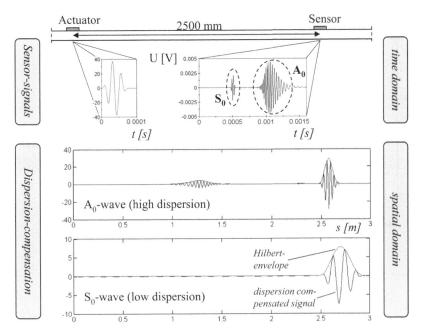

Figure 5.   Results of the dispersion-compensation (example).

5   DAMAGE DETECTION WITH A 6 MM STEEL PLATE

Experiments were conducted with a quadratic 6 mm steel plate with an edge length of 1500 mm and piezoceramic discs Ø 25 mm, d = 0,5 mm (Figs. 2, 6). The piezo array consisting of 5 single piezos that can switch arbitrarily to actuator or sensor function is located centrally on the plate. The structural change that is to be detected and localized by means of beamforming [Sundararaman et al. 2003, Christensen et al. 2004] consists of a small steel cylinder with a mass of 9 g. This mass is bonded to the steel plate with Cyanoacrylate-glue at several positions.

Before glueing the small mass to the plate a baseline measurement with the undistorted system is made. Subsequently the mass is bonded to the plate and the measurements with this distortion are conducted. The signals are digitally filtered by an appropriate IIR-bandpass-filter twice eliminating a phase-shift. Due to the inherent high-pass filtering the signals are set to zero-offset.

In the first experiments only the center piezo of the 5 piezo-array was used as an actuator while the other piezos worked as sensors. A big improvement of the damage-detection sensitivity was achieved by also using all other piezos as actuators (but only one at a time).

For better detection of the waves reflected at the 9 g-mass in the measured data the differential signal is calculated (measurement with and without mass). The procedure features the essential benefit of a nearly total elimination of the reflections at the plate edges. For the same reason inherent distortions (e.g. welded seams) are no limitations of the presented method for damage detection.

Subsequently to the calculation of the differential signals the dispersion-compensation based on the A_0 as well as on the S_0-mode is applied. The results are two datasets in the spatial domain that are interpreted by means of beamforming.

5.1   *Principle beamforming*

From the antenna technology the principle of phased-array-antennas is known for the directional radiation of signals. By a phase-shifted excitation of single transducers and their interaction as an

313

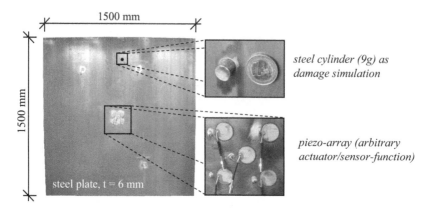

Figure 6.   Steel plate 6 mm with piezo-array and applied mass 9 g.

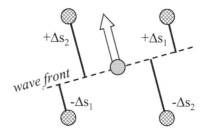

Figure 7.   Sensor-beamforming.

array not only the signal-range compared to a single transducer is increased. Also directivity with main lobes and side lobes of the emitted signals is obtained.

Within the scope of this work this principle of phased-array-antennas is swapped, i.e. the phase-shift is applied to the measured sensor-signals instead of the actuators. Thus the array works as a directional microphone. The actual beamforming is realized by accounting for an individual distance shift for every single signal in dependence on the focused angle (Fig. 7). A superposition of all processed signals is the final step.

5.2   *Interpretation and results*

The described procedure of beamforming is applied to the quadratic steel plate with the central piezo-array shown in Figure 6. The excitation-signal is a short windowed toneburst with a center frequency of 80 kHz and the measured data is digitally filtered and dispersion-compensated. As shown in Figure 6 the small steel cylinder is located 850 mm horizontally and 200 mm vertically from the upper left corner of the plate (see also Figure 8).

Figure 8 shows the beamforming-results of the damage detection with regard to above-mentioned dispersion-compensation. The darker the spot the more wave-energy is reflected. The position of the damage is detected excellently although is has to be mentioned again that the damage consisting only of a small 9 g-mass bonded to the plate surface is a very weak distortion. A different damage e.g. a hole or crack would cause locally a total reflection, therefore considerably stronger signals and thus an even better detection.

Although the interpretations in Figure 8 are based on differential signals a significant reflection from the plate borders can be noticed. The reason is a very small and inevitable time shift between the single measurements (caused by the measurement hardware) that creates small remaining signals in the region of the plate edges. For a subsequent calculational elimination of this problem

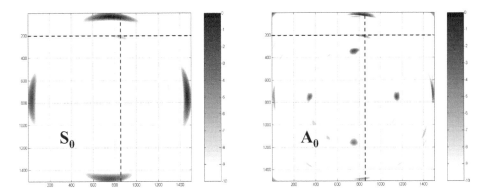

Figure 8.   Results based on S_0 and A_0 dispersion-relations.

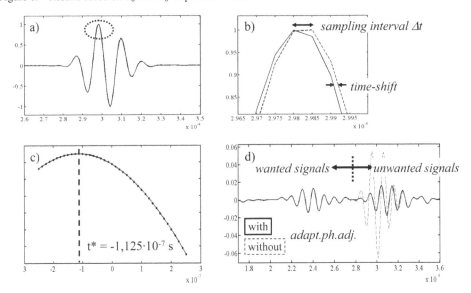

Figure 9.   Adaptive-phase-adjustment, example.

a procedure called adaptive-phase-adjustment was developed. The principal idea of this procedure is the numerical new sampling of the relevant signal-region with shorter sampling intervals. With these newly sampled signals a cross-correlation $\rho(t)$ of both signals (eq.3) is performed to get an ideal time-shift $t^* < \Delta t$ (with $\Delta t =$ original sampling interval) for the maximum reduction of the amplitude of the unwanted signal. Finally the signal is interpolated with respect to t^* and with this modified signal the differential signal is calculated again.

$$\rho(\tau) = \int_{-\Delta t}^{+\Delta t} U_{\text{Reference}}(\tau) \cdot U_{\text{Damage}}(t + \tau)\, d\tau \qquad (\textit{Cross-correlation}) \qquad (3)$$

Figure 9a shows an example of two signals (with and without distortion) in the region of the arrival of the strong signal-reflection from the plate edges. Only by zooming-in the small time-shift t^* is identifiable (Fig. 9b). The discrete calculation of the cross-correlation-integral (eq. 3) leads to Figure 9c where the maximum can be noticed at $t^* = 1,125 \cdot 10^{-7}$ s (the sampling interval $\Delta t = 5,0 \cdot 10^{-7}$ s is more than 4 times larger). Figure 9d shows the improvement of the differential signal quality and the influence of the adaptive-phase-adjustment respectively. The amplitude of the distortion is significantly reduced while the influence on the wanted signal is nearly neutral.

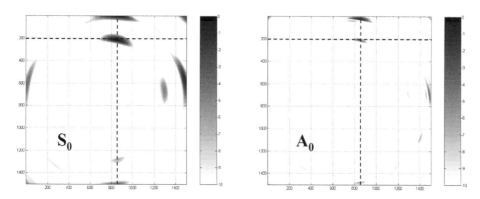

Figure 10.   S$_0$ and A$_0$ interpretation allowing for the adaptive phase-adjustment.

By implementing the adaptive-phase-adjustment before the dispersion-compensation is performed the sensitivity of the whole method for damage detection can be increased significantly, see Fig. 10. Compared to Fig. 8 (S$_0$) the detection of the damage is much stronger, compared to Fig. 8 (A$_0$) the unwanted spots disappear. Thus the use of the adaptive-phase-adjustment leads to a considerable improvement of the damage detection by means of beamforming.

6   CONCLUSION AND OUTLOOK

The results of the present work show that a damage detection and precise localization is possible by means of simple and affordable piezoceramic discs. The used methods of signal processing are partly self-developed and their results are easy-to-interpret-diagrams.

Future work will focus on the variation of different parameters e.g. thickness, geometry or damage. Furthermore alternative procedures for damage detection will be developed (also based on simple piezoceramic elements). First experiments were carried out with single transducers bonded as a raster to a plate with the single piezos monitoring the inner region of the raster.

REFERENCES

Achenbach, J.D. 1975. Wave propagation in elastic solids. Elsevier Science Publishers 1975
Ansys Inc. 2005: Ansys Release 10 Documentation
CeramTec AG 2004. Piezoelektrische Bauteile (literature of the company)
Christensen, J.J. & Hald, J. 2004. Beamforming. Technical Review No.1 – 2004. Brüel&Kjær Sound & Vibration Measurement A/S, Nærum Denmark, 2004
Doerlemann, C., Muss, P., Schugt, M. & Uhlenbrock, R. 2004. Improved active vibration control using the linear properties of piezoelectric actuators. Actuator 2004, 9th International Conference on new Actuators, Bremen 14.–16. Juni 2004
Guided Ultrasonics Ltd. Wavemaker rapid pipe screening system (product information)
Keats Wilkie, W. et al. 2000. Low-Cost Piezocomposite Actuator for Structural Control Application. SPIE's 7th Annual International Symposium on Smart Structures and Materials, Newport Beach 5.–9. march 2000
Oppenheim, A.V. & Schafer, R.W. 1999. Zeitdiskrete Signalverarbeitung. 3rd edition, R.Oldenbourg Verlag Muenchen Wien 1999
PI Ceramic GmbH 2004: Piezokeramische Materialien und Bauelemente (literature of the company)
Sundararaman, S., Adams, D.E. & Rigas, E.J. 2003. Structural Damage Characterization through Beamforming with Phased Arrays. Proceedings of the 4th International Workshop on Structural Health Monitoring, Stanford USA, September 2003
Viktorov, I.A. 1967. Rayleigh and Lamb Waves. Plenum Press, New York 196
Wilcox, P.D. 2003. A Rapid Signal Processing Technique to Remove the Effect of Dispersion from Guided Wave Signals. IEEE Transactions on Ultrasonics, Ferroelectrics, and Frequency Control, Vol.50, April 2003

Author index